Advances in Intelligent Systems and Computing

Volume 532

Series editor

Janusz Kacprzyk, Polish Academy of Sciences, Warsaw, Poland
e-mail: kacprzyk@ibspan.waw.pl

About this Series

The series "Advances in Intelligent Systems and Computing" contains publications on theory, applications, and design methods of Intelligent Systems and Intelligent Computing. Virtually all disciplines such as engineering, natural sciences, computer and information science, ICT, economics, business, e-commerce, environment, healthcare, life science are covered. The list of topics spans all the areas of modern intelligent systems and computing.

The publications within "Advances in Intelligent Systems and Computing" are primarily textbooks and proceedings of important conferences, symposia and congresses. They cover significant recent developments in the field, both of a foundational and applicable character. An important characteristic feature of the series is the short publication time and world-wide distribution. This permits a rapid and broad dissemination of research results.

Advisory Board

Chairman

Nikhil R. Pal, Indian Statistical Institute, Kolkata, India
e-mail: nikhil@isical.ac.in

Members

Rafael Bello, Universidad Central "Marta Abreu" de Las Villas, Santa Clara, Cuba
e-mail: rbellop@uclv.edu.cu

Emilio S. Corchado, University of Salamanca, Salamanca, Spain
e-mail: escorchado@usal.es

Hani Hagras, University of Essex, Colchester, UK
e-mail: hani@essex.ac.uk

László T. Kóczy, Széchenyi István University, Győr, Hungary
e-mail: koczy@sze.hu

Vladik Kreinovich, University of Texas at El Paso, El Paso, USA
e-mail: vladik@utep.edu

Chin-Teng Lin, National Chiao Tung University, Hsinchu, Taiwan
e-mail: ctlin@mail.nctu.edu.tw

Jie Lu, University of Technology, Sydney, Australia
e-mail: Jie.Lu@uts.edu.au

Patricia Melin, Tijuana Institute of Technology, Tijuana, Mexico
e-mail: epmelin@hafsamx.org

Nadia Nedjah, State University of Rio de Janeiro, Rio de Janeiro, Brazil
e-mail: nadia@eng.uerj.br

Ngoc Thanh Nguyen, Wroclaw University of Technology, Wroclaw, Poland
e-mail: Ngoc-Thanh.Nguyen@pwr.edu.pl

Jun Wang, The Chinese University of Hong Kong, Shatin, Hong Kong
e-mail: jwang@mae.cuhk.edu.hk

More information about this series at http://www.springer.com/series/11156

Somnuk Phon-Amnuaisuk
Thien-Wan Au · Saiful Omar
Editors

Computational Intelligence in Information Systems

Proceedings of the Computational Intelligence
in Information Systems Conference
(CIIS 2016)

 Springer

Editors
Somnuk Phon-Amnuaisuk
School of Computing and Informatics
Universiti Teknologi Brunei
Gadong
Brunei Darussalam

Saiful Omar
School of Computing and Informatics
Universiti Teknologi Brunei
Gadong
Brunei Darussalam

Thien-Wan Au
School of Computing and Informatics
Universiti Teknologi Brunei
Gadong
Brunei Darussalam

ISSN 2194-5357   ISSN 2194-5365 (electronic)
Advances in Intelligent Systems and Computing
ISBN 978-3-319-48516-4   ISBN 978-3-319-48517-1 (eBook)
DOI 10.1007/978-3-319-48517-1

Library of Congress Control Number: 2016954698

Printed on acid-free paper

This Springer imprint is published by Springer Nature
The registered company is Springer International Publishing AG
The registered company address is: Gewerbestrasse 11, 6330 Cham, Switzerland

Preface

On behalf of the organizing committee, it is an honor and a great pleasure to welcome all of you to Brunei and to the Computational Intelligence in Information Systems (CIIS 2016) Conference.

CIIS Conference is initiated from the INNS-CIIS 2014, with help from the International Neural Network Society (INNS). CIIS Conference aims to bring together researchers from countries in the Asian Pacific Rim to exchange ideas, present recent results and discuss possible collaborations in general areas related to computational intelligence and their applications in various domains.

This year, the international program committee constitutes 80 researchers from 19 different countries. CIIS 2016 has attracted a total of 62 submissions from 21 countries. These submissions underwent a rigorous double-blind peer-review process. Of those 62 submissions, 26 submissions (42 %) have been selected to be included in this book.

First and foremost, we would like to thank the keynote speaker, the invited speakers and all the authors who have spent the time and effort to contribute significantly to this event. We would like to thank members of the technical committee who have provided their expert evaluation of the submitted papers; Janusz Kacprzyk and Thomas Ditzinger from AISC series, Springer; members of the local organizing committee for their contributions to this event; Pg. Hj Mohd. Esa Al-Islam Bin Pg Hj Md. Yunus and all members of the steering committee for their useful advice; and last but not least, Hjh Zohrah Binti Haji Sulaiman, our Vice Chancellor.

We would also like to acknowledge the following organizations: Universiti Teknologi Brunei for its institutional and financial support, and for providing the venues and administrative assistance; Brain Science Research Center, International Neural Network Society, Asia Pacific Neural Network Society, Cisco, Google Developer Group Brunei and BAG networks for their technical support.

Finally, we thank all the participants of CIIS 2016 and hope that you will continue to support us in the future.

September 2016

Somnuk Phon-Amnuaisuk
Thien-Wan Au
Saiful Omar

Organization

Honorary Chair/Advisor

Hjh Zohrah binti Haji Sulaiman   Vice Chancellor,
Universiti Teknologi Brunei

Steering Committee

Chairperson

Pg Hj Mohd. Esa Al-Islam bin Pg   Assistant Vice Chancellor Academic,
    Hj Md. Yunus   Universiti Teknologi Brunei

Members

Nawaf Hazim Saeid	Assistant Vice Chancellor Research, UTB
Dyg Jennifer Hiew Lim	Registrar and Secretary, UTB
Awang Hamdani bin Hj Ibrahim	Finance, UTB
Thien-Wan Au	Chairman CIIS 2016, UTB
Mohamad Saiful bin Haji Omar	Co-chairman CIIS 2016, UTB

International Advisory Board

Akira Hirose	University of Tokyo, Japan
Derong Liu	University of Illinois at Chicago, USA
Eran Edirisinghe	Loughborough University, UK
Fushuan Wen	Universiti Teknologi Brunei
Hussein A. Abbass	University of New South Wales, Australia

Irwin King	Chinese University of Hong Kong, China
Jain, Lakhmi C.	University of Canberra, Australia
Kay-Chen Tan	National University of Singapore, Singapore
Laszlo T. Koczy	Budapest University of Technology and Economics, Hungary
Paul Chung	Loughborough University, UK
Soo-Young Lee	Korea Advanced Institute of Science and Technology, Korea
Yew-Soon Ong	Nanyang Technological University, Singapore

Working Committees

Chairman and Co-chairs

| Thien-Wan Au | Universiti Teknologi Brunei |
| Mohd Saiful Haji Omar | Universiti Teknologi Brunei |

Secretariat

Hajah Nor Zainah Haji Siau	Universiti Teknologi Brunei
Haji Rudy Erwan Haji Ramlie, Rahman	Universiti Teknologi Brunei
Haji Irwan Mashadi Haji Mashud	Universiti Teknologi Brunei

Technical

Somnuk Phon-Amnuaisuk	Universiti Teknologi Brunei
Sheung-Hung Poon	Universiti Teknologi Brunei
Haji Afzaal H. Seyal	Universiti Teknologi Brunei
Wida Susanty	Universiti Teknologi Brunei

Finance

| Hamdani bin Hj Ibrahim | Universiti Teknologi Brunei |

Ceremony and Logistics

Hj Idham Maswadi Haji Mashud	Universiti Teknologi Brunei
Siti Noorfatimah Haji Awg Safar	Universiti Teknologi Brunei
Hj Ady Syarmin bin Hj Md Taib	Universiti Teknologi Brunei
Hj Morsidi bin Haji Kasim	Universiti Teknologi Brunei

Welfare and Accommodation

Hj Sharul Tazrajiman Hj Tajuddin	Universiti Teknologi Brunei

Web Master

Ak Hj Azhan Hj Pg Ahmad	Universiti Teknologi Brunei

Sponsorship, Promotion and Publicity

Jennifer Voon	Universiti Teknologi Brunei
Sey Mey	Universiti Teknologi Brunei
Siti Noorfatimah Haji Awg Safar	Universiti Teknologi Brunei
Hj Ady Syarmin bin Hj Md Taib	Universiti Teknologi Brunei

Publishing

Mohd Peter D. Shannon	Universiti Teknologi Brunei

Invitation and Protocol

Ibrahim Edris	Universiti Teknologi Brunei

Refreshment

Norhuraizah Haji Md Jaafar	Universiti Teknologi Brunei

Car and Traffic

Haji Mohd Noah Haji   Universiti Teknologi Brunei
   Abd. Rahman

Souvenir and Certificate

Siti Noorfatimah Haji Awg Safar   Universiti Teknologi Brunei

Special Session Organizers

SoCCA, SoMET   Sheung-Hung Poon, UTB
SoDM   Kok-Chin Khor, Keng-Hoong Ng, MMU
SoIoT   Thien-Wan Au, Saiful Omar, UTB
SoITS   Afzaal Seyal, Sharul Tajuddin, UTB
SoNIC   Somnuk Phon-Amnuaisuk, UTB

International Technical Committee

Abdelrahman Osman Elfaki   Tabuk University,
                                      Kingdom of Saudi Arabia
Abdollah Dehzangi   Griffith University, Australia
Adham Atyabi   Yale University, USA
Ahmad Rafi   Multimedia University, Malaysia
Ak Hj Azhan bin Pg Ahmad   Universiti Teknologi Brunei, Brunei
Ali Selamat   Universiti Teknologi Malaysia, Malaysia
Alex Sim   Universiti Teknologi Malaysia
Atikom Ruekbutra   Mahanakorn University of Technology,
                                      Thailand
Bok-Min Goi   Universiti Tunku Abdul Rahman
Chuan-Min Lee   Ming Chuan University, Taiwan
Chuan-Kang Ting   National Chung Cheng University, Taiwan
Chun-Cheng Lin   National Chiao Tung University, Taiwan
David Geraint Hassell   Universiti Teknologi Brunei
Daphne Lai Teck Ching   Universiti Brunei Darussalam
Deenina Salleh   Universiti Teknologi Brunei, Brunei
Deshi Ye   Zhejiang University
Eran Edirisinghe   Loughborough University
Florence Choong Chiao Mei   Taylor's University, Malaysia
Greg Aloupis   Tufts University, USA
Lee-Kien Foo   Multimedia University, Malaysia

Shahid Anjum   Universiti Teknologi Brunei
Simon Colton   Goldsmiths College, University of London,
                                   UK
Simon Fong   University of Macau
Sung-Bae Cho   Yonsei University, Korea
Suresh Sankaranarayanan   SRM University, India
Tossapon Boongern   Royal Thai Air Force Academy, Thailand
Truong Ton Hien   Universiti Brunei Darussalam
Tutut Herewan   University of Malaya
Ukrit Watchareeruetai   King Mongkut Institute of Technology,
                                   Ladkrabang, Thailand
Werasak Kurutach   Mahanakorn University of Technology,
                                   Thailand
Wen-Chieh Lin   National Chiao Tung University
Weng-Kin Lai   TAR University College, Malaysia
Wida Susanty Suhaili   Universiti Teknologi Brunei
Worrawat Engchuan   Centre for Applied Genomics, Canada
Xujin Che   Chinese Academy of Sciences, Beijing,
                                   China
Yota Otachi   Japan Advanced Institute of Science
                                   and Technology, Japan
Yo-Sub Han   Yonsei University, Korea
Yuan-Hsuan Lee   National Taichung University of Education
Yun-Li Lee   Sunway University, Malaysia

Organizer

School of Computing and Informatics, Universiti Teknologi Brunei

Technical Sponsors

International Neural Network Society (INNS)
Brain Science Research Center (BRSC), KAIST
Asia Pacific Neural Network Society (APNNS)
CISCO
Google Developer Group (GDG) Brunei
BAG networks

Contents

Data Mining and Its Applications

Internetworking, Security and Internet of Things

Intelligent Systems and their Applications

On Using Genetic Algorithm for Initialising Semi-supervised Fuzzy c-Means Clustering

Daphne Teck Ching Lai[1(✉)] and Jonathan M. Garibaldi[2]

[1] Faculty of Science, Universiti Brunei Darussalam, Gadong BE1410, Brunei
daphne.lai@ubd.edu.bn
[2] School of Computer Science, University of Nottingham,
Nottingham NG8 1BB, UK
jon.garibaldi@nottingham.ac.uk

Abstract. In a previous work, suitable initialisation techniques were incorporated with semi-supervised Fuzzy c-Means clustering (ssFCM) to improve clustering results on a trial and error basis. In this work, we present a single fully-automatic version of an existing semi-supervised Fuzzy c-means clustering framework which uses genetically-modified prototypes (ssFCMGA). Initial prototypes are generated by GA to initialise the ssFCM algorithm without experimentation of different initialisation techniques. The framework is tested on a real, biomedical dataset NTBC and on the Arrhythmia UCI dataset, using varying amounts of labelled data from 10 % to 60 % of the total data patterns. Different ssFCM threshold values and fitness functions for ssFCMGA are also investigated (sGAs). We used accuracy and NMI to measure class-label agreement and internal measures WSS, BSS, CH, CWB, DB and DU to evaluate cluster quality of the clustering algorithms. Results are compared with those produced by the existing ssFCM. While ssFCMGA and sGAs produced slightly lower agreement level than ssFCM with known class labels based on accuracy and NMI, the other six measurements showed improvement in the results in terms of compactness and well-separatedness (cluster quality), particularly when labelled data are low at 10 %. Furthermore, the cluster quality are shown to further improve using ssFCMGA with a more complex fitness function (sGA2). This demonstrates the application of GA in ssFCM improves cluster quality without exploration of different initialisation techniques.

Keywords: Semi-supervised · Genetic algorithms · Fuzzy clustering

1 Introduction

Clustering is a pattern recognition approach for discovering natural and hidden groupings of similar data patterns, defined by a distance metric. One challenge in an unsupervised learning task such as clustering is that the solution varies according to the initial prototypes (cluster centres) chosen, often at random. One way is to use labelled data patterns as examples for guiding the clustering

© Springer International Publishing AG 2017
S. Phon-Amnuaisuk et al. (eds.), *Computational Intelligence in Information Systems,*
Advances in Intelligent Systems and Computing 532, DOI 10.1007/978-3-319-48517-1_1

of unlabelled ones, this is known as semi-supervision. Pedrycz and Waletzky [1] has applied semi-supervision in fuzzy clustering by using labels as known membership to clusters. As opposed to a binary approach, fuzzy clustering allows data points to belong to more than one cluster. This is considered a more realistic representation. Thus, semi-supervised fuzzy clustering takes the advantages of learning from available labels and using fuzzy membership to represent the degree of belongingness of data patterns to all clusters.

Initial prototypes affect the classification accuracy of clustering algorithms and thus, there are initialisation techniques available to improve accuracy [2–4]. Previously, ssFCM was investigated as an automatic (post-initialisation) technique [5] to improve accuracy rates where initialisation techniques were applied prior to ssFCM. However, the search for good initial prototypes was manual involving separate experimentation with different techniques; simple cluster seeking [2], KKZ [3] and subtractive clustering [4]. In this paper, our aim is to investigate the use of genetic algorithm (GA) in a fully automatic ssFCM framework to improve clustering results, investigating agreement level with known class labels and cluster quality in terms of compactness and well-separatedness. The idea is to use simple, existing GA operators to produce initial prototypes that will lead to better clustering results together with ssFCM rather than using ssFCM alone. As the clustering algorithm used is semi-supervised, the number of clusters is assumed to be the number of classes provided by known labels. Thus, this paper focuses on investigating the performance (accuracy rate and cluster quality) of ssFCM with the application of GA.

Färber et al. [6] warned against using class labels for evaluating clustering algorithms or using class labels in designing clustering algorithm to learn class structure instead of the internal structure of data. The class labels used in our study are for guiding the learning of internal structure of the dataset, and the labelled data patterns themselves undergo learning and gets updated at each iteration. For this reason, we use internal measures to evaluate cluster quality and thus, the internal structure, in addition to using external measures to evaluate class-label agreement in this investigation.

Hruschka et al. [7] wrote an extensive piece on the design of evolutionary algorithms for clustering, mostly applied in an unsupervised setting to improve Kmeans or FCM clustering [8]. Liu and Huang [9] proposed an evolutionary ssFCM algorithm applied to textual data, which is different to our study where an evolutionary technique was applied externally of ssFCM on numerical data. As part of continuing work [5], we explore external improvements to ssFCM because ssFCM is simple and performs well with suitable distance metrics.

The paper is organised as follows: We discuss ssFCM and GA methodologies used in the framework in Sect. 2. The experiments are presented Sect. 3, followed by results and discussion in Sect. 4 and conclusions in Sect. 5.

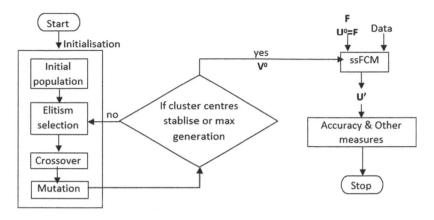

Fig. 1. Flowchart of ssFCMGA.

Algorithm 1. Semi-supervised fuzzy c-means [1].

1: Initialise c, labelled data membership matrix $\mathbf{F}$ and initial membership matrix $\mathbf{U^0}$
2: Calculate prototypes using

$$\mathbf{v}_i = \frac{\sum_{j=1}^{N} u_{ij}^2 \mathbf{x}_j}{\sum_{k=1}^{N} u_{ij}^2}, \ 1 \leq i \leq c. \tag{1}$$

3: Compute squared Euclidean dist. d_{ij}^2 between prototype $\mathbf{v}_i$ and data pattern $\mathbf{x}_j$.
4: Update partition matrix, $\mathbf{U}$ using equation :

$$u_{ij} = \frac{1}{1+\alpha} \left\{ \frac{1 + \alpha(1 - b_j \sum_{l=1}^{c} f_{lj})}{\sum_{l=1}^{c} \left(\frac{d_{ij}}{d_{lj}}\right)^2} + \alpha f_{ij} b_j \right\} \tag{2}$$

5: If $||\mathbf{U'} - \mathbf{U}|| < \epsilon$, stop. Else, go to Line 2 with $\mathbf{U} = \mathbf{U'}$

---

2 Methodology

In this section, a ssFCM framework using Genetic Algorithm (ssFCMGA) generated prototypes is presented. Figure 1 shows the flowchart of ssFCMGA where GA is run to supply the initial prototypes, $\mathbf{V^0}$ for ssFCM.

2.1 Semi-supervised Fuzzy C-Means

The objective function of ssFCM proposed by Pedrycz and Waletzky [1] contains unsupervised learning in the first term and supervised learning in the second term as follows:

$$J = \sum_{i=1}^{c} \sum_{j=1}^{N} u_{ij}^p d_{ij}^2 + \alpha \sum_{i=1}^{c} \sum_{j=1}^{N} (u_{ij} - f_{ij} b_j)^p d_{ij}^2, \tag{3}$$

where u_{ij} is the membership value of data pattern j in cluster i, c is the number of clusters, d_{ij} the distance (Euclidean) between data pattern j and prototype

P1	C1	1.5	1.3	1.2	2.0	0.5
	C2	0.2	1.7	1.1	0.8	0.3
P2	C1	1.7	1.8	2.1	2.2	1.9
	C2	1.4	1.1	1.3	0.9	0.1

Fig. 2. Matrix-based real encoding representing prototypes as solutions.

P1	C1	1.5	1.3	1.2	2.0	0.5
	C2	0.2	1.7	1.1	0.8	0.3
P2	C1	1.7	1.8	2.1	2.2	1.9
	C2	1.4	1.1	1.3	0.9	0.1

$\downarrow$

P1′	C1	1.5	1.3	2.1	2.2	1.9
	C2	0.2	1.7	1.3	0.9	0.1
P2′	C1	1.7	1.8	1.2	2.0	0.5
	C2	1.4	1.1	1.1	0.8	0.3

Fig. 3. One-point crossover in matrix-based real encoding.

v_i, f_{ij} the membership value of labelled data pattern j in cluster i, b_j indicates if data pattern j is labelled, p is the fuzzifier parameter (commonly 2) and α is a scaling parameter for maintaining balance between supervised and unsupervised learning components such that supervised learning does not dominate. The authors recommend α to be proportional to N/M, where M is the number of labelled data. The algorithm is summarised in Algorithm 1.

2.2   The Genetic Algorithm

Encoding Scheme. The potential solutions (initial prototypes) are represented with a matrix-based real encoding. Figure 2 shows an example of two solutions, P1 and P2 containing two prototypes each, C1 and C2, for a dataset with 5 features. The matrix-based encoding is chosen to ensure that the feature representation is not lost [7] when parts of the prototypes are swapped during crossover. To allow for slight mutation of values directly, the real encoding scheme is used. The genetic algorithm steps are as follow:

Step 1: Population Initialisation. At the first generation of the GA, the initial population of 50 cluster solutions are chosen randomly from the dataset. Smaller population sizes of 10 and 20 were tried in preliminary experiments but poor results were obtained.

Step 2: Selection Criteria. Each solution is evaluated using fitness function (3). Using elitism scheme, two solutions with the minimum J values in a pool of parent and children solutions are chosen for crossover and mutation.

Table 1. Dataset specifications showing number of data patterns (N), number of features (n) and number of classes (c).

Dataset	N	n	c
Nottingham Tenovus Breast Cancer (NTBC) [11]	663	25	6
Arrhythmia (Arrhy)	420	277	3

Table 2. Parameter settings for the experiment

Parameter	Setting
Population size	50
Max generation	20
Parameter threshold	10
Probability of crossover	1
Mutation frequency	Random number of times between 1 to n/10 for each data pattern where n is number of features
Number of runs	100
Percentage of labelled data	{10 %, 20 %, 30 %, 40 %, 50 %, 60 %}

Step 3: Crossover. A one-point crossover scheme is used, as shown in the example in Fig. 3. Feature values in column 3 to 5 in P1 has now been swapped into P2' and vice versa for P2. As the feature values in the original column stays, the context (feature representation) is not lost as explained earlier.

Step 4: Mutation. The prototypes are modified using an approach based on Maulik and Bandyopadhyay's work [10]. Rather than changing gene positions, we slightly change the values. In our case, every data patterns will undergo a random variable rate of mutation depending on the number of features, n, from 1 to n/10 times per data pattern. For each data pattern, the randomly chosen feature to mutate is based on the equation: $x'_l = x_l \pm \lambda$ where $0 < \lambda < 1$ and there is equal probability for the mutation to be an addition or subtraction.

Step 5: Termination Criteria The selected parents undergo crossover and mutation to produce children. Together, they form a new population. The best parent is selected from this population for the next generation, repeating Step 2 to 4. The algorithm terminates when the best parent has converged based on criteria: $||\mathbf{P}' - \mathbf{P}''|| < \delta$, where $\mathbf{P}'$ and $\mathbf{P}''$ are the best solutions found in the previous and current generations respectively. So far, δ was set to 10.

3   Experiments

The experiments are run 600 times for each data set, 100 times for each amount of labelled data setting. The datasets, Nottingham Tenovus Breast Cancer (NTBC) and Arrhythmia from UCI repository, are used and summarised in Table 1. Experimental settings used are summarised in Table 2. Normalised Mutual Index (NMI) [12] and accuracy (rate of true positives) were calculated based on matches between ssFCM clusters with class labels to measure agreement level with known labels. Internal measures within-sum-of-squares (WSS), between-sum-of-squares (BSS), Calinski-Harabasz index (CH) [13], Davies-Bouldin index (DB) [14], Compose Within and Between scattering validity index (CWB) [15] and Dunn index (DU) [16] were used to evaluate cluster quality in terms of compactness and well-separatedness.

The threshold ϵ value (Step 5 in Algorithm 1) for ssFCM and ssFCMGA is set to 0.05. ssFCMGA with ϵ values of 0.5 (sGA0.5) and 0.8 (sGA0.8) are explored to study their effects on class-label agreement and cluster quality. An experiment was also run using ssFCMGA with a fitness function that adds three functions, (3), DB and CWB with 0.05 ϵ value, referred as sGA2.

For WSS, CWB and DB, a smaller value is more favourable while for accuracy rate, NMI, BSS, CH and DU, a larger value is. Apart from accuracy rate and NMI, the range of the other measures are data-dependent and they serve as relative measures for cluster quality comparison between algorithms.

4   Results and Discussion

Figure 4 shows graphs of respective scores against varying amounts of labelled data for NTBC. ssFCMGA and ssFCM produced competitive accuracy and NMI values. Slight improvement is found in six measures (Fig. 4(c)–(h)), particularly at low amounts of labelled data 10 % and 20 %. At above 30 % labelled data, both ssFCMGA and ssFCM produced competitive values for these six measures. These trends were also found in Arrhythmia, as shown in Fig. 5. This shows that where labelled data is scarce at below 30 %, GA in ssFCMGA is able to find better initial clusters improves clustering result of existing ssFCM.

We increase ϵ value from 0.05 to 0.5 and 0.8 to study the effects on cluster quality and class-label agreement. In Fig. 6(a) and (b), we observed slight increase in class-label agreement but the cluster quality deteriorate drastically (see (c), (d) and (g)) for NTBC. However, both class-agreement and cluster quality improves with increasing amount of labelled data using ssFCM and all ssFCMGAs. In Fig. 7, no visible improvement in class-label agreement and cluster quality is observed to worsen (see (c) to (h)). With higher ϵ values, the algorithms sGA0.5 and sGA0.8 may not have converged despite achieving similar NMI and accuracy values as ssFCMGA.

Using the three combined function as a fitness function in sGA2, we observed further improved cluster quality for both datasets (the green dashed-dotted line with + in Figs. 6 and 7(c)–(h)) from using 10 % to 60 % labelled data. By combining the three functions, the different definitions of compactness and well-separatedness used in these functions are considered, further pushing the GA to search for cluster centres that will produce more compact and well-separated clusters. Table 3 showed significant improvement for sGA2, particularly with

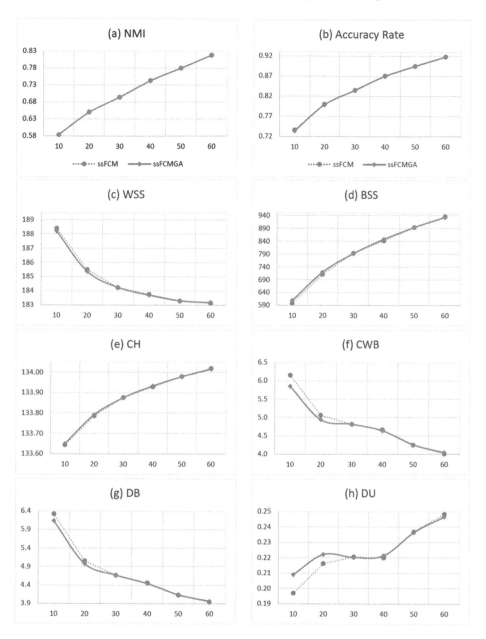

Fig. 4. Graphs of respective scores against percentage of labelled data for NTBC.

small amounts of labelled data, demonstrating the importance of GA to find good initial clusters where label data is scarce. Results are compared with Kmeans using applicable measures, shown in Figs. 6 and 7. While Kmeans show better cluster quality than sGA2 (see WSS, BSS, CH, DB and DU values) the agreeable

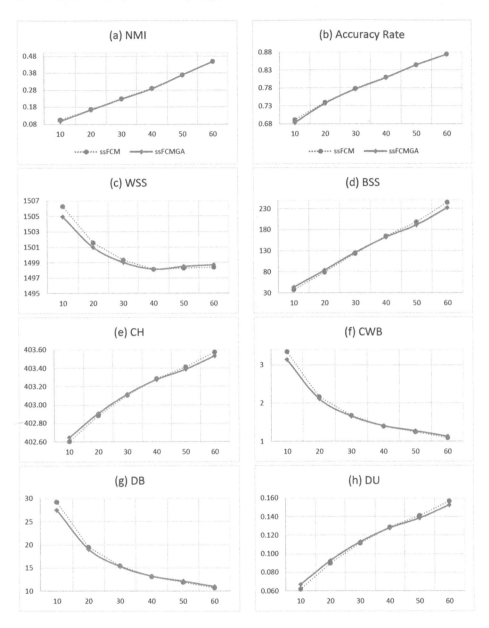

Fig. 5. Graphs of respective scores against percentage of labelled data for Arrhythmia.

level (NMI) is worse than ssFCM. sGA2, on the other hand, maintains competitive agreement levels with ssFCM while producing clusters of better quality than ssFCM. A poor agreement level may mean producing clusters that may not be meaningful or have relevance to the class structure.

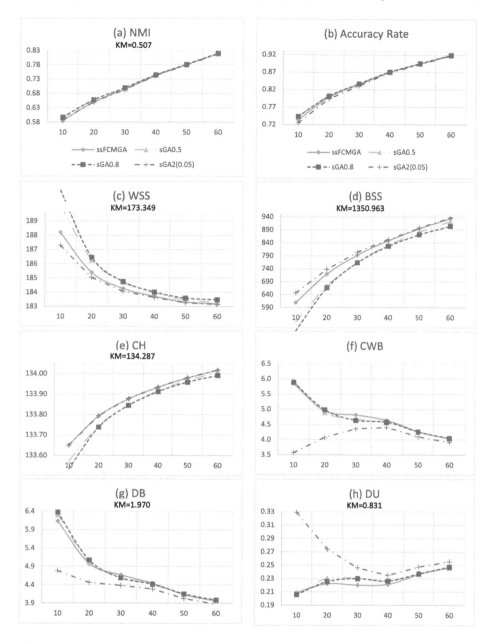

Fig. 6. Graphs comparing different ssFCMGAs for NTBC. Results from Kmeans (KM) are reported under graph title.

Table 4 shows the run time in seconds of the different ssFCM algorithms. It is obvious using more functions as fitness function requires more computational time, shown in sGA2.

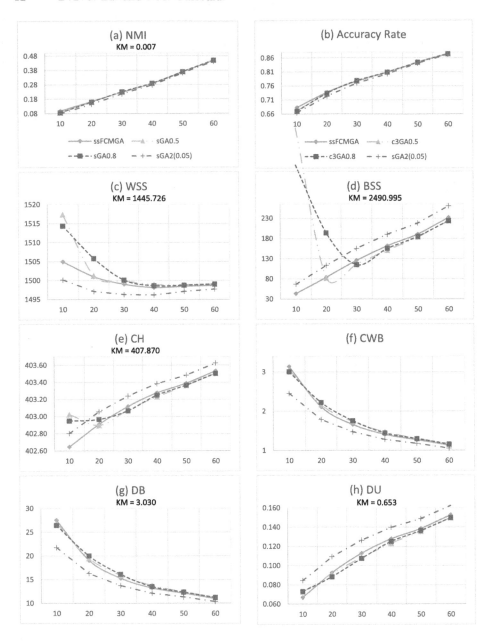

Fig. 7. Graphs comparing different ssFCMGAs for Arrhythmia. Results from Kmeans(KM) are reported under graph title.

Table 3. p-values obtained using Mann-Whitney test between ssFCM and the respective ssFCMGAs on NTBC dataset with varying amount of labelled data using WSS, BSS, CH, CWB, DB and DU measures.

NTBC							Arrhythmia					
ssFCMGAs	10 %	20 %	30 %	40 %	50 %	60 %	10 %	20 %	30 %	40 %	50 %	60 %
WSS												
ssFCMGA	0.001	+	−	+	−	−	0.997	0.938	0.834	0.911	0.951	0.809
sGA2(0.05)	0.001	+	+	+	+	−	0.440	0.506	0.481	0.554	0.583	0.539
BSS												
ssFCMGA	0.058	0.023	0.547	0.063	0.902	0.322	0.049	0.304	0.647	0.557	0.183	0.016
sGA2(0.05)	+	+	0.002	0.002	0.307	0.952	+	+	+	+	0.001	0.009
CH												
ssFCMGA	0.204	0.116	0.618	0.076	0.625	0.080	0.051	0.299	0.673	0.540	0.180	0.015
sGA2(0.05)	0.223	0.022	0.454	0.035	0.932	0.363	+	+	+	+	0.001	0.010
CWB												
ssFCMGA	0.197	0.583	0.997	0.919	0.892	0.605	0.053	0.296	0.663	0.539	0.179	0.015
sGA2(0.05)	+	+	0.002	0.032	0.070	0.163	+	+	+	+	0.001	0.010
DB												
ssFCMGA	0.173	0.577	0.943	0.877	0.850	0.547	0.052	0.312	0.635	0.560	0.187	0.016
sGA2(0.05)	+	+	+	0.013	0.030	0.125	+	+	+	+	0.001	0.008
DU												
ssFCMGA	0.169	0.510	0.989	0.882	0.879	0.673	0.041	0.286	0.582	0.663	0.185	0.026
sGA2(0.05)	+	+	+	0.008	0.024	0.083	+	+	+	+	0.001	0.006

− No improvement with p<0.001
+ Significant improvement with p<0.001

Table 4. Running time in seconds for ssFCM and the respective ssFCMGAs on NTBC dataset with varying amount of labelled data.

NTBC							Arrhythmia					
ssFCMs	10 %	20 %	30 %	40 %	50 %	60 %	10 %	20 %	30 %	40 %	50 %	60 %
ssFCM	1.6	1.2	1.2	1.0	1.0	0.9	0.2	0.2	0.2	0.2	0.2	0.2
ssFCMGA	420.1	417.8	423.1	425.2	426.5	426.0	64.1	57.8	62.0	63.5	68.0	67.3
sGA0.5	412.7	419.3	422.1	423.1	427.4	425.5	63.3	65.5	69.9	64.9	62.0	64.4
sGA0.8	425.5	428.5	429.3	431.5	435.5	435.4	63.4	62.1	71.8	71.3	68.9	67.6
sGA2(0.05)	3694.8	3094.1	2546.8	2148.5	2050.5	1987.3	334.7	309.5	319.6	322.9	312.3	284.1

5 Conclusion

In this study, two datasets were experimented in detail using ssFCMGA. ssFCMGA, particularly sGA2 generated good initial cluster centres to produce clusters of higher quality than ssFCM, particularly when labelled data are scarce (below 30 %). This was achieved automatically without having to experiment using different initialisation techniques. Furthermore, sGA2 perform as competitively well as ssFCM in producing clusters that agree with class labels. While Kmeans produced clusters of better quality than the ssFCMGAs, the agreement levels with class labels are lower which may signify good quality clusters that may less meaningful or have relevance.

One limitation in our study is that only two datasets are studied. As future work, we would study more datasets. In addition, we would like to explore multi-objective evolutionary methods for semi-supervised clustering.

Acknowledgements. This work was supported by the Universiti Brunei Darussalam under Grant UBD/PNC2/2/RG/1(311).

References

1. Pedrycz, W., Waletzky, J.: Fuzzy clustering with partial supervision. IEEE Trans. Syst. Man Cybern. **27**(5), 787–795 (1997)
2. Tou, J., Gonzales, R.: Pattern Recognition Principles. Addison-Wesley, Reading (1974)
3. Katsavounidis, I., Kuo, C.C.J., Zhang, Z.: A new initialization technique for generalized lloyd iteration. Sig. Process. Lett. **1**(10), 144–146 (1994)
4. Chiu, S.: Fuzzy model identification based on cluster estimation. J. Intell. Fuzzy Syst. **2**, 267–278 (1994)
5. Lai, D.T.C., Garibaldi, J.M.: Investigating distance metrics in semi-supervised fuzzy c-means for breast cancer classification. In: Peterson, L.E., Masulli, F., Russo, G. (eds.) CIBB 2012. LNCS, vol. 7845, pp. 147–157. Springer, Heidelberg (2013). doi:10.1007/978-3-642-38342-7_13
6. Färber, I., Günnemann, S., Kriegel, H.P., Kröger, P., Müller, E., Schubert, E., Seidl, T., Zimek, A.: On using class-labels in evaluation of clusterings. In: MultiClust: 1st International Workshop on Discovering, Summarizing and Using Multiple Clusterings held in Conjunction with KDD, p. 1 (2010)
7. Hruschka, E.R., Campello, R.J., Freitas, A.A., De Carvalho, A.C., et al.: A survey of evolutionary algorithms for clustering. IEEE Trans. Syst. Man Cybern. Part C: Appl. Rev. **39**(2), 133–155 (2009)
8. Wikaisuksakul, S.: A multi-objective genetic algorithm with fuzzy c-means for automatic data clustering. Appl. Soft Comput. **24**, 679–691 (2014)
9. Liu, H., Huang, S.T.: Evolutionary semi-supervised fuzzy clustering. Pattern Recogn. Lett. **24**, 3105–3113 (2003)
10. Maulik, U., Bandyopadhyay, S.: Genetic algorithm-based clustering technique. Pattern Recogn. **33**(9), 1455–1465 (2000)
11. Soria, D., Garibaldi, J.M., Ambrogi, F., Green, A.R., Powe, D., Rakha, E., Macmillan, R.D., Blamey, R.W., Ball, G., Lisboa, P.J., Etchells, T.A., Boracchi, P., Biganzoli, E., Ellis, I.O.: A methodology to identify consensus classes from clustering algorithms applied to immunohistochemical data from breast cancer patients. Comput. Biol. Med. **40**(3), 318–330 (2010)
12. Strehl, A., Ghosh, J.: Cluster ensembles - a knowledge reuse framework for combining multiple partitions. J. Mach. Learn. Res. **3**, 583–617 (2003)
13. Caliński, T., Harabasz, J.: A dendrite method for cluster analysis. Commun. Stat. **3**(1), 1–27 (1974)
14. Davies, D.L., Bouldin, D.W.: A cluster separation measure. IEEE Trans. Pattern Anal. Mach. Intell. **1**, 224–227 (1979)
15. Rezaee, M.R., Lelieveldt, B., Reiber, J.: A new cluster validity index for the fuzzy c-mean. Pattern Recogn. Lett. **19**(3–4), 237–246 (1998)
16. Dunn, J.C.: A fuzzy relative of the isodata process and its use in detecting compact well-separated clusters. J. Cybern. **3**(3), 32–57 (1973)

Estimation of Confidence-Interval for Yearly Electricity Load Consumption Based on Fuzzy Random Auto-Regression Model

Riswan Efendi[1,2(✉)], Nureize Arbaiy[1], and Mustafa Mat Deris[1]

[1] Faculty of Computer Science and Information Technology,
University Tun Hussein Onn Malaysia, 86400 Batu Pahat, Malaysia
{riswan,nureize,mmustafa}@uthm.edu.my
[2] Mathematics Department, UIN Sultan Syarif Kasim Riau,
Pekanbaru, Indonesia

Abstract. Many models have been implemented in the energy sectors, especially in the electricity load consumption ranging from the statistical to the artificial intelligence models. However, most of these models do not consider the factors of uncertainty, the randomness and the probability of the time series data into the forecasting model. These factors give impact to the estimated model's coefficients and also the forecasting accuracy. In this paper, the fuzzy random auto-regression model is suggested to solve three conditions above. The best confidence interval estimation and the forecasting accuracy are improved through adjusting of the left-right spreads of triangular fuzzy numbers. The yearly electricity load consumption of North-Taiwan from 1981 to 2000 are examined in evaluating the performance of three different left-right spreads of fuzzy random auto-regression models and some existing models, respectively. The result indicates that the smaller left-right spread of triangular fuzzy number provides the better forecast values if compared with based line models.

Keywords: Fuzzy random variable · Auto-regression model · Left-right spread · Triangular fuzzy number · Forecasting error · Electricity

1 Introduction

The decision makers and researchers should pay attention seriously to enhance the studies in organizing and managing the electricity load demand and consumption, respectively. The output of these studies is very determinative for energy planning and power management. Additionally, load forecasting helps an electric utility to make important decisions including decisions on purchasing and generating electric power, load switching, and infrastructure development [1].

Forecasting is a predictive analytical technique that deals with estimation the future, generally by considering the past data sets and models. It can be applied in various domains of management, finance-economic, energy, engineering, computer science, and others. In electricity forecasting, among the models frequently used for electricity forecasting are autoregressive integrated moving average (ARIMA), regression time series, time series, genetic algorithm (GA), artificial neural network (ANN), and

© Springer International Publishing AG 2017
S. Phon-Amnuaisuk et al. (eds.), *Computational Intelligence in Information Systems*,
Advances in Intelligent Systems and Computing 532, DOI 10.1007/978-3-319-48517-1_2

particle swarm optimization (PSO) [1]. In this decade, the implementation of fuzzy theories with regression and time series are frequently used to forecast the electricity load consumptions by researchers [2–8].

In the electricity forecasting models, the accuracy of forecasted values is still in issue and very important. Because, not easy to get the historical data accurately and many factors may influence the behavior of electricity load data. Moreover, the randomness and fuzziness of these data play the important role. To solve both conditions, the fuzzy random regression and auto-regression models and its applications have been introduced [9, 10, 15, 16].

From [9, 10], we are interested to modify some aspects such as the formatting of fuzzy data and the left-right spreads (LRS) of TFN in this paper. Both aspects are very essential to be considered in improving of the estimated confidence interval (CI) performance and the forecasting accuracy of fuzzy random auto-regression (FR-AR) model. The rest of paper is organized as follows: In Sect. 2, the theories of fuzzy random variable (FRV) and fuzzy random auto-regression (FR-AR) are described. The proposed ideas are presented in Sect. 3. In Sect. 4, the empirical analysis of electricity load consumption are discussed. In the end of this paper, the conclusion is mentioned briefly.

2 Fundamental Theories of Fuzzy Random Variable and Fuzzy Random Auto-Regression Model

In this section, there are two fundamental theories, namely, fuzzy random variable and fuzzy random auto-regression. Both theories are very important in building the proposed procedure of LRS of TFN for FR-AR model as described in Sects. 2.1 and 2.2.

2.1 Fuzzy Random Variables

Suppose some universe Γ, let Pos be a possibility measure that is defined on the power set $P(\Gamma)$ of Γ. Let R be the set of real numbers. A function $Y : \Gamma \rightarrow R$ is said to be a fuzzy variable defined on Γ [11]. The possibility distribution μ_Y of Y is defined by $\mu_Y(t) = Pos\{Y = t\}, t \in R$, which is the possibility of event $\{Y = t\}$. For fuzzy variable Y, with possibility distribution μ_Y, the possibility, necessity, and credibility of event $Y \leqslant r$ are given as follows:

$$Pos\{Y \leq r\} = \sup \mu_Y(t), \ t \leq r, \tag{1}$$

$$Nec\{Y \leq r\} = 1 - \sup \mu_Y(t), \ t \geq r, \tag{2}$$

$$Cr\{Y \leq r\} = \frac{1}{2}(1 + \sup t \leq r \, \mu_Y(t) - \sup t \geq r \, \mu_Y(t)). \tag{3}$$

The credibility measure is an average of the possibility and the necessity measures from Eq. (3), i.e., $Cr\{.\} = \frac{Pos\{.\} + Nec\{.\}}{2}$. The motivation behind the introduction of the credibility measure is to develop a certain measure, which is a sound aggregate of the two extreme cases, such as the possibility (which expresses a level of overlap and highly optimistic in this sense) and necessity (that articulates a degree of inclusion and is pessimistic in its nature). Based on credibility measure, the expected value of fuzzy variable is presented as follows.

Definition 1. Expected value of fuzzy variable [12]
Let Y be a fuzzy variable. The expected value of Y is defined as:

$$E(Y) = \int Cr\{Y \geq r\}dr - \int Cr\{Y \leq r\}dr, \tag{4}$$

under the condition that the two integral are finite. Assume that $Y = [a^l, c, a^r]_T$ is triangular fuzzy variable (TFV = TFN) whose possibility distribution is given by

$$\mu_Y(t) = \begin{cases} \frac{x-a^l}{c-a^l}, & if\ a^l \leq x \leq c \\ \frac{a^r-x}{a^r-c}, & if\ c \leq x \leq a^r, \\ 0, & otherwise \end{cases} \tag{5}$$

Making use of Eq. (4), the expected value of Y can be written as

$$E(Y) = \frac{(a^l + 2c + a^r)}{4}. \tag{6}$$

Definition 2. Fuzzy random variable [13]
Suppose that (Ω, Σ, Pr) is a probability space and F_v is a collection of fuzzy variables defined on possibility space $(\Gamma, P(\Gamma), Pos)$. A fuzzy random variable is a mapping $X : \Omega \to F_v$ such that for any Borel subset B of R, $Pos\{X(\omega) \in B\}$ is a measurable function of ω.

Let X be a fuzzy random variable on Ω. From the previous definition, we know, for each $\omega \in \Omega$, that $X(\omega)$ is a fuzzy variable. Moreover, a fuzzy random variable X is said to be positive if, for almost every ω, fuzzy variable $X(\omega)$ is positive almost surely. For any fuzzy random variable X on Ω, for each $\omega \in \Omega$, the expected value of the fuzzy variable $X(\omega)$ is denoted by $E(X(\omega))$, which has been proved to be a measurable function of ω [13], i.e., it is random variable. Given this, the expected value of the fuzzy random variable X is defined as the mathematical expectation of the random variable $E(X(\omega))$.

Definition 3. Expected value of fuzzy random variable [13]
Let X be a fuzzy random variable defined on probability space (Ω, Σ, Pr). Then, the expected value of X and variance of X are defined as

$$E(X) = \int \Omega[\int Cr\{\xi(\omega) \geq r\}dr - \int Cr\{\xi(\omega) \leq r\}dr] Pr(\omega), \tag{7}$$

$$Var(X) = E(X-e)^2, \tag{8}$$

where $e = E(X)$ is given by Eq. (7).

2.2   Fuzzy Random Auto-Regression (*FR-AR*) Model

In time series, autoregressive or AR(p) model can be written as [14]:

$$Y_t = \emptyset_1 Y_{t-1} + \emptyset_2 Y_{t-2} + \cdots + \emptyset_p Y_{t-p} + e_t, \tag{9}$$

where $\emptyset_1, \ldots, \emptyset_p$ are coefficients of $Y_{t-1}, \ldots, Y_{t-p}$, respectively, e_t is an error models at time-t. From [10], the fuzzy random auto-regression (FR-AR) model can be defined as input and output data Y_{t-p} for all $p = 0, 1, 2, \ldots, n$ are fuzzy random variables, which are written as:

$$Y_t = \bigcup_{i=1}^{n} \left[\left(Y_{it}^l, Y_{it}^c, Y_{it}^r \right)_T, P_{it} \right], \tag{10}$$

where Y_t is a time series data at time-t and its formatted as a triangular fuzzy number [left, l; center, c; right, r]. From Eq. (10), all values given as fuzzy numbers with probabilities, P_{it}. These data, Y_t also can be presented in Table 1.

Table 1.   Fuzzy random input-output time series data

Time/Sample	Output	Input
0	Y_t	Y_{t-1} Y_{t-2} ... Y_{t-k}
1	Y_{t-1}	Y_{t-2} Y_{t-3} ... $Y_{t-(k+1)}$
2	Y_{t-2}	Y_{t-3} Y_{t-4} ... $Y_{t-(k+2)}$
...	...	... ...
n	Y_{t-n}	$Y_{t-(n+1)}$ $Y_{t-(n+2)}$ ... $Y_{t-(k+n)}$

Let a simple FR-AR model with coefficients $\left[\emptyset_1^l, \emptyset_1^r \right]$ and $\left[\emptyset_2^l, \emptyset_2^r \right]$ can be written as:

$$Y_t = \left[\emptyset_1^l, \emptyset_1^r \right] Y_{t-1} + \left[\emptyset_2^l, \emptyset_2^r \right] Y_{t-2} + \left[e_t^l, e_t^r \right], \tag{11}$$

To estimate CI of both coefficients in Eq. (11) can be derived by following steps:

Step 1: Provide the real time series data in the fuzzy data format [min, max] per interval time-t, such as, per one week, per one month, etc. For example, week-1; [3020, 3050], week-2; [3000, 3057], etc.

Step 2: Divide the fuzzy data into the fuzzy random data [min, center, right] with probabilities. For example, week-1; FRD1 = [3020, 3030, 3040], Pr1 = 0.4 and FRD2 = [3030, 3040, 3050], Pr2 = 0.6.

Step 3: Calculate the expected value (*EV*) and standard deviation (*Std.Dev*) of fuzzy random data (FRD) in Step 2, respectively.

$$EV = E(Y) = (\text{Center of FRD1} \times \text{Pr1}) + (\text{Center of FRD2} \times \text{Pr2})$$
$$= (3030 \times 0.4) + (3040 \times 0.6)$$
$$= 3036$$

$$\text{Variance}(Y) = E(Y - e)^2$$

Standard deviation $(Std.Dev) = s(Y) = \sqrt{\text{Variance}(Y)} = 7.4$

Step 4:   Determine the confidence interval (CI) of FRD. For example,
Week -1 : $[(EV - Std.Dev), (EV + Std.Dev)] = [3028.6, 3043.4]$.

Step 5:   Estimate CI for each coefficient model by using linear programming (LP) approach.

Objective function: $\min J(\emptyset) = \sum_{i=1}^{n} (\emptyset_i^r - \emptyset_i^l)$,

Subject to
$\emptyset_i^r \geq \emptyset_i^l$

$$a_1\emptyset_{11}^l + \left(a_1 + \frac{1}{3}l\right)\emptyset_{12}^l \leq E_1(Y) - Std.Dev_1(Y)$$

$$a_2\emptyset_{21}^l + \left(a_2 + \frac{1}{3}l\right)\emptyset_{22}^l \leq E_2(Y) - Std.Dev_2(Y)$$

$$\vdots$$

$$a_n\emptyset_{n1}^l + \left(a_n + \frac{1}{3}l\right)\emptyset_{n2}^l \leq E_n(Y) - Std.Dev_n(Y)$$

and

$$\left(a_1 + \frac{2}{3}l\right)\emptyset_{11}^r + (b_1)\emptyset_{12}^r \geq E_1(Y) + Std.Dev_1(Y)$$

$$\left(a_2 + \frac{2}{3}l\right)\emptyset_{21}^r + (b_2)\emptyset_{22}^r \geq E_2(Y) + Std.Dev_2(Y)$$

$$\vdots$$

$$\left(a_n + \frac{2}{3}l\right)\emptyset_{n1}^r + (b_n)\emptyset_{n2}^r \geq E_n(Y) + Std.Dev_n(Y)$$

Step 6:   From Step 5, define the estimated confidence-interval (CI) for each coefficient model.

$$\hat{Y}_t = \left[\hat{\emptyset}_1^l, \hat{\emptyset}_1^r\right]Y_{t-1} + \left[\hat{\emptyset}_2^l, \hat{\emptyset}_2^r\right]Y_{t-2}$$

3   Proposed LRS of TFN in Estimating Confidence-Interval of FR-AR Model

In fuzzy random auto-regression model, the left-right spreads (LRS) of TFN are very important to be considered, because their contributions are very significant in reducing the length of confidence-interval (CI) and the forecasting error. In this paper, the main motivation is to investigate the effect of various LRS in achieving the high forecasting accuracy and to introduce a new formatting of fuzzy data which not clearly described in the previous studies. The forecasting procedure can be derived by following steps:

Step 1: Define the new data format. We suggest to transform the real data into TFN by using various L-R spreads ($\pm k$).
Real data $\rightarrow$ TFN : $Y_t \rightarrow [Y_t - k, Y_t, Y_t + k]$, $k = 5, \ldots, 10$
$3000 \rightarrow [2990, 3000, 3010]$, if $k = 10$

Step 2: Define the real data in Step 1 as new fuzzy data $[Y_t - k, Y_t + k]$.
Year-t: $3000 \rightarrow [2990, 3010]$

Step 3: Divide fuzzy data (FD) into FRD1 and FRD2 as described in Sect. 2.
FRD1: [2990, 2996.66, 3003.33], FRD2: [2996.66, 3003.33, 3010]

Step 4: Calculate EV and $Std.Dev$ of FRD.

Step 5: Determine CI of FRD.

Step 6: Estimate coefficients FR-AR(p) model using LP.

Step 7: Determine the estimated CI for each coefficient model.

Step 8: Change $k = 6, 7,..., 10$ and repeat Steps 1–7.

Step 9: Find and state the best coefficients model based on various k.

The effect of LRS (k) to the forecasting accuracy can be explained as follows: Since $k_1 < k_2 < k_3 < \ldots < k_n$. Thus, the area of triangles can be written as:

$$A_1 < A_2 < \cdots < A_n, \tag{12}$$

By using Eq. (12), $E(Y)$ and $Var(Y)$, of the fuzzy random variables can be written as:

$$E_1(Y) < E_2(Y) < \cdots < E_n Y, \tag{13}$$

and

$$Var_1(Y) < Var_2(Y) < \cdots < Var_n(Y), \tag{14}$$

Thus, the confidence intervals of fuzzy random variables (FRDs) can be written as:

$$\left[\left(E_1(Y) - \sqrt{Var_1(Y)}, E_1(Y) + \sqrt{Var_1(Y),}\right)\right], \ldots, \left[\left(E_n(Y) - \sqrt{Var_n(Y)}, E_1(Y) + \sqrt{Var_n(Y),}\right)\right], \tag{15}$$

From Eq. (15), the range of FRDs can be denoted as:

$$R_1(FRD) < R_2(FRD) < \cdots < R_n(FRD), \tag{16}$$

By using (16), the range of FRD decrease gradually by following values of k (LRS). Therefore, the smaller k will produces the better coefficients of FR-AR model. From this equation, we can claim that the adjusting of LRS is very important in improving of forecasting accuracy.

Table 2. Actual and TFN electricity load data with $k = 10$

Year	Actual data	TFN data
1981	3388	[3378, 3388, 3398]
1982	3523	[3513, 3523, 3533]
1983	3752	[3742, 3752, 3762]
...	...	...
2000	12924	[12914, 12924, 12934]

Table 3. Yearly electricity load of fuzzy data

Year	Fuzzy data
1981	[3378, 3398]
1982	[3513, 3533]
...	...
2000	[12914, 12934]

Table 4. FRD of electricity load consumption

Year	FRD-1	Pr-1	FRD-2	Pr-2
1981	[3378, 3384.6, 3391.3]	0.4	[3384.6, 3391.3, 3398]	0.6
1982	[3513, 3519.6, 3526.3]	0.2	[3519.6, 3526.3, 3533]	0.8
...	...	...	...	...
2000	[12914, 12920.6, 12927.3]	0.1	[12920.6, 12927.3, 12934]	0.9

Table 5. EV and SD of FRD-1 and FRD-2

Year	Expected value	Standard deviation
1981	3386.67	4.3
1982	3523.00	4.6
...	...	...
2000	12926.67	3.6

Table 6. CI of FRD1 and FRD2

Year	Confidence intervals
1981	[3382.4, 3390.9]
1982	[3518.4, 3527.6]
...	...
2000	[12923, 12930]

4 Empirical Analysis

In this section, the various LRS of TFN are examined to investigate the best CI of the yearly electricity load consumption of North-Taiwan, the period 1981 to 2000 [6, 7] which are used as model building. By using the proposed algorithm given in Sect. 3, the estimated CI for model's coefficients can be calculated as follows:

Step 1:   Transform the yearly electricity load consumption into TFN format as shown in Table 2. In this paper, we examine $k = 10, 8, 5$.

Step 2:   Define the fuzzy data using TFN in Step 1 as shown in Table 3.

Step 3:   Divide fuzzy data (FD) in Step 2 into FRD1 and FRD2 with probabilities as shown in Table 4.

Step 4:   Calculate EV and SD of FRD of electricity load consumption as shown in Table 5.

Step 5:   Determine CI of FRD as presented in Table 6.

Step 6:   Estimate coefficients FR-AR(p) model using LP.
$$\text{Min} = ((\alpha_t)_{T, r} - (\alpha_t)_{T, 1}) + ((\delta_t)_{T, r} - (\delta_t)_{T, 1}), (\alpha_t)_{T, r} \geq (\alpha_t)_{T, 1}, (\delta_t)_{T, r} \geq (\delta_t)_{T, 1}.$$
Subject to
Inequalities of Left-LP:
$$3378(\alpha_t)_{T, 1} + 3384.6(\delta_t)_{T, 1} \leq 3382.4$$
$$3513(\alpha_t)_{T, 1} + 3519.0(\delta_t)_{T, 1} \leq 3518.4$$
… … …
$$12914(\alpha_t)_{T, 1} + 12920(\delta_t)_{T, 1} \leq 12923$$
Inequalities of Right-LP:
$$3391.3(\alpha_t)_{T, r} + 3398(\delta_t)_{T, r} \leq 3390.9$$
$$3526.3(\alpha_t)_{T, r} + 3533(\delta_t)_{T, r} \leq 3527.6$$
… … …
$$12927.3(\alpha_t)_{T, r} + 12934(\delta_t)_{T, r} \leq 12930$$
$$(\alpha_t)_{T, 1} \geq 0, (\alpha_t)_{T, r} \geq 0, (\delta_t)_{T, 1} \geq 0, (\delta_t)_{T, r} \geq 0$$

Step 7:   Write the estimated CI for each model with $k = 5, 8, 10$ in Table 7.

Table 7. The estimated of model coefficients

k	Model-1
10	$\alpha = (\alpha_t)_{T, 1} = (\alpha_t)_{T, r} = 0.598$
	$\delta = (\delta_t)_{T, 1} = (\delta_t)_{T, r} = 0.401$
k	Model-2
8	$\alpha = (\alpha_t)_{T, 1} = (\alpha_t)_{T, r} = 0.615$
	$\delta = (\delta_t)_{T, 1} = (\delta_t)_{T, r} = 0.384$
k	Model-3
5	$\alpha = (\alpha_t)_{T, 1} = (\alpha_t)_{T, r} = 0.624$
	$\delta = (\delta_t)_{T, 1} = (\delta_t)_{T, r} = 0.375$

Remark: $(\alpha_t)_{T, 1} = (\alpha_t)_{T, r} = \left[\emptyset_1^l, \emptyset_1^r\right]$, $(\delta_t)_{T, 1} = (\delta_t)_{T, r} = \left[\emptyset_2^l, \emptyset_2^r\right]$

Table 8. Actual, Forecasted Values and MSE using M1 – M3 models

Year	North	M1	M2	M3
1981	3388	3384.0	3384.0	3384.2
1982	3523	3518.8	3518.9	3519.1
1983	3752	3747.6	3747.6	3747.8
1984	4296	4291.0	4291.1	4291.3
1985	4250	4245.1	4245.1	4245.3
1986	5013	5007.3	5007.4	5007.6
1987	5745	5738.6	5738.6	5738.8
1988	6320	6313.0	6313.1	6313.3
1989	6844	6836.5	6836.5	6836.7
1990	7613	7604.7	7604.8	7605.0
1991	7551	7542.8	7542.8	7543.0
1992	8352	8343.0	8343.0	8343.2
1993	8781	8771.6	8771.6	8771.8
1994	9400	9389.9	9390.0	9390.2
1995	10254	10243.1	10243.1	10243.3
1996	11222	11210.1	11210.2	11210.4
1997	10719	10707.6	10707.7	10707.9
1998	11642	11629.7	11629.7	11629.9
1999	11981	11968.4	11968.4	11968.6
2000	12924	12910.4	12910.5	12910.7
MSE		78.6	77.9	74.6

Mathematically, the predicted Models 1–3 can be written as:

$$Y_{1t} = 0.598Y_{t-1} + 0.401Y_{t-2}, \qquad (17)$$

$$Y_{2t} = 0.615Y_{t-1} + 0.384Y_{t-2}, \qquad (18)$$

$$Y_{3t} = 0.624Y_{t-1} + 0.305Y_{t-2}, \qquad (19)$$

Step 9:   Find and state the best coefficients model based on various k. By using Eqs. (17–19), the comparison of forecasting errors are measured using

Table 9. Comparison MAPE between FR-AR and the Existing Models

Model	MAPE (%)
Support Vector Regression (SVR-CAS)	1.30
SVR-CGA	1.35
SVR-CPSO	1.31
Artificial Neural Network (ANN)	1.06
Regression	2.46
Fuzzy Time Series (FTS)	1.42
FR-AR (Proposed LRS with $k = 5$)	0.10[a]

[a]Smallest MAPE.

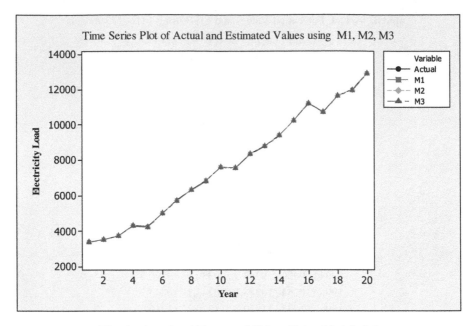

Fig. 1. Actual and Forecasted Values Using Models 1-3

mean square error (MSE) from three different models can be shown in Table 8.

From Table 8, Model-3 (M3) indicates the smaller MSE as compared with M1 and M2 in term of forecasting accuracy. Through this model, the estimated of CI with $k = 5$ is better than $k = 10$ and $k = 8$, respectively. The decreasing of k contributes to reduce the forecasting error, thus, the forecasting accuracy can be improved significantly. Moreover, the time series plot between actual electricity load consumption and its forecasted values are also illustrated in Fig. 1 by using Models 1–3.

Figure 1 shows the forecasted values which derived by M1, M2 and M3 are not too much different. Thus, the graphs of actual and models look like similar in this figure.

Furthermore, the comparison of mean absolute percentage error (MAPE) is also presented with the existing models in Table 9.

Table 9 indicates the proposed LRS with $k = 5$ has smaller MAPE as compared with existing models. Our proposed LRS is able to achieve the higher level forecasting significantly. From this table, the contribution of smaller LRS is very satisfactory in reducing the forecasting error of FR-AR model.

5 Conclusion

The new formatting of the real time series data into the fuzzy data has been introduced in this paper clearly. Moreover, in achieving the higher forecasting accuracy of FR-AR model, we adjusted the left-right spreads of TFN. The smaller of LRS in TFN is a promising procedure to achieve the best estimated confidence-interval (CI) which shown by MSE of three different models (M1, M2, M3). Furthermore, the comparison MAPE with existing models is also done in this paper, the result indicates the forecasting error which obtained by proposed LRS is better than others. From this study, the increasing of LRS in TFN will increase the forecasting error also. Finally, the further study should be completely investigated with various k and others time series data in determining the smaller LRS of TFN.

Acknowledgment. The authors are grateful to Research and Innovation Fund, UTHM for their financial support.

References

1. Ucal, I., Oztaysi, B.: Forecasting energy demand using fuzzy seasonal time series. In: Kahraman, C. (ed.) Computational Intelligence Systems in Industrial Engineering. Atlantis Computational Intelligence Systems, pp. 251–269. Atlantis Press, Amsterdam (2012)
2. Pai, P.F.: Hybrid ellipsoidal fuzzy systems in forecasting regional electricity load. Energ. Convers. Manag. **47**(15–16), 2283–2289 (2006)
3. Bolturuk, E., Oztayzi, B., Sari, I.U.: Electricity consumption forecasting using fuzzy time series. In: Proceeding 13th IEEE International Symposium on Computer Intelligence and Informatics, Istanbul, Turkey, pp. 245–249. IEEE Press (2012)
4. Azadeh, A., Saberi, M., Gitiforouz, A.: An integrated simulation-based fuzzy regression-time series algorithm for electricity consumption estimation with non-stationary data. J. Chin. Inst. Eng. **34**(8), 1047–1066 (2012)
5. Efendi, R., Ismail, Z., Deris, M.M.: New linguistic out-sample approach of fuzzy time series for daily forecasting of Malaysian electricity load demand. App. Soft. Comput. **28**, 422–430 (2015)
6. Ismail, Z., Efendi, R., Deris, M.M.: Application of fuzzy time series approach in electric load forecasting. New Math. Nat. Comput. **11**(3), 229–248 (2015)
7. Efendi, R., Ismail, Z., Deris, M.M.: A reversal model of fuzzy time series in regional load forecasting. Int. J. Energ. Stat. **3**(1), 1–17 (2015)
8. Efendi, R., Ismail, Z., Deris, M.M.: Implementation of fuzzy time series in forecasting of the non-stationary data. Int. J. Comput. Intell. Appl. **15**(2), 1–10 (2016)
9. Watada, J., Wang, S., Pedrycz, W.: Building confidence-interval-based fuzzy random regression models. IEEE Trans. Fuzzy Syst. **17**(6), 1273–1283 (2009)
10. Shou, L., Tsai, Y.-H., Watada, J., Wang, S.: Building fuzzy random auto-regression model and its application. In: Watada, J., Watanabe, T., Phillips-Wren, G., Howlett, R.J., Jain, L.C., et al. (eds.) Intelligent Decision Technologies. SIST, vol. 68, pp. 155–164. Springer, Heidelberg (2012)
11. Nahmias, S.: Fuzzy variables. Fuzzy Sets Syst. **1**(2), 97–111 (1978)

12. Liu, B., Liu, Y.K.: Expected value of fuzzy variable and fuzzy expected value models. IEEE Trans. Fuzzy Syst. **10**(4), 445–450 (2002)
13. Liu, Y.K., Liu, B.: Fuzzy random variable: a scalar expected value operator. Fuzzy Optim. Decis. Making **2**(2), 143–160 (2003)
14. Hanke, J.E., Wichern, D.W.: Business Forecasting. Prentice Hall, London (2009)
15. Nureize, A., Watada, J.: Linear fractional programming for fuzzy random based possibilistic programming problem. Int. J. Simul. Syst. Sci. Technol. **14**(2), 24–30 (2014)
16. Nureize, A., Watada, J.: Constructing fuzzy random goal constraints for stochastic goal programming. In: Huynh, V.-N., et al. (eds.) Integrated Uncertainty Management and Applications. AISC, vol. 68, pp. 293–304. Springer, Heidelberg (2010)

Improved Discrete Bacterial Memetic Evolutionary Algorithm for the Traveling Salesman Problem

Boldizsár Tüű-Szabó[1(✉)], Péter Földesi[2], and László T. Kóczy[1,3]

[1] Department of Information Technology,
Széchenyi István University, Győr, Hungary
tszboldi@gmail.com, koczy@sze.hu
[2] Department of Logistics, Széchenyi István University, Győr, Hungary
foldesi@sze.hu
[3] Department of Telecommunications and Media Informatics,
Budapest University of Technology and Economics, Budapest, Hungary

Abstract. In recent years a large number of evolutionary and other population based heuristics were proposed in the literature for solving NP-hard optimization problems. In 2015 we presented a Discrete Bacterial Memetic Evolutionary Algorithm (DBMEA) for The Traveling Salesman Problem. It provided results tested on series of TSP problems. In this paper we present an improved version of the DBMEA algorithm, where the local search is accelerated, which is the most time consuming part of the original DBMEA algorithm. This modification led to a significant improvement, the runtime of the improved DBMEA was 5–20 times shorter than the original DBMEA algorithm. Our DBMEA algorithms calculate real value costs better than integer ones, so we modified the Concorde algorithm be comparable with our results. The improved DBMEA was tested on several TSPLIB benchmark problems and other VLSI benchmark problems and the following values were compared: - optima found by the improved DBMEA heuristic and by the modified Concorde algorithm with real cost values - runtimes of original DBMEA, improved DBMEA and modified Concorde algorithm. Based on the test results we suggest the use of the improved DBMEA heuristic for the more efficient solution of TSP problems.

Keywords: Traveling Salesman Problem · Discrete optimization · Memetic algorithm

1 Introduction

1.1 The Traveling Salesman Problem

The Traveling Salesman Problem (TSP) was first formulated in 1930, and also nowadays is one of the most widely researched combinatorial optimization problems (which also has wide practical applications).

The original problem involves a salesman who starts the journey from the company's headquarters, visits each city at least (and preferable at most) once, and then he

© Springer International Publishing AG 2017
S. Phon-Amnuaisuk et al. (eds.), *Computational Intelligence in Information Systems*,
Advances in Intelligent Systems and Computing 532, DOI 10.1007/978-3-319-48517-1_3

returns to the starting place. The task is to find the route which allows the salesman visit all cities with the minimum time spent or distance travelled [1, 2].

This problem has several application areas, such as logistics, planning, manufacture of microchips and genome sequencing. In the case of genome sequencing the cities represent DNA fragments and the distance is the similarity value between these DNA fragments [1].

1.2   The TSP as an NP-Hard Task

The Traveling Salesman Problem can be defined as a graph search problem where weights are assigned to each edge (cost between the endpoints of the edge) (1):

$$
\begin{aligned}
G_{TSP} &= (V_{cities}, E_{conn}) \\
V_{cities} &= \{v_1, v_2, \ldots, v_n\}, E_{conn} \subseteq \{(v_i, v_j) | i \neq j\} \\
C &: V_{cities} \times V_{cities} \rightarrow R, C = (c_{ij})_{n \times n}
\end{aligned}
\tag{1}
$$

C is called cost matrix, where c_{ij} the cost of going from city i to city j.

The goal is to find the cycle with the lowest cost that visits every vertex (the directed Hamiltonian cycle with minimal total length). Otherwise, the goal is to find a permutation of vertices $(p_1, p_2, p_3, \ldots, p_n)$ that minimalizes the total cost (2).

$$
C(i) = \left(\sum_{i=1}^{n-1} c_{p_i, p_{i+1}} \right) + c_{p_n, p_1}
\tag{2}
$$

Depending on the properties of the cost matrix TSPs are divided into 2 classes. If the cost matrix is symmetric ($c_{ij} = c_{ji}$ for all i and j) then the TSP is called symmetric, otherwise the problem is asymmetric.

The corresponding optimization problem (the decidable question whether the given directed Hamiltonian cycle has the lowest possible cost) is known to be NP-hard (Non deterministic Polynomial-time hard) problem. A problem is NP-hard if every problem in NP can be reduced to this problem, meaning it is at least as hard or harder than any problem in NP [3].

To the set of our present knowledge NP-hard problems cannot be solved in polynomial time with exact algorithms, their time complexity is in worst case exponential. Algorithms that generate all possible tours and search for the shortest one cannot deal with large-sized problems in real time in the case of TSP.

Many discrete optimization problems are NP-hard to be solved optimally. With implementation of an efficient algorithm for the TSP, also a generally effective method to solve other NP-hard optimization problems may be obtained (other logistics optimization problems, bin-packing etc.).

1.3   Our Previous Work

In recent years we did extensive research on the comparison of various population based algorithms (genetic algorithm [4], bacterial evolutionary algorithm [5], particle swarm algorithm [6] and their memetic versions [7, 8]). We analyzed the properties (especially the speed of convergence) of these algorithms by applying them on several numerical optimization benchmark functions [7].

In 2005 based on the concept of Moscato [9] we proposed a bacterial memetic algorithm which combines the bacterial evolutionary algorithm with a second order gradient based method (Levenberg-Marquardt method) [8].

While a gradient type local search is not possible for discrete problems, we proposed the use of bounded complexity local search, and our investigations showed that the discrete version of bacterial memetic algorithm (Discrete Bacterial Memetic Evolutionary Algorithm DBMEA) is effective for small-sized Bin Packing [10] and Traveling Salesman problems [11].

Considering the above, we came to the conclusion that it would be worth while to investigate the method deeper, and we started carry out further research to compare the properties (tour length, runtime) of our DBMEA with the most efficient algorithm, Concorde algorithm [1, 12] for several TSP benchmark problems [13]. The algorithm showed good properties, the runtime was more predictable than in the case of Concorde algorithm. The local search was the weakest, most time consuming part of the algorithm, so we decided to try the acceleration of it.

2   The Discrete Bacterial Memetic Evolutionary Algorithm

The improved DBMEA is a memetic algorithm, an extension of the bacterial evolutionary algorithm with accelerated 2-opt and 3-opt local search.

Memetic algorithms are the combination of global search evolutionary algorithms and local search methods. In each iteration for the individuals a local search step is applied [9]. Memetic algorithms eliminate the disadvantages of both methods. Evolutionary algorithms search globally, however, in most cases they don't find the optimal solution due to their slow convergence speed. Gradient based methods use only local information in their search process, so they converge always to the nearest local optimum, however they converge to it much faster. As a result, in many cases the addition of local search approaches usually can improve significantly the performance of the classical evolutionary algorithms, so the memetic algorithms can be used efficiently for solving TSP and other NP-hard optimization problems [14].

2.1   Bacterial Evolutionary Algorithm

The development of Bacterial Evolutionary Algorithm (BEA) [5] was inspired by the biological phenomenon of microbial evolution. The process of bacterial recombination that inspired BEA is the following: Bacteria can transfer DNA to recipient cells through mating. Male cells transfer strands of genes to female cells. After that, those female cells get features of the male cells and transform themselves into male cells. By these means, the features of one bacterium can be spread among the entire population [5].

The Bacterial Evolutionary Algorithm contains two special operations for evolving its population, the bacterial mutation and the gene transfer operations. The process of BEA consists of three steps as illustrated in Fig. 1.

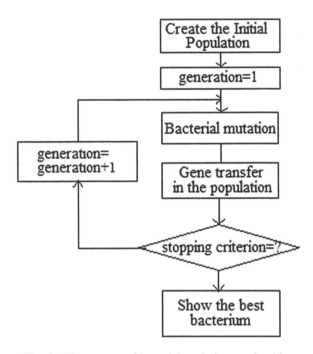

Fig. 1. The process of bacterial evolutionary algorithm

In the first step the algorithm creates a random initial population with *Nind* individuals. Each individual means a solution for the original problem. Then the algorithm improves the individuals of the population using the two operators (bacterial mutation and gene transfer). The bacterial mutation improve the bacteria individually. The gene transfer operation allows the transfer of information in the population between bacteria in the hope that it can produce better bacteria. This process is repeated until the stopping criterion is fulfilled.

Encoding the Individuals. One of the most critical problem in applying an evolutionary algorithm is to find a suitable encoding of the individuals in the population. A good choice of representation will limit the search space, so it will make the search easier.

In DBMEA we use permutation encoding. Each individual represent a possible tour for the Traveling Salesman Problem. We assigned an index *(0 ... n − 1)* to every vertex in the graph, where n is the number of the cities. So a possible tour can be described as a sequence of indices.

The starting point is the city with index 0 in every tour, therefore the start city is not present in the encoded representation of the tours. If each city is visited only once, then each index (excepting index 0) appears once in the code, therefore the length of the string is $n - 1$. In this work the examined TSP benchmark problems are Euclidean. The Euclidean problems are metric, which means that they satisfy the triangle inequality ($c_{ik} \leq c_{ij} + c_{jk}$ for all i, j and k). As a result each encoded tour contains each point only once because the optimal tour visits all points only once in the metric case (Fig. 2).

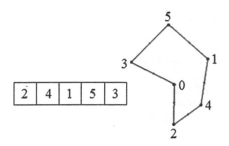

Fig. 2. The encoding of the tour

Creating the Initial Population. The algorithm begins by creating the initial population. The creation of the initial population is a crucial part because it can affect the convergence speed. In order to achieve this goal we generate random and deterministic individuals in the population. Our former test results showed that the use of deterministic individuals can be effective in solving TSP [15].

Random creation: Most of the individuals (excepting the 3 deterministic individuals) are created randomly. The length of the random individuals equals to the number of vertices in the graph (each city is visited only once). Random creation guarantees the uniform distribution of the population in the search space.

We generate three deterministic individuals with the following three eugenic heuristics.

Nearest neighbour (NN) heuristic: The Nearest neighbour heuristic represents a deterministic method, which represent a tour in which always the nearest unvisited city is visited. The advantage of this heuristic is that it is easy to implement and executes quickly.

Secondary nearest neighbour (SNN) heuristic: The Secondary nearest neighbour heuristic is also a deterministic method, which visits always the second nearest unvisited city in the tour.

Alternating nearest neighbour (ANN) heuristic: The combination of above two methods. It represents a tour in which the nearest and second nearest unvisited cities are visited in alternating order.

Bacterial Mutation. Bacterial mutation optimizes the bacteria individually. The process of the bacterial mutation can be seen in Fig. 3. For every bacterium of the population, the operator does the following.

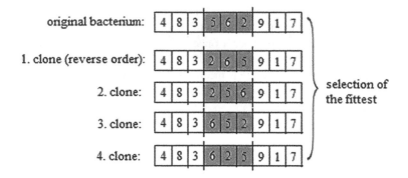

original bacterium:

1. clone (reverse order):

2. clone:

3. clone:

4. clone:

selection of the fittest

Fig. 3. Bacterial mutation

Initially N_{clones} clones are created from the original bacterium. The chromosomes are divided into fixed length (I_{seg}) but not necessary coherent segments (genes). In our algorithm we use both loose and coherent segment mutations.

Next it chooses randomly one from the genes of the bacterium, and it randomly modifies the value of the selected segment in the clones but the same gene in the original bacterium remains unchanged. Beside the randomly mutated clones in the DBMEA algorithm also a deterministic clone is generated, which has a reverse ordering segment.

The next step is the evaluation of each clone bacterium including the original. If one of the clones proved to be better than the original bacterium, the new value of the gene is copied back to the original bacterium and to all the clones. This process is consecutively applied until all the genes of the original bacteria are mutated.

At the end of the mutation the best (most fit) clone is selected from the clones, all the others are deleted. As a result of the bacterial mutation the cloned bacteria are more or equally fit than the original bacteria.

Coherent segment mutation: In this case the elements of the segments are consecutive in the chromosome. It is easy to execute: the chromosome is cut into segments with equal length as it can be seen in Fig. 4.

Loose segment mutation: As opposed to the coherent segment mutation, the elements of the segments are not necessarily adjacent, it may come from different parts of the bacterium. An easy way to create loose segments is the permutation of indices that points to alleles in the chromosome as it can be seen in Fig. 5.

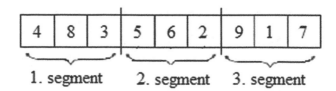

Fig. 4. Coherent segments

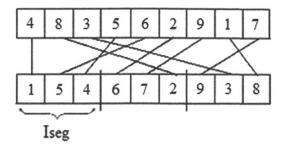

Fig. 5. Loose segments

The time complexity of the bacterial mutation is in one generation $O(N_{ind}N_{clones}n^2)$, and the space requirement is $O(N_{ind}N_{clones}n)$ [15].

Gene Transfer. The gene transfer operation allows the transfer of information between the bacteria in the population in the hope that the bacteria become better and better.

In the gene transfer operation first the population is sorted in a descending order according to their fitness values sorted and divided into two parts (a superior and an inferior half). Next, the operator repeats N_{inf} times the following: it chooses randomly one bacterium (source bacterium) from the superior part and another bacterium (destination bacterium) from the inferior part. After that, it copies some randomly selected segment with pre-defined length ($I_{transfer}$) of the source bacterium into the other one.

Gene Transfer in the DBMEA algorithm is the following: A source segment with pre-defined length is selected randomly from the source bacterium and this segment is transferred to the destination bacterium. The destination offset is set randomly, not necessary equal with the source offset.

In Fig. 6, the source segment is (6, 2, 9), and the destination bacterium get this segment (between 5 and 7). The length of the destination bacterium remains unchanged, the elements in the destination bacterium, which are identical with the elements in the transferred segment, must be deleted (thus eliminating double occurrence). (In the example the first element of the destination bacterium, 2 is located in the transferred segment, so it is deleted, and the first element of the infected destination bacterium becomes 5 etc.).

At the beginning of the gene transfer operation the fitness value of the bacteria are calculated $O(N_{ind}n)$, and then the bacteria are sorted in a descending order $O(N_{ind}logN_{ind})$. After each gene transfer the new fitness value of the destination bacterium is calculated, and the infected destination bacterium is added to the sorted population. The time complexity of this calculation is $O(N_{inf}(n+N_{ind}))$. Summarizing the required operations the total time complexity of the gene transfer operation is $C_{GT} = O(N_{ind}(n+logN_{ind}) + N_{inf}(n+N_{ind}))$ [15].

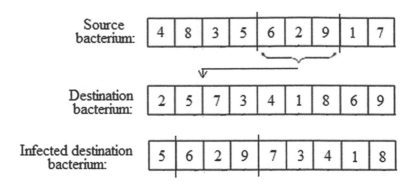

Fig. 6. Gene transfer

2.2 Local Search

Local search algorithms are important parts of the memetic algorithms because iteratively improve the candidate solutions in the population by moving to neighbouring solutions. In many optimization problems local search methods can be much more efficient than genetic search, but usually they need to be combined with metaheuristics (genetic algorithms, bacterial evolutionary algorithms etc.) to find the optimal solution [16].

We applied a 2-opt and 3-opt local search technique in the new DBMEA algorithm.

The local search is accelerated in the following way. A candidate list is registered for all vertices. The candidate list contains the indices of the closest vertices in ascending order. The local search doesn't examine the whole graph searching for improvements, only the pre-defined number of closest vertices (candidate lists contain them) for each vertices.

2-opt Local Search. 2-opt local search replaces two edge pairs in the original graph to reduce the length of the tour. The algorithm works on metric TSP instances.

Edge pairs (AB, CD) are iteratively replaced with AC and BD edges and the following inequality is examined: $|AB| + |CD| > |AC| + |BD|$.

The whole graph isn't examined searching for vertex C, only the members of candidate list of vertex A.

If the inequality holds then edge pairs are exchanged; the deleted AB and CD edges are replaced with AC and BD edges (Fig. 7). One of the sub-tours between the original edges is reversed in the improved tour. This is continued till no further improvement is possible.

3-opt Local Search. The 3-opt local search replaces edge triples. The deleting of three edges results three sub-tours. There are two possible new edge orders to reconnect these sub-tours (Fig. 7). The output of the 3-opt step is always the less costly tour.

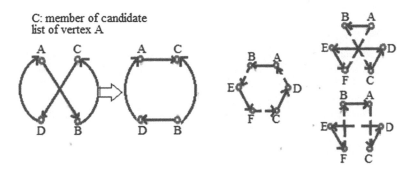

Fig. 7. 2-opt and 3-opt local steps

Computational Results

The DBMEA algorithm was tested on 10 TSPLIB benchmark problems [17] and other VLSI benchmark problems [18]. More parameter settings were tested, and the following configuration gave the fastest convergence:

- the number of bacteria in the population (N_{ind} = 100)
- the number of infections in the gene transfer (N_{inf} = 40)
- the length of the chromosomes (I_{seg} = 10)
- the length of the transferred segment ($I_{transfer} = n_{cities}/10$)
- length of the candidate lists (square root of the number of cities)
- the number of clones in the bacterial mutation (N_{clones} = 20)

Each benchmark problem was tested 10 times with the above mentioned parameter setting. Our results are compared with an other efficient TSP solver algorithm (Concorde) in terms of the optimal tour and the run time, that is considered in the literature as the most efficient approach.

In these instances the distances between the points are Euclidean distances, so the problems are always symmetrical.

2.3   Comparison of Optimal Tour Lengths

We compare the best tour found by our improved DBMEA algorithm with the optimum found by the modified Concorde algorithm. Table 1 also contains the average tour lengths in the case of the improved DBMEA algorithm. As mentioned before, we modified the Concorde algorithm to calculate the distances between the cities with real values.

As it can be seen in Table 1, for the majority of the investigated benchmark problems (8 out of the 10 tested problems) we achieved identical results. In the remaining few (two) cases DBMEA produced near-optimal solutions.

Table 1. Comparison of minimal tour lengths

Name	Best tour lengths found by our improved DBMEA algorithm	Average tours lengths found by our improved DBMEA (10 tests)	Optimal tour lengths found by the modified Concorde algorithm (real values)	Gap (best tour DBMEA and optimal tour) [%]
pr107	44 301.68	44 301.68	44 301.68	0
XQF131	566.42	566.42	566.42	0
XQG237	1 029.18	1 029.18	1 029.18	0
LIN318	42 024.54	42 024.54	42 024.54	0
PMA343	1 387.51	1 387.51	1 387.51	0
PKA379	1 344.41	1 344.41	1 344.41	0
XQL662	2 550.84	2 550.84	2 550.84	0
DKG813	3 243.41	3 244.56	3 243.41	0
DKA1376	4 743.6	4 745.13	4 736.87	0.14
DCA1389	5 155	5 157.26	5 152.69	0.04

2.4   Comparison of Runtimes

In the following the runtimes used for the achieved results on the benchmark problems will be examined. The comparison of runtimes is difficult and relative because it depends on the performance of the computer.

In Table 2, the comparison of runtimes averaging 10 tests can be seen. The calculation of the optimal tours was on an Intel Pentium Dual CPU T2390 1,86 GHz, 2 GB RAM workstation.

According to the test results the structure of the TSP instance has a big effect on the run time of the Concorde algorithm. For example, the run time of DCA1389 is more

Table 2. Comparison of runtimes

Name	Average run times of our improved DBMEA algorithms (10 tests)	Average run times of the modified Concorde algorithm (10 tests)	Ratio between original and improved DBMEA algorithms
pr107	3.07 s	0.41 s	22.19
XQF131	3.79 s	0.85 s	29.98
XQG237	96.11 s	4.46 s	12.79
LIN318	125.14 s	3.92 s	10.59
PMA343	147.68 s	9.65 s	15.04
PKA379	160.94 s	15.08 s	14.69
XQL662	871.157 s	157.27 s	7.63
DKG813	2 504.74 s	2 031.11 s	7.31
DKA1376	18 610.15 s	807.11 s	5.58
DCA1389	34 380.12 s	380 155.1 s	10.71

than 471 times longer than the run time of DKA1376, although the size of the problem is almost equal. In the case of the DBMEA algorithm the difference between the run time of DCA1389 and DKA1376 was much smaller, the structure of the problem has much smaller impact. In the case of DCA1389 our algorithm was faster than in the case of Concorde.

The ratio between the run time of the original and the improved DBMEA algorithm is between 5 and 20, so we achieved significant improvement.

According to the above we propose that the new DBMEA algorithm is efficient for solving TSP, especially for large-sized problems with complicated structure.

3 Conclusions

It can be concluded that the improved DBMEA algorithm produces optimal or near-optimal tour lengths on TSP benchmark problems. The run time is more predictable in the case of our DBMEA algorithm because the structure of the TSP instance has much smaller impact on the run time than in the case of the Concorde algorithm. The runtime was 5–20 times shorter than our original DBMEA algorithm on the tested benchmark problems.

In our further work we will examine other population based techniques for solving TSP and other NP-hard optimization problems. We plan to test DBMEA algorithm on other NP-hard optimization benchmark problems (e.g. bin packing).

References

1. Applegate, D.L., Bixby, R.E., Chvátal, V., Cook, W.J.: The Traveling Salesman Problem: A Computational Study, pp. 1–81. Princeton University Press, Princeton (2006)
2. Gutin, G., Punnen, A.P.: The Traveling Salesman Problem and Its Variations, pp. 1–28. Springer, New York (2007)
3. Karp, R.M.: Reducibility among combinatorial problems. In: Miller, R.E., Thatcher, J.W., Bohlinger, J.D. (eds.) Complexity of Computer Computations, pp. 85–103. Springer, New York (1972)
4. Holland, J.H.: Adaption in Natural and Artificial Systems. The MIT Press, Cambridge (1992)
5. Nawa, N.E., Furuhashi, T.: Fuzzy system parameters discovery by bacterial evolutionary algorithm. IEEE Trans. Fuzzy Syst. 7, 608–616 (1999)
6. Kennedy, J., Eberhart, R.: Particle swarm optimization. In: Proceedings of the IEEE International Conference on Neural Networks, ICNN 1995, Perth, WA, Australia, vol. 4, pp. 1942–1948 (1995)
7. Balázs, K., Botzheim, J., Kóczy, T.L.: Comparison of various evolutionary and memetic algorithms. In: Huynh, V.-N., Nakamori, Y., Lawry, J., Inuiguchi, M. (eds.) Integrated Uncertainty Management and Applications, vol. 68, pp. 431–442. Springer, Heidelberg (2010)
8. Botzheim, J., Cabrita, C., Kóczy, L.T., Ruano, A.E.: Fuzzy rule extraction by bacterial memetic algorithms. In: Proceedings of the 11th World Congress of International Fuzzy Systems Association, IFSA 2005, Beijing, China, pp. 1563–1568 (2005)

9. Moscato, P.: On evolution, search, optimization, genetic algorithms and martial arts-towards memetic algorithms, Technical Report Caltech Concurrent Computation Program, Report 826, California Institute of Technology, Pasadena, USA (1989)

10. Dányádi, Zs., Földesi, P., Kóczy, T.L.: A fuzzy bacterial evolutionary solution for crisp three-dimensional bin packing problems. In: IEEE World Congress on Computational Intelligence, Brisbane, Australia, WCCI 2012 (2012)

11. Farkas, M., Földesi, P., Botzheim, J., Kóczy, T.L.: Approximation of a modified Traveling Salesman Problem using Bacterial Memetic Algorithms. In: Rudas, I.J., Fodor, J., Kacprzyk, J. (eds.) Towards Intelligent Engineering and Information Technology, SCI 243, pp. 607–625. Springer, Heidelberg (2009)

12. Applegate, D.L., Bixby, R.E., Chvátal, V., Cook, W.J., Espinoza, D., Goycoolea, M., Helsgaun, K.: Certification of an optimal tour through 85,900 cities. Oper. Res. Lett. 37(1), 11–15 (2009)

13. Kóczy, L.T., Földesi, P., Tüű-Szabó, B.: A Discrete Bacterial Memetic Evolutionary Algorithm for the Traveling Salesman Problem. In: The Congress on Information Technology, Computational and Experimental Physics (CITCEP 2015), Cracow, Poland, pp. 57–63 (2015)

14. Tang, M., Yao, X.: A memetic algorithm for VLSI floorplanning. Syst. Man Cybernet. 37 (1), 62–69 (2007)

15. Földesi, P., Botzheim, J.: Modeling of loss aversion in solving fuzzy road transport traveling salesman problem using eugenic bacterial memetic algorithm. Memetic Comput. 2(4), 259–271 (2010)

16. Hoos, H.H., Stutzle, T.: Stochastic Local Search: Foundations and Applications. Morgan Kaufmann, San Francisco (2005)

17. Reinelt, G.: TSPLIB – A traveling salesman problem library. ORSA J. Comput. 3, 376–385 (1991)

18. VLSI TSP dataset, April 2015. http://www.math.uwaterloo.ca/tsp/vlsi/index.html

Improved Stampede Prediction Model on Context-Awareness Framework Using Machine Learning Techniques

Fatai Idowu Sadiq[2,3], Ali Selamat[1,2,4(✉)], and Roliana Ibrahim[2]

[1] UTM-IRDA Digital Media Center of Excellence,
University Teknologi Malaysia, 81310 Skudai, Johor Bahru, Malaysia
[2] Faculty of Computing, University Teknologi Malaysia,
81310 Skudai, Johor Bahru, Malaysia
sfatai2011@gmail.com, {aselamat,roliana}@utm.my
[3] Ambrose Alli University, P. M. B 14, Ekpoma, Edo State, Nigeria
[4] Faculty of Informatics and Management,
Center for Basic and Applied Research, University of Hradec Kralove,
Rokitanskeho 62, 500 03 Hradec Kralove, Czech Republic

Abstract. The determination of stampede occurrence through abnormal behaviors is an important research in context-awareness using individual activity recognition (IAR). An application such as an intelligent smartphone for crowd monitoring using inbuilt sensors is used. Meanwhile, there are few algorithms to recognize abnormal behaviors that can lead to a stampede for mitigation of crowd disasters. This study proposed an improved stampede prediction model which can facilitate abnormal detection with k-means. It can identify cluster areas among a group of people to know susceptible places that can help to predict stampede occurrence using IAR with the help of geographical positioning system (GPS) and accelerometer sensor data. To achieve this, two research questions were formulated and answered in this paper. (i) How to determine crowd of people in an area? (ii) How to know when stampede will occur in the identified area? The experimental results on the proposed model with decision tree (DT) algorithm shows an improved performance of 98.6 %, 97.7 % and 10.9 % over 94.4 %, 95 % and 18 % in the baselines for specificity, accuracy and false-negative rate (FNR) respectively thereby reducing high false negative alarm.

Keywords: Individual activity recognition · Context-awareness framework · Stampede prediction model · Participant identification node · Specificity

1 Introduction

Context sensing, context acquisition, context-awareness are increasing problems towards contextual information in a ubiquitous society [1, 2]. Most importantly context play a significant role in physical, computing, personal and environmental entities especially with the interactions between smartphones and other handheld devices. These problems are important to human lives and our immediate environment with relevant

© Springer International Publishing AG 2017
S. Phon-Amnuaisuk et al. (eds.), *Computational Intelligence in Information Systems*,
Advances in Intelligent Systems and Computing 532, DOI 10.1007/978-3-319-48517-1_4

interaction factors. The origin of crowd monitoring system (CMS) were traced back to 1995 when Close Circuit Televisions was used with pattern recognition for CMS [3]. This technique has no feedback, it involves physical efforts, and requires regular participations of security officers. This result to the work in [4], where Wireless Sensor Networks using wireless communication technique for better situational awareness was proposed. The study suffered a high rate of false negative alarm (FNA) [5], as a result of redundant features and inadequate preprocess of the activity recognition data from the mobile sensor. Consequently, the use of low power sensors, specifically temperature and acoustic sensors were less reliable as well. The limitations above led to emerging technology of mobile phone sensing for monitoring activity of people for possible safety control in case of any unforeseen situation in crowd related scenario [5]. Automatic recognition of user activities using different contextual data for enhancement of pervasive system using context-awareness application is still in its infancy [6, 7], considering its potential to offer more adaptable, flexible and user-friendly services [11]. Despite the improvement in the work of [6, 7]. The study has some limitations, which drew the attention of [5] to improve the work by introducing context-aware computing and wireless sensor network (CAC-WSN) to investigate activity recognition accuracy; inadequate real-time information dissemination and high rate of FNA for efficient stampede prediction to mitigate crowd disaster. Given the shortcomings found in [5], this ongoing research work intends to develop an enhanced context-awareness framework for mitigation of crowd disaster (MCD) capable of providing efficient stampede prediction with most relevant features for reducing time complexity with improving recognition accuracy for effective MCD. The proposed stampede prediction model in this paper is one of the possible solutions in the ongoing research to the limitations found in the baseline [5].

1.1 Purpose of the Study

Mitigation of crowd disaster is paramount [5, 8, 9]. Reason to reduce the risk and dangers to human lives in everyday activities around us. The causes of crowd disasters were traced back to stampede in a crowded area refer to [8]. Interestingly, the menace and causes of the crowd disaster can be addressed using pervasive computing precisely CAC-WSN technology, together with individual activity recognition and machine learning as the technique [5]. The motivation lies in the fact that the solution proposed in the baseline; as observed in [5]; assumed high-density crowd without criteria to ascertain how to determine it with 20 people used in their study. This in the actual sense is necessary to know the extent to which the crowd is prone to stampede in real life scenario. More so, the density of a particular place remains constant, as only the number of people will vary depending on the event of activity. On this premises, this paper proposes an improved stampede prediction model that will help to predict stampede in a crowd scenario using machine learning technique with individual activity recognition to extend the work of [5], for effective MCD. The ultimate goal is to reduce the false negative alarm as shown in our experiment. To achieve this, two (2) research questions (RQs) are formulated and answer in this paper. The following are the RQs:

1. How to determine crowd of people in an area?
2. How to know when stampede will occur in the identify crowd?

2 Related Works

According to Okoli and Nnorom [9], the human stampede has recently happened in places like China, Russia, Nigeria and another part of the world. The implication of this danger represents critical challenges to the national security of any nation. In the study of [8], the leading causes of mass death and injury worldwide were traced to human stampede [10]. Early frameworks [11, 12]; does not look into monitoring activity of people for possible occurrences of dangers and does not consider any activity scenario using the IAR to investigate AR_{ac} and FNA related to crowd disaster, unlike those listed in Table 1. Nearly all the existing framework focuses on the utilization of standard context interaction (SCI) [13]. In [14] a framework that specifically catered to the need of users using the mobile device was proposed, its feasibility was demonstrated based on the scenario. A Java based having core design principles, runtime infrastructure, and a programming model was introduced in [15]. Ontology-based for the ubiquitous environment is proposed in [16], context modeling, its extensibility and open nature of system with dynamic pluggability of component distributed over several nodes was suggested in [17], and a framework for the provision of intelligent network embedded systems were presented [18].

Table 1. Context-awareness frameworks with justifications of outstanding issues

Reference (R)	MLT	ATSI	FNA	SCI	AR_{ac}	MPS
R [13]				✓		
R [14]				✓		
R [15]				✓		
R [16]				✓		
R [17]				✓		
R [18]				✓		
R [23]				✓		
R [12]	DT					
R [24]	DT					
R [5][a]	DT	✓	High		Low 92 %	
R [25]	NN					
Proposed framework	DT others	✓	low		improved	✓
Key						

[a]Baseline. MLT – Machine Learning Technique.
DT – Decision tree algorithm
NN – Neural Network algorithm
SCI – Standard context interaction.
ATSI –Accurateness and timeliness of sensed information support for critical situation (real time).
AR_{ac} – Activity Recognition accuracy.
MPS – Mobile Phone Sensing
SPM – Stampede prediction model

According to Ravindran et al. [12], context-aware computing is a peculiar kind of application that can sense the physical environment and reacts accordingly. To facilitate the quick and efficient development of the useful application, it combines context-aware services and MLT using Mumbai as a case study in Context-Aware and Pattern-Oriented Machine Learning Framework (CAPOMF). The essence was to help commuters to avoid potholes and saves the repair costs of vehicles in Mumbai. It shows the potential of context-awareness application to ease the problem faced by people of Mumbai [12]. As proposed in this paper MLT is the choice to achieve the proposed model. In the previous context-awareness research, MLT is rarely used [11, 19–21] to realize context-aware computing and application goal. In [12], it was remarked that many context-awareness systems store contextual information, but no one uses MLT to provide context-awareness services proactively [11, 19–21]. A stampede prediction model was investigated in (Zhang et al. 2013) using a video camera with MLT, but the work does not use IAR, context acquisition, real-time management of the mobile sensors, context modeling, and inference. At the same time, feedback to a potential victim of disaster in an emergency may be difficult due to static nature of video camera. According to Oscar et al. [22], recognition of human activities has been somehow individualized, it can be taken a step further. This observation was investigated by [5] using the scenario such as standing, fall, peak-shake-whst, still, climbing-up, climbing-down and jogging for both individual and group recognition. The baseline utilized activity recognition accuracy of 92 % for onset stampede prediction to mitigate crowd disaster in a crowded area. The study used a smartphone as a participatory sensing node with embedded tri-axial accelerometer sensor using 20 students to acquired data. However, the baseline has the following limitations. (i) The result of 92 % AR_{ac} employed with stampede algorithm may be inadequate for effective MCD. (ii) The algorithm used for the prediction which utilized high-density does not consider how the participant clustered in a group based on the area of the location which can facilitate the stampede detection. Table 1 shows the summary of existing context-awareness framework.

3   Proposed Work

In this section, the details of the methods used for the improved stampede prediction model are discussed. These include decision tree (DT) and K-means algorithms.

3.1   Decision Tree

In classification decision tree (DT) has been reported to give higher accuracy result in a simple body motion in activity recognition (AR) [23]. It has delivered the best results in the derived feature vector for AR in the study of [26]. The evidence from the literature on this algorithm has confirmed the reason for its choice in this paper. DT is essentially a hierarchical decomposition of the (training) data space, in which a predicate or a condition on the attribute value is used to divide the data space hierarchically [27]. It has been used for various learning tasks including classification, regression, and survival analysis [28]. It has the following advantages: (i) Easy to understand (ii) Useful for data exploration (iii) fewer data cleaning is required (iv) No constraints on data types and (v) High performance with less effort. However, it uses a straightforward idea to resolve classification problem [29]. This method was chosen to compare the result of the proposed method in this paper with the previous work on activity recognition on machine learning by [30], and the baseline on crowd abnormality monitor (CAM) by [5]. However, Fig. 1 outlines the design flow of the proposed model starting from the data collection using the smartphone and activities used for individual activity recognition (IAR) in the first phase of the study. After that, interactions take place among the individuals' label for group activity recognition (GAR). The second step employed the context-awareness application with the said smartphone used to monitor the scenario. The mobile sensor data acquired from GAR are then utilized to build, compare and report the best model following preprocessing and extraction methods not detailed in this paper.

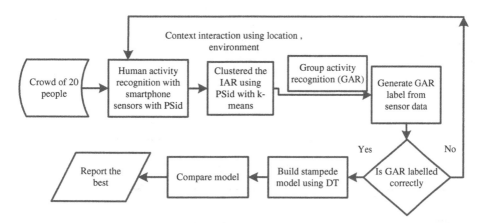

Fig. 1. Proposed model for group activity recognition

3.2 K-Means

To answer research question RQs 1. K-means is adapted in this paper to help us clustered the activity of the IAR in the group using the PS_{id} nodes to know the various sub-area with clusters of the participant which belongs to each subarea list in a group. Each participant node, PS_{id} is assumed to fall into a particular cluster of a set of a participant in the crowd often refer to as a group. The PS_{id} which is a node relating to a participant at one time is closer to the cluster's centroid k than any centroid far from k, the centroid. We assume that variation between monitoring participant whose monitor's device is represented by PS_{id} and another participant $P \in K_i$ and k_i (centroid), is the representation of group cluster, which is measured by dist (p, k_i). The dist (x, y) is the Euclidean distance. In this case x, y represents longitude and latitude between two PS_{id} nodes in a particular crowd scenario given in Eq. 1. The cluster is initialized by randomly chosen points as centers among a group of participant. We utilized the sci-kit learn default in python program for the employed k-means, where average complexity is given by O (K n T), and k is the number of clusters, n is number of samples and T is the number of iteration.

Let $x = (x_1, x_2)$ and $y = (y_1, y_2)$ then the distance between the two nodes is given by:

$$dist(\text{x,y}) = \sqrt{(x_1 - y_1)^2 + \left((x_2 - y_2)^2\right)} \tag{1}$$

Equation 1 represents the Euclidean distance. The quality of the cluster K_i can be measured using the cluster difference. This is done with sum of the square error between all nodes PS_{id} using geographical position system (GPS) sensor data in K_i and the centroid k_i defined as follows in Eq. 2.

$$E = \sum_{i=1}^{c} \sum_{P \in K_i} dist(p, k_i)^2 \tag{2}$$

In Eq. 2, E represents the sum of the square error (SSE). SSE is determined by using the node of participant that is nearest to each pair of node PS_{id} in the monitoring group and subsequent ones in the group. P is the sensor data from GPS in subarealist for any group and k_i is the cluster K_i centroid. The longitude and latitude sensor for each PS_{id} in each cluster group, is the distance from cluster center to every PS_{id} node. The squared and, the distance are summed up as shown in Eq. 2. The algorithm adapted is presented in Algorithm 1. It depicts an example of k-means clustering of points in subarea, with k = 8 using the number of activities scenario. The main advantage of k-means: (i) it has been successful for clustering of mobile sensor data [31]. (ii) It is better for large data with time complexity of O(KT) [32], particularly with crowd of people it varies in number and it will make the data to increase based on the number of people in the scene.

Algorithm 1. K-Means algorithm for clustering

Require: K, numbers of clusters; D, a dataset of T points

Ensure: J set of K clusters

1. Initialization.
2. **repeat**
3.   **for** each points a in D **do**
4.   find the nearest center and assign t to the

        corresponding cluster.

5.   **end for**
6.   Update clusters by calculating new centers
7.   using mean of the participants' node.
   until stop-iteration criteria met
8. **return** clustering result.

3.3   Participant Nodes Behavior for Group Clustering

To answer RQs. 2. The crowd behavior monitoring was done with the use of sensor data for each participant identification node PS_{id} on an individual basis. Individual sensor analyses with context recognition performed on human activities are illustrated in Fig. 3. The mapping between the program sensors and activities considered were learn using MLT. The individual participant id node is represented as PS_{id}. The algorithm for individual and group behavior participants' node clustering is shown in Fig. 2 with the help of Algorithm 1. The crowd formation distribution is divided into sets of Subarealist using the crowd controller station (CCS). When a new participant PS_{id} node is noticed, the context-awareness application senses and notifies the CCS to automatically add the new node into the specific subarealist1 based on the present location. The procedures for grouping a participant in Subarealist is equally achieved using line 5 module in Fig. 2. It takes care of the mobility of the participant PS_{id} node whenever the distance traversed by the participant is greater than a threshold value. The threshold is about 30 m from the wifi-hotspot for the indoor condition which is of greater value to the outdoor condition. Figure 2 also takes notes of any nearby participant and distance between two participants PS_{id} nodes in the subareas unit. In case the distance is less than 2 m then the new PS_{id} nodes is added to the Subarealist 2 through CCS. In order to achieve the clustering of the participant in all the Subarealist 1-2 k-means algorithm described in Sect. 3.2 is utilized to clusters all the nodes PS_{id}. The region identification and grouping of participants are actualized by using the smartphone of the participant with the PS_{id} as shown in Fig. 2.

4   Experimental Results

The details about data collection, preprocessing, feature extraction and evaluation metrics formulas for accuracy (AC), precision (P), recall(R) and f-scores (F) are not included in this paper, refer to page 17 of [33]. The confusion matrix based on the proposed model is summarized as follows using eight classes of IAR in Fig. 1. Detail is as shown in Table 2 and performance evaluation is shown in Fig. 3.

4.1   Performance Evaluation Criteria

In addressing research question 2. In addition to metrics refer to in Sect. 4, two others specifically specificity (SP) and macro-average (M_{mav}) are chosen to evaluate the performance of the classifier. SP, this is used to determine the negative activity prediction that is correctly recognized as negative. The mathematical definition of the metrics are defined in Eq. 3, it was not considered in the previous studies.

$$SP = \frac{TN}{TN + FP}, \quad M_{mav} = \frac{1}{N}\sum_{i=1}^{N} \frac{\sum_{c=1}^{C} n_{ci}M_{ci}}{\sum_{c=1}^{C} n_{ci}} \tag{3}$$

Require: S participant node PS$_{id}$. Lat: Latitude. Long: Longitude. T: Time

Ensure: SA: SubArea SA. SA1: SubArealist1. SA2: SubArealist2.

SAn: SubArealistn. Dist: distance

1.	Start Input S, Output (Lat, Long, T)
2.	Input SubAreaList (SA, SA1, SA2…SAn, Lat, Long, T)
3.	Output S grouped in a SubArea, SA
4.	*While S is ready do*
5.	*for each S of participant SubAreaList do*
6.	Set locationUpdateWindowSize
7.	SetminTime for location manager minimum power consumption
8.	With mint milliseconds between location update to reserve power
9.	SetminDist = locationbroadcasting in case device moves using the

minDistance meters

10.	TDeviation = Location.getT( ) – currentbestlocation.getT( )
11.	*If* TDeviation > TWindow then participant have moved and transit to
12.	new location by crowd controller station (CCS) using timestamp
13.	*elseif* (Lat,Long) as SA location context clusters using Dist between nodes S, group S into
14.	SA, SA1 and Crowdcount (S+1)
15.	*endif*
16.	*endfor*
17.	*endwhile*
18.	*End*

Fig. 2. Enhanced algorithm for group clustering using participant smartphone as nodes

Table 2. Confusion matrix summary in 2 × 2 for group clustering classification

Class	Actual	Predicted climb-down			
		TP	FN	TN	FP
P:Climb_down	1225	1112	113	10548	128
Q:Climb_up	1225	1096	129	10542	134
R:Stampede	1225	560	665	10671	5
S:Jogging	594	577	17	11292	15
T:Fight	2553	2504	49	8672	704
U:Standing	1480	1440	40	10381	40
V:Panic	1377	1344	33	10489	35
W:Walking	2222	2154	68	9626	53

To evaluate the overall effectiveness of activity recognition for mitigation of crowd disaster, macro-averaged is used. Recognition is basically a classification task in content recognition and another domain to determine the overall interaction between

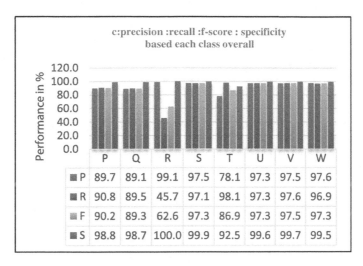

Fig. 3. Evaluation based on precision, recall, F-score and specificity based on class

classes macro–average is necessary [34, 35]. It shows the overall average of the entire class. The macro-average represent M_{mav}, N is type of validation (split or cross validation), C represent no of clsases, n_c is number of instances that belong to a class c, M_{ci} is the value of metric M for class c in i[th] type of classification. C = 8, N: 10-fold. Now using Eq. 3, with DT algorithm the following results are obtained to answer RQs 2.

5   Discussion of Results

Table 2 explains the details of AR classes and their instances in the classification model by summarizing the confusion matrix. It shows that out of the 1225 stampede recognized 560 (TP) were predicted correctly and 665 were predicted to belong to fight (663) and walking (2) known as FN. Out of the 2553 fights, 2504 (TP) predicted correctly and 49 were predicted to belong to climb-down, climb-up, stampede and walking, 10, 9, 2, and 28 as (FN) respectively. In the case of the panic of 1377, 1344 (TP) predicted correctly and 33 were predicted to belong to jogging 5 and standing 28 as (FN). In this study all other classes were regarded as normal i.e. P, Q, S, U and W amounted to 6746 instances Table 2, because participant can be in any of these stage in crowd before any of the R, T and V of 5,155 instances can occur since they are the major anomalies in the crowd that can lead to stampede known as crowd disasters (refer to Sect. 4.1 page 4) of [5]. The evaluation for each activity class is as shown in Fig. 3.

Figure 4 using the same dataset shows that the specificity and accuracy (AC) of the proposed model 98.6 %, 97.7 % outperformed 94.4 %, 95 % for baseline (BS) [5] features when used in the proposed model and 94.4 %, 92 % of baseline1 (BS1) [30]. Note that the specificity and macro-average were not used in the work of the baselines, we derived this through evaluation using the same features and dataset in our study, because the measurement can be evaluated through the 4 metric parameters in Table 2. The SP is necessary when any danger which can result to panic or fight is sensed by the

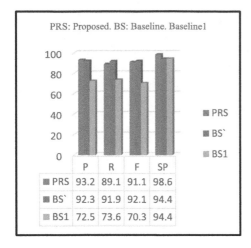
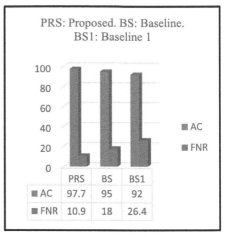

Fig. 4. Performance evaluation using macro-average based on precision, recall, f-score, specificity and accuracy for the proposed model and the baselines

monitoring device, predicting the activity correctly will help to save the participant and know what measure to take or security personnel. This is evident from our validation result with low risk of reduced FNR 10.9 % in proposed model and high risk of FNR 18 % [5] and 26.4 % [30] respectively with same features and methods.

The proposed model is the benchmark against the state-of-the-art for human activity recognition [30] and that of context-aware ad-hoc network for mitigation of crowd disasters using activity recognition in [5]. The comparisons are summarized in Table 3. In the proposed model the chance of stampede occurrence is very low with reduced FNA using false negative rate FNR 10.9 %, against the baseline (BS) and (BS1) with FNR 18 % and FNR 26.4 % respectively using the same dataset and methods. The proposed model outperforms the baselines as shown above. However [25] uses video sensors and different methods which makes it unsuitable for

Table 3. Benchmark of the proposed model with similar model

Reference learning technique	Dataset	Number of sample/device	Method	Framework/metric
BS [5]	Individual activity recognition/group	20:Smartphone sensing	Decision-tree/ Stampede algorithm	CAM: AC 95 %, SP 94.4 % FNR 18 %
BS1 [30]	Individual activity recognition on Heterogeneities	30:Smartphone sensing	Decision tree	AC 92 %,SP 94.4 %, FNR 26.4 %
Proposed approach improved stampede model	**Individual activity recognition/group**	**20: smartphone sensing**	**Decision-tree/K-means for clustering**	**CAM, AC 97.7 %, SP 98.6 %, FNR 10.9 %**

benchmarking hence the replacement with [30] a public dataset of human activity recognition from machine learning repository.

6 Conclusion

The study is evaluated using machine learning algorithms for effective mitigation of crowd disaster with an improved stampede prediction model to extend the work of baseline. Data collected in real time from 20 students on individual and group activity recognition were obtained using smartphone sensing and analyzed offline using decision tree algorithm. The clustering of the participant using the smartphone as participant nodes PS_{id} was achieved using K-means algorithm to determine subarealist with various clusters to know the area that is more susceptible to stampede which can lead to crowd disasters. Experimental results show that the proposed model achieve better classification result using decision tree algorithm and can facilitate safety control measure in a disaster-prone area in crowded area.

References

1. Fereshteh, F.C., Lilly, S.A.: Impact of mobile context-aware applications on human computer interaction. J. Theoret. Appl. Inf. Technol. **62**(1), 1–7 (2014)
2. Jeong, H.-Y., Xiao, Y., Chen, Y.-S., Peng, X.: Ubiquitous context-awareness and wireless sensor networks. Int. J. Distrib. Sensor Netw. **2014**, 1–2 (2014)
3. Davies, A.C., Yin, J.H., Velastin, S.A.: Crowd monitoring using image processing. Electron. Commun. Eng. J. **7**(1), 37–47 (1995)
4. Gomez, L., Laube, A., Ulmer, C.: Secure sensor networks for public safety command and control system. In: 2009 IEEE Conference on Technologies for Homeland Security, HST 2009, pp. 59–66 (2009)
5. Ramesh, M.V., Shanmughan, A., Prabha, R.: Context aware ad hoc network for mitigation of crowd disasters. Ad Hoc Netw. **18**, 55–70 (2014)
6. Wirz, M., Roggen, D., Troster, G.: Decentralized detection of group formations from wearable acceleration sensors. In: 2009 International Conference on Computational Science and Engineering, CSE 2009, vol. 4, pp. 952–959 IEEE (2009)
7. Roggen, D., Wirz, M., Troster, G.: Recognition of crowd behavior from mobile sensors with pattern analysis and graph clustering methods. Netw. Heterogen. Media 6(3), 521–544 (2011). arXiv preprint arXiv:1109.1664
8. Fruin, J.J.: The causes and prevention of crowd disasters. In: Engineering for Crowd Safety, pp. 99–108 (1993)
9. Okoli, A.C., Nnorom, K.C.: Disaster Risks in Crowded Situations: Contemporary Manifestations and Implications of Human Stampede in Nigeria. Risk, p. 62 (2007)
10. Still, K.: Crowd Risk Analysis (2014). Retrieved from http://gkstill.com Accessed Sept 2015
11. Schilit, B.N., Theimer, M.M.: Disseminating active map information to mobile hosts. IEEE Netw. **8**(5), 22–32 (1994)
12. Ravindran, R., Suchdev, R. Tanna, Y., Swamy, S.: Context aware and pattern oriented machine learning framework (CAPOMF) for android. In: 2014 International Conference on Advances in Engineering and Technology Research (ICAETR), pp. 1–7. IEEE (2014)

13. Dey, A.K., Abowd, G.D., Salber, D.: A conceptual framework and a toolkit for supporting the rapid prototyping of context-aware applications. Hum. Comput. Interact. **16**(2), 97–166 (2001)
14. Hofer, T., Schwinger, W., Pichler, M., Leonhartsberger, G., Altmann, J., Retschitzegger, W.: Context-awareness on mobile devices-the hydrogen approach. In: 2003 Proceedings of the 36th Annual Hawaii International Conference on System Sciences, pp. 1–10. IEEE (2003)
15. Bardram, J.E.: The java context awareness framework (JCAF) – a service infrastructure and programming framework for context-aware applications. In: Gellersen, H.-W., Want, R., Schmidt, A. (eds.) PERVASIVE 2005. LNCS, vol. 3468, pp. 98–115. Springer, Heidelberg (2005)
16. Lee, K.-C., Kim, J-H., Lee, J-H, Lee, K-M.: Implementation of ontology based context-awareness framework for ubiquitous environment. In: 2007 International Conference on Multimedia and Ubiquitous Engineering, MUE 2007, pp. 278–282. IEEE (2007)
17. Fides-Valero, Á., Freddi, M., Furfari, F., Tazari, M.-R.: The PERSONA framework for supporting context-awareness in open distributed systems. In: Aarts, E., Crowley, J.L., Ruyter, B., Gerhäuser, H., Pflaum, A., Schmidt, J., Wichert, R. (eds.) AmI 2008. LNCS, vol. 5355, pp. 91–108. Springer, Heidelberg (2008)
18. Badii, A., Crouch, M., Lallah, C.: A context-awareness framework for intelligent networked embedded systems. In: 2010 Third International Conference on Advances in Human-Oriented and Personalized Mechanisms, Technologies and Services (CENTRIC), pp. 105–110. IEEE (2010)
19. Baldauf, M., Dustdar, S., Rosenberg, F.: A survey on context-aware systems. Int. J. Ad Hoc Ubiquitous Comput. **2**(4), 263–277 (2007)
20. Chen, H.: An intelligent broker architecture for context-aware systems. Ph.D. proposal in Computer Science. University of Maryland, Baltimore, USA, pp. 1–129 (2003)
21. Chen, H., Finin, T., Joshi, A.: An ontology for context-aware pervasive computing environments. Knowl. Eng. Rev. **18**(03), 197–207 (2003)
22. Lara, O.D., Labrador, M.A.: A survey on human activity recognition using wearable sensors. IEEE Commun. Surv. Tutor. **15**(3), 1192–1209 (2013)
23. Saguna, S., Zaslavsky, A., Chakraborty, D.: Complex activity recognition using context-driven activity theory and activity signatures. ACM Trans. Comput. Hum. Interact. (TOCHI) **20**(6), 32 (2013)
24. Guinness, R.E.: Beyond where to how: a machine learning approach for sensing mobility contexts using smartphone sensors. Sensors **15**(5), 9962–9985 (2015)
25. Zhang, D., Peng, H., Haibin, Y., Lu, Y.: Crowd abnormal behavior detection based on machine learning. Inf. Technol. J. **12**(6), 1199–1205 (2013)
26. Saeedi, S., El-Sheimy, N.: Activity recognition using fusion of low-cost sensors on a smartphone for mobile navigation application. Micromachines **6**(8), 1100–1134 (2015)
27. Aggarwal, C.C.: Mining text data. In: Aggarwal, C.C. (ed.) Data Mining, pp. 429–455. Springer, Heidelberg (2015)
28. Rokach, L.: Decision forest: twenty years of research. Inf. Fusion **27**, 111–125 (2016)
29. Devi, B.R., Rao, K.N., Setty, S.P.: Towards better classification using improved particle swarm optimization algorithm and decision tree for dengue datasets. Int. J. Soft Comput. **11**(1), 18–25 (2016)
30. Stisen, A., Blunck, H., Bhattacharya, S., Prentow, T.S., Kjaergaard, M.B, Dey, A., Sonne, T., Jensen, M.M.: Smart devices are different: assessing and mitigating mobile sensing heterogeneities for activity recognition. In: Proceedings of the 13th ACM Conference on Embedded Networked Sensor Systems, pp. 127–140. ACM (2015)
31. Wawrzyniak, S., Niemiro, W.: Clustering approach to the problem of human activity recognition using motion data. In: Proceedings of FedCSIS, vol. 5, pp. 411–416 (2015)

32. Celebi, M.E., Kingravi, H.A., Vela, P.A.: A comparative study of efficient initialization methods for the k-means clustering algorithm. Expert Syst. Appl. **40**(1), 200–210 (2013)
33. Sadiq, F.I., Selamat, A., Ibrahim, R.: Performance evaluation of classifiers on activity recognition for disasters mitigation using smartphone sensing. Jurnal Teknologi **77**(13), 11–19 (2015)
34. Bhatia, S., Biyani, P., Mitra, P.: Identifying the role of individual user messages in an online discussion and its use in thread retrieval. J. Assoc. Inf. Sci. Technol. **67**(2), 1–13 (2015)
35. Van Asch, V.: Macro-and micro-averaged evaluation measures, pp. 1–14 (2013)

Image Classification for Snake Species Using Machine Learning Techniques

Amiza Amir[✉], Nik Adilah Hanin Zahri, Naimah Yaakob,
and R. Badlishah Ahmad

School of Computer and Communication Engineering,
Universiti Malaysia Perlis (UniMAP), 02600 Arau, Perlis, Malaysia
amizaamir@unimap.edu.my

Abstract. This paper investigates the accuracy of five state-of-the-art machine learning techniques — decision tree J48, nearest neighbors, k-nearest neighbors (k-NN), backpropagation neural network, and naive Bayes — for image-based snake species identification problem. Conventionally, snake species identification is conducted manually based on the observation of the characteristics such head shape, body pattern, body color, and eyes shape. Images of 22 species of snakes that can be found in Malaysia were collected into a database, namely the Snakes of Perlis Corpus. Then, an intelligent approach is proposed to automatically identify a snake species based on an image which is useful for content retrieval purpose where a snake species can be predicted whenever a snake image is given as input. Our experiment shows that backpropagation neural network and nearest neighbour are highly accurate with greater than 87 % accuracy on CEDD descriptor in this problem.

1 Introduction

There are approximately 3,000 snake species has been identified so far with 600 of them are categorized as venomous [13]. For many countries with vast agricultural sectors especially in South East Asia Region, snake-bite is a common occupational hazard to farmers and plantation workers [19]. Snakebite is a severe medical emergency which requires fast first-aid treatment and typically leads to hospital admission. Globally, at least 421,000 envenoming and 20,000 deaths occur each year due to snakebite [13]. In Malaysia, according to a five-year review of snakebite patients admitted to a tertiary university hospital, there were 260 cases of snakebites reported, and 52.9 % of them were bitten by unidentified species [4].

To perform optimal clinical treatment, the diagnosis of the snake species responsible for the bite is vital. The slightest delay might result in severe morbidity and mortality [4]. Therefore, it is important to precisely and concisely determine the type or species of the snakes. This information is important to identify if the snake is venomous or not, thus helps physicians to determine suitable anti-venom therapy and further treatment plan. Since different species

© Springer International Publishing AG 2017
S. Phon-Amnuaisuk et al. (eds.), *Computational Intelligence in Information Systems*,
Advances in Intelligent Systems and Computing 532, DOI 10.1007/978-3-319-48517-1_5

inhabits many different regions, the dynamic approach is needed and developed for diagnosing the type of snakes in various parts of the region.

Other than medical purposes, the snakes species identification is also useful for wildlife monitoring purpose [9]. Image captured by camera traps in national parks and forest can be used to assess the movement patterns of the corresponding reptiles [15]. Wildlife monitoring is essential to track the changes of the population size and the condition, as well as to detect new species inhabiting a particular area. Nature disaster and environmental pollution can affect the balance of ecosystem if not monitor adequately.

Nonetheless, considering the difficulty faced by most people in identifying the snake species, the main aim of this work is to perform species recognition by the input of a snake image. Satisfying this objective may serve various purposes in medical treatment and environmental management. Conventionally, snake species are distinguished manually based visual features such as head shape, skin color, eye shape, and body shape. This process requires knowledge of snakes features which is not common for most people where only a handful experts have this useful knowledge.

Therefore, in this paper, we present textural image classification of snake species from South East Asia using multiple supervised machine learning techniques to assist people (not necessarily experts) in recognizing the snake species. Our work compares and analyzes a few techniques to obtain the most precise and accurate outcome of the species classification. We manually gathered and created snake species corpus of 22 species from Perlis Snake Park in Malaysia. Then, we extracted textural features from each images using Color and Edge Directivity (CEDD) [3] descriptor. Finally, we performed species classification using five different state-of-the-art classifiers which are naive Bayes [8], decision tree J48 [17], k-nearest neighbors (k-NN) [6], nearest neighbors (k-NN with $k = 1$) [6], and backpropagation neural network [1,14]. The overview of the existing works regarding textural image classification is presented in Sect. 2. Section 3 describes database of snake images used in this research. In Sect. 4, we report experimental results and conclude our discussion with some direction for further works in Sect. 5.

2   Related Works

Numbers of automated image detection and recognition techniques were proposed several decades ago. Recently, the image classification and detection of animal has been explored to protect and monitor wildlife [5,16,20], and also vice versa, to protect human from animal threat [18]. Machine learning algorithms have been used widely in many studies to achieve this purposes.

Image-based wood species recognition was proposed by Zhao et al. [21–23]. In [22], they considered color, texture, and spectral features to discern the wood species visually by using the backpropagation neural network. This technique compares the classification output with desired output, and the errors are propagated back to the previous layer and adjusted until it meets the predetermined threshold. Zamri et al. [21] applied Improved-Basic Gray Level Aura Matrix (I-BGLAM) technique to extract 136 features from each wood image. Backpropagation neural network classifier then was used to classify 52 wood species based on the extracted features. In 2012, backpropagation neural network was used to recognize butterfly species in [12]. Features used were calculated by using branch length similarity (BLS) entropies from the boundary pixels of a butterfly shape. While in [23], color and texture features were used with the k-nearest neighbour (k-NN) as the classifier. kNN algorithm classifies wood species according to similarities of extracted features. Another work, Faria et al. [7] also proposed an image-based plant species recognition by using k-nearest neighbour. The research analyzed the RGB information channels to extract the suitable features for plant species identification problem.

Christiansen et al. proposed a detection of wildlife animal using thermal cameras to promote wildlife-friendly farming to reduce injury and death of animals due to agricultural machinery [5]. This work uses k-NN classifier to discriminate animal and non-animal based on heat characteristics of objects. In the work of Yu et al., sparse coding spatial pyramid matching (ScSPM) has been used to extract features and classify image captured by camera traps using Support Vector Machine (SVMs) [20]. They successfully used the method to identify over 57 animal species and achieved an average classification accuracy of 82 %. Parikh et al. presented multiple object detection techniques such as object matching, edge based matching and skeleton extraction for real-time application [16]. In contrast, Rangdal et al. proposed animal detection of targeted animal and prevent animal intrusion in the residential area. Their system uses Haar of Oriented Gradient to extract shape and texture features and use SVMs as classifier [18].

These developments have shown that the growing interest in nature and wildlife conservation increases the needs of utilizing the available technological breakthrough—such as sensors, digital technology, and artificial intelligence—for species identification. Snake is one of the animals with the highest number of species. Thus, there is a demand for an automatic snake species identification. To our knowledge, the closest work to our research can be found in [10]. In [10], an automatic snake species identification from snake images was proposed by using machine learning algorithms. In contrast to our work which applies texture based approach as features, James et al. [10] used features describing top, side and body views of snake images. However, these features were manually converted into feature vectors whereas, in our study, the texture-based features are extracted automatically from the snake images.

3   The Snakes of Perlis Corpus

The Snakes of Perlis Corpus consists of snake images which were collected from the management of Taman Ular and Reptilia, Perlis, Malaysia. The corpus contains 349 samples of 22 different snakes species. The highest number of samples with 33 samples is of the William Ratsnake species while the lowest number of samples with three samples is of the Bungarus candidus species. Figure 1 shows

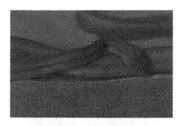

(a) Yellow Ratsnake

(b) William Ratsnake

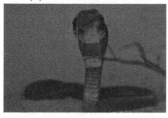

(c) Naja Tripudians

(d) Green Tree Python

Fig. 1. Examples of snake images in the snake of Perlis corpus

Table 1. The snakes of Perlis corpus: samples distribution

No.	Snake species	Number of samples	No.	Snake Species	Number of samples
1	Albino Burmese python	12	12	Reticulated Python	32
2	Green tree python	10	13	Malayan Krait	3
3	Albino naja sputatrix	31	14	Savu Python	12
4	Leucistic Texas rat snake	10	15	Malayan Pit Viper	9
5	Ball python	12	16	Mangrove Pit Viper	8
6	Naja tripudians	29	17	Dog-toothed Cat Snake	25
7	Black naja sputatrix	11	18	Asian Rat Snake	33
8	Ophiohagus Hannah	28	19	Copperhead Rat Snake	7
9	Boa constrictor	29	20	Yello Naja Sputatrix	8
10	Oriental rat snake	10	21	Javelin Sand Boa	4
11	Mangrove snake	7	22	Yellow Rat Snake	19

several examples of snake images in this experiment from the species of Yellow Ratsnake, William Ratsnake, Naja Tripudians and Green Tree Python (Table 1).

3.1 Feature Extraction

Feature extraction is a process to transform raw images into particular forms that are useful for machine learning tools to perform classification. As previously mentioned, we used a texture-based descriptor, Color and Edge

Directivity Descriptor (CEDD) [3] in this project since a snake species can usually be identified by human eyes based on its colour and skin texture. The CEDD descriptor uses colour and texture characteristics represented in histogram form. The features are created in two steps. Firstly, the histogram is divided into six regions based on texture [3]. Secondly, 24 regions are calculated from each of these regions by using colour characteristics. Hence, 144 features (6 × 24) are generated. The texture-based descriptor was selected to distinguish snake species based on the patterns of the snake skin. JFeaturelib package [11] was used to generate feature vectors for this descriptor. Hence, this procedure produces a dataset called Snake-CEDD dataset where the total number of attributes is 144.

4   Experiment

We conducted an extensive series of experiments to compare the accuracy of five state-of-the-art machine learning algorithms on Snake-CEDD data set (as described in Sect. 3. The Weka package [2], was used to implement the five machine learning techniques (nearest neighbours, k-nearest neighbours (k-NN), naive Bayes, backpropagation neural network, and decision tree J48) in these experiments. For k-NN, we randomly selected $k = 7$ in this experiment. While for backpropagation neural network, we used three-layer network where the hidden layer consists of $\frac{c+d}{2}$ nodes where c is the number of species and d is the number of attributes. In all experiments, a standard machine learning evaluation technique, the 10-fold cross-validation, was performed for ten repetitions to ensure the correctness of the experimental results. The ten repetitions ensure that all classes including classes with small sample size were included in the training and testing process.

4.1   Performance Metrics

The snake species identification problem is presented as a multi-class problem. We present the accuracy of five of the state-of-the-art algorithms in Sect. 4.2. The accuracy is measured by four metrics: the percentage of the correct predictions, precision, recall and F-measure. Correct predictions give the percentage of the images which were correctly classified. The precision explains the fraction of test images classified as a class x that are truly from the class x whereas recall is the fraction of test images from a class x that are correctly identified as the class x. For instance, given a class label Green Tree Python, the precision gives the fraction of the images classified as Green Tree Python that are correctly from the Green Tree Python species. The recall explains the fraction of the Green Tree Python images which are correctly identified as the Green Tree Python species. F-measure is the harmonic means of precision and recall. These performance metrics are calculated by using Eqs. 1, 2, 3 and 4.

$$\text{Correct predictions} = \frac{TP + FP}{TP + FP + TN + FN} \tag{1}$$

where TP is the number of true positive cases, TN is the number of true negative cases, FP is the number of false positive cases, and FN is the number of false negative cases.

$$\text{Precision} = \frac{TP}{TP + FP} \tag{2}$$

$$\text{Recall} = \frac{TP}{TP + FN} \tag{3}$$

$$\text{F-measure} = 2 \cdot \frac{precision \cdot recall}{precision + recall} \tag{4}$$

The average accuracy for each set of 10 cross-validation folds is calculated resulting one result is produced for each run. Hence, for each algorithm, 100 estimates (10 folds times 10 runs) are used to calculate the accuracy metrics in these experiments.

4.2   Results and Discussion

Table 2 shows the accuracy of all algorithms on Snake-CEDD dataset. The correct predictions value for each of the five algorithms: 75.64 % for naive Bayes, 87.93 % for backpropagation neural network, 89.22 % for nearest neighbour, 80.34 % for k-NN, and 71.29 % for J48. The correct predictions score of backpropagation neural network and nearest neighbours are significantly better than naive Bayes at the 5 % level of statistical significant. Naive bayes, however, has the correct predictions score better than the accuracy of J48 but this maybe due to chance since we cannot rule it out at 5 % significant level. All techniques perform worse than nearest neighbour where the correct predictions score of J48, k-NN, and naive Bayes are significantly worse than nearest neighbours (at 5 % significance level). Nearest neighbour gives better prediction since all snake images of a particular species are usually similar in texture and colour.

The recall, precision and F-measure for backpropagation neural network, nearest neighbour and k-NN are significantly higher than naive Bayes and J48 algorithm. They perform almost perfect predictions with 0.99, 0.96, and 0.96 precision for backpropagation neural network, nearest neighbour and k-NN, respectively. For recall, these algorithms (backpropagation neural network,

Table 2. Accuracy result on Snake-CEDD dataset

Classifier	Correct predictions	Recall	Precision	F-measure
Naive Bayes	75.64	0.92	0.94	0.93
Backpropagation neural network	87.93	1	0.99	0.99
Nearest neighbour	89.22	1	0.96	0.97
k-NN ($k = 7$)	80.34	1	0.96	0.97
Decision tree J48	71.29	0.79	0.71	0.72

nearest neighbour and k-NN) obtain perfect prediction (score 1 for recall). Backpropagation neural network obtains slightly higher F-measure score compared to nearest neighbours and k-NN, hence, it is not significantly better than nearest neighbours and k-NN. The results demonstrate the high probability of backpropagation neural network, nearest neighbour and k-NN to provide correct and consistent predictions.

5   Conclusion

This work introduced a snake images corpus, namely Snake of Perlis Corpus which consists of snake images from Taman Ular & Reptilia, Perlis, Malaysia. The corpus was used to experiment the effectiveness of a snake species identification by using machine learning techniques. The experiment shows that the nearest neighbour algorithm perform and the backpropagation neural network perform the best on the dataset extracted from the images by using texture-based feature descriptor (CEDD descriptor). To our knowledge, this is the first time texture-based features are used for automatic snake species identification. In future work, we plan to apply other image-based features (such as Haralick, shape, and spectral features) and to apply segmentation process for snake species recognition. We will also conduct further test on the accuracy of backpropagation neural networks and k-NN for this problem with different parameters.

Acknowledgments. Authors would like to thank the Taman Ular & Reptilia, Perlis and the School of Computer and Communication Engineering for the facilities provided in conducting this research.

References

1. Anuar, S., Selamat, A., Sallehuddin, R.: Hybrid artificial neural network with artificial bee colony algorithm for crime classification. In: Phon-Amnuaisuk, S., Au, T.W. (eds.) Computational Intelligence in Information Systems. AISC, vol. 331, pp. 31–40. Springer, Heidelberg (2015). doi:10.1007/978-3-319-13153-5_4
2. Bouckaert, R.R., Frank, E., Hall, M., Kirkby, R., Reutemann, P., Seewald, A.,Scuse, D.: WEKA Manual for Version 3-7-11 (April 2014). http://www.cs.waikato.ac.nz/ml/weka/documentation.html
3. Chatzichristofis, S.A., Boutalis, Y.S.: CEDD: color and edge directivity descriptor: a compact descriptor for image indexing and retrieval. In: Gasteratos, A., Vincze, M., Tsotsos, J.K. (eds.) ICVS 2008. LNCS, vol. 5008, pp. 312–322. Springer, Heidelberg (2008). doi:10.1007/978-3-540-79547-6_30
4. Chew, K.S., Khor, H.W., Ahmad, R., Rahman, N.A.H.N.: A five-year retrospective review of snakebite patients admitted to a tertiary university hospital in malaysia. Int. J. Emerg. Med. **4**(1), 1–6 (2011)
5. Christiansen, P., Steen, K.A., Jrgensen, R.N., Karstoft, H.: Automated detection and recognition of wildlife using thermal cameras. Sensors **14**(8), 13778 (2014)
6. Cover, T., Hart, P.: Nearest neighbor pattern classification. IEEE Trans. Inf. Theor. **13**(1), 21–27 (2006). http://dx.doi.org/10.1109/TIT.1967.1053964

7. Faria, F.A., Almeida, J., Alberton, B., Morellato, L.P.C., Rocha, A., da Torres, R.S.: Time series-based classifier fusion forfine-grained plant species recognition. Pattern Recogn. Lett. **81**, 101–109 (2015)
8. Friedman, N., Geiger, D., Goldszmidt, M.: Bayesian network classifiers. Mach. Learn. **29**(2–3), 131–163 (1997)
9. Gray, M.J., Chamberlain, M.J., Buehler, D.A., Sutton, W.B.: Wetlandwildlife monitoring and assessment. In: Anderson, T.J., Davis, A.C. (eds.) Wetland Techniques: Volume 2: Organisms, pp. 265–318. Springer, Dordrecht (2013)
10. James, A.P., Mathews, B., Sugathan, S., Raveendran, D.K.: Discriminative histogram taxonomy features for snake species identification. Human-Centric Comput. Inf. Sci. **4**(1), 1–11 (2014)
11. JFeatureLib: JFeatureLib: A free java library containing feature descriptorsand detectors, April 2016. http://code.google.com/p/jfeaturelib/. Accessed 6 Apr 2015
12. Kang, S.H., Song, S.H., Lee, S.H.: Identification of butterfly species with a single neural network system. J. Asia-Pacific Entomol. **15**(3), 431–435 (2012)
13. Kasturiratne, A., Wickremasinghe, A.R., de Silva, N., Gunawardena, N.K., Pathmeswaran, A., Premaratna, R., Savioli, L., Lalloo, D.G., de Silva, H.J.: The global burden of snakebite: a literature analysis and modelling based on regional estimates of envenoming and deaths. PLoS Med. **5**(11), 1–14 (2008)
14. Li, J., Cheng, J.H., Shi, J.Y., Huang, F.: Brief introduction of back propagation (bp) neural network algorithm and its improvement. In: Jin, D., Lin, S. (eds.) Advances in Computer Science and Information Engineering. AISC, vol. 2, pp. 553–558. Springer, Heidelberg (2012)
15. Meek, P.D., Ballard, G.A., Fleming, P.J.S.: The pitfalls of wildlife camera trapping as a survey tool in australia. Aust. Mammal. **37**, 13–22 (2015)
16. Parikh, M., Patel, M., Bhatt, D.: Animal detection using template matching algorithm. Int. J. Res. Mod. Eng. Emerg. Technol. **1**(3), 26–32 (2013)
17. Quinlan, J.R.: C4.5: Programs for Machine Learning. Morgan Kaufmann Publishers Inc., San Francisco (1993)
18. Rangdal, M.B., Hanchate, D.B.: Animal detection using histogram oriented gradient. Int. J. Recent Innov. Trends Comput. Commun. **2**(2), 178–183 (2014)
19. Warrel, D.: Guidelines for the management of Snake-Bites. World Health Organization (2010)
20. Yu, X., Wang, J., Kays, R., Jansen, P.A., Wang, T., Huang, T.: Automated identification of animal species in camera trap images. EURASIP J. Image Video Process. **2013**(1), 1–10 (2013)
21. Zamri, M.I.P., Cordova, F., Khairuddin, A.S.M., Mokhtar, N., Yusof, R.: Tree species classification based on image analysis using improved-basic gray level aura matrix. Comput. Electron. Agric. **124**, 227–233 (2016)
22. Zhao, P., Dou, G., Chen, G.S.: Wood species identification using feature-level fusion scheme. Optik Int. J. Light Electron Optics **125**(3), 1144–1148 (2014)
23. Zhao, P., Dou, G., Chen, G.S.: Wood species identification using improved active shape model. Optik Int. J. Light Electron Optics **125**(18), 5212–5217 (2014)

Rides for Rewards (R4R): A Mobile Application to Sustain an Incentive Scheme for Public Bus Transport

Nyuk Hiong Voon[1(✉)], Siti Noora'zam Haji Abdul Kadir[1],
Matius Anak Belayan[2], Sheung Hung Poon[1],
and El-Said Mamdouh Mahmoud Zahran[2]

[1] School of Computing and Informatics,
Universiti Teknologi Brunei, Gadong, Brunei Darussalam
{jennifer.voon, sheunghung.poon}@utb.edu.bn,
sitinoorazam.itb@gmail.com
[2] Faculty of Engineering, Universiti Teknologi Brunei,
Gadong, Brunei Darussalam
matiusbels98@gmail.com, elsaid.zahran@utb.edu.bn

Abstract. Bus transit is not popular in Brunei partly due to high ownership of private cars and this will lead to severe traffic congestion in the future. This paper discusses the design of a mobile application to sustain an incentive scheme for public bus transportation in Brunei. It supports a case study on whether an incentive scheme has impact to increase bus transit ridership under the prevailing transport conditions. The application has a front-end mobile app to collect dates and time stamps of each ride and a back-end infrastructure to manage stakeholder details, incentive schemes, bus routes and a repository for travel time stamps.

Keywords: Public transportation · Incentive scheme · Bus system · Bus ridership · BaaS · Rides for rewards

1 Introduction

Transportation is among the more vital economic activities to move goods and people on demand from source to destination. As population increases, the cost of transportation manifests itself in many forms such as higher costs of sustaining land use for transportation, and health, safety and environmental effects from pollution and traffic congestions. There is urgent need to strive for optimal use of land for roads and highways, to address traffic congestion and air pollution contributed mostly by petroleum powered vehicles. Hence, an efficient transportation system is critical to balance the needs of essential road users and the environmental impacts from transport.

The two main forms of transportation, private and public differ largely on their usage. Public transportation such as buses and trains generally convey a larger number of people and goods, and available for public use. Normally, it runs on fixed routes and charges set fares. Private transportation is on-demand and not available for public use. Public transportation such as buses and rapid transits are effective for their high

© Springer International Publishing AG 2017
S. Phon-Amnuaisuk et al. (eds.), *Computational Intelligence in Information Systems*,
Advances in Intelligent Systems and Computing 532, DOI 10.1007/978-3-319-48517-1_6

capacity modes to move people and goods. Findings argue that effective mobility management resulting in high quality public transportation and transit oriented development reduces vehicle miles travel, thereby reduces pollution emissions and traffic congestion, resulting in better consumer savings and economic development [1]. Such findings place emphasis for countries to develop effective public transportation strategies to encourage its citizens to peruse of public transportation. In Brunei, this emphasis has yet to be realised nor experienced by the populace due to the ease of owning private vehicles, generous fuel subsidies and relative small population, contributing to low ridership in public transportation. Such issues are highlighted in a white paper on the need to formulate a masterplan to sustain a land transportation system for the country [2]. The current low ridership is a concern. To alleviate this issue, a case study is conducted on whether an incentive scheme has impact to increase ridership under the prevailing bus transit conditions. This paper discusses the design of a mobile application to support the incentive scheme. The application has a front-end mobile app focusing on building the interfaces for the respective stakeholder roles and collects the dates and time stamps of rides, while the back-end supports the administration and database infrastructure.

The rest of the paper is organized into the following sections: Sect. 2 discusses the prevailing bus ridership problem; Sect. 3 discusses the background of incentive programs on public transit; Sect. 4 discusses our approach in designing and developing the R4R incentive application; Sect. 5 provides a critical discussion of the results of the pilot study; and the conclusion and further research are presented in Sect. 6.

2 Problem Statement

The current low ridership of the public bus transport is a concern and a reflection of the high ownership of private vehicles in Brunei. Results of the Transport Attitude Survey conducted by the Centre for Strategic and Policy Studies (CSPS) confirm high car dependency across the surveyed population and a large proportion of those who use the bus transport regularly have no access to cars [2]. It is evident that public transportation, particularly buses, is not effectively utilized by the local populace. Despite efforts by the Land Transport Department to improve the bus transport system such as enhancements in route maps, schedules and interior comfort [3], the number of bus riders remain low. Meanwhile ownership of private vehicles increased by 19 % over two years (2011–2013) and the car ownership rate at 2.65 people per vehicle is high [4]. Clearly the enhancement efforts have not motivated more local populace to use the public bus transport. This may be due to the common attractions of private car, such as diverse activity locations, abundant parking lots and relative lower levels of traffic jams [5].

3 Related Literature

Incentive schemes have been applied on public transport to reduce commuter congestions at peak times or to encourage more commuters to switch from private to public travel mode. One such scheme is the Travel Smart Rewards (TSR) in Singapore.

The TSR is introduced to optimize public transport capacity by reducing morning peak period congestion on the Mass Rapid Transit (MRT) or Light Rail Transit (LRT) [6]. TSR uses the accumulation of points for cash rebates. Points are earned per trip on the MRT or LRT with the perk of extra points in the morning off-peak hours. Further motivation to induce off-peak riders is the use of "gamification" where riders login to play "Spin to Win" for more point accumulation or gain cash rebate. In 2014, the TSR extended to corporate-tier rewards to encourage companies to create supportive environments for flexi-travel for their employees, i.e. to be able to travel on public transport during off-peak periods [7]. TSR was launched in January 2012 and within two years of implementation, the results are encouraging. On average about 12 % employees (pilot study of 12 companies) shifted out of the morning peak periods [8] and an estimated 7–8 % decrease in commuters number in the morning peak periods since 2013 [9].

The INSTANT project is another example of a pilot program to use an incentive mechanism to decongest roads in Bangalore, India, by encouraging commuters to travel at less congested periods. The focus group is the employees of Infosys Technologies, Bangalore. Qualified commuters are awarded credits daily based on arrival times, and weekly, an algorithm is applied to qualify winning commuters for cash rewards. Winners are randomly selected based on different levels, and non-winners at higher levels qualify for lower levels. After the weekly draw, credit deduction is applied to all winners and non-winners. The credit deduction feature influences the behavior of commuters to arrive early to maintain credit balances for the weekly draws and previous winners would need a longer period to build up credit balances [10]. The project ran for a period of six months and recorded results over three time segments (before 8 AM, 8–8.30 AM, 8.30–9 AM). The results show that number of commuters in each segment has roughly doubled over the study period [10].

By contrast, our R4R incentive scheme approach focusses on increasing bus ridership overall rather than shifting commuter travel to off peak times. TSR and INSTANT incorporated "gamification" and employers participation to entice travel at less congested periods while R4R is based on specific routes of a bus company and a group of vendors to provide rewards points accumulated per ride, mainly due to the reasons that the Brunei bus transport system has not matured to an extent with convenient routes (bus on-off boarding is within walking distance) and ticketing is manual. Hence the R4R design would need to account for these issues. In addition, for the same public transport patronage scheme, outcomes are different in different countries [11]. This adds up to the novelty of the R4R scheme, introduced in this paper, since no similar schemes have ever been implemented in Brunei.

4   Methodology of the Rides for Rewards (R4R) Application

Figure 1 shows the methodology applied to develop the R4R mobile application. It has five stages, namely pre-initiation survey, collect requirements, assess requirements, develop prototype and conduct a pilot study. The pre-initiation survey is a pre-development stage conducted early to gauge the viability to develop the application.

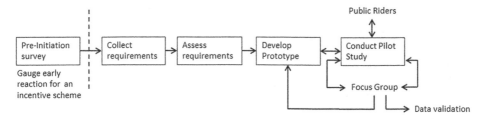

Fig. 1. Methodology applied to develop the R4R application.

4.1 Pre-Initiation Survey

Prior to the development of the application, an initial survey is conducted to gauge early opinions on whether the surveyed group would switch to bus transport mode if a mobile based incentive scheme is in place and to get some insights on their preferences on type of rewards. The target group is students from the same university, the rationale being that university students are perceived to be low income earners, dependent on others for transportation and of an age where they are independent travelers. A total of 102 responses are received and 84 % are students. While 59 % has indicated that they would switch to bus transport if the incentive scheme is introduced, however, only about 21 % rate that the scheme would be successful to increase bus ridership. In terms of preferences for rewards, the top three choices are free movie tickets (59 %), food vouchers (51 %) and entertainment discounts (48 %). However, when the respondents are asked to list local businesses by name for rewards redemption, the top business choices are food and beverages related at about 33 %. These results show some evidence that an incentive scheme would be supported and preferred rewards are geared towards food and beverages.

4.2 Collect Requirements

Collect requirements is in the form of a survey conducted over a period. A landing page (http://ride4rewards.tumblr.com/) is created to provide information on the R4R scheme and direct respondents to one of two surveys, based on age category: 15–29 years and over 30 years. Link to the survey is propagated on buses and social media. The objectives are to gain insights on the demographic details of bus commuters, frequencies of bus rides, further insights on inducements to convert non-commuters to commuters and the choice of mobile technology in common use. Table 1 is a summary of the responses.

A total of 226 responses were recorded, 191 responses from the 15–29 years category and 35 responses from the over 30 years. In the 15–19 years category, respondents are mostly predominantly students (87 %) with 85 % in low income level. The over 30 years category are mostly working in the public sector (66 %) and 57 % are in the considerable high income level. In frequency of using public bus, both categories show similar results i.e. 63 % have never taken the bus. This is reflected in their daily mode of travel of very high usage of private vehicles at 90–100 %. There is

Table 1. Summary of responses from the Collect Requirements stage

	Category: 15–29 years	Category: over 30 years
Number of responses	191	35
Demographics		
Age group	43 % in the 20–24 age group	40 % in the 30–34 age group
Nationality.	87 % citizens; 8 % permanent residents	91 % citizens; 6 % permanent residents
Occupation	4 % government sector; 87 % students	66 % government sector; 2 % students
Average monthly income	85 % in the $0–$499 range	57 % in the above $3000 range
Bus ridership		
Permission to ride bus	61 % have parents who allow them to use the bus on their own	63 % would allow their young children, 15 years and above, to ride the bus on their own
Frequency of rides	63 % have never used the bus; 3 % use the bus daily	63 % have never used the bus; 3 % use bus at least monthly
Incentives		
R4R incentive scheme (claim travel behavior)	51 % stated R4R will encourage them to ride the bus	29 % stated R4R will encourage them to ride the bus
Rewards	15 % towards free bus rides; 26 % towards gadgets, clothes and food/beverage vouchers	26 % towards free bus rides; 20 % towards gadgets, clothes and food/beverage vouchers
Others		
Mobile technology	66 % used Android phones; 30 % used iPhone	63 % used Android phones; 37 % used iPhone
Daily transport mode	90 % private vehicle; others 10 %	100 % private vehicle

some evidence that a bus ridership incentive scheme would encourage more riders in both categories; 51 % from the 15–24 years and 29 % in the over 30 indicated that they would take the bus with an incentive scheme in place. The surveyed results also show Android device is predominantly high for both categories at over 60 %, though there is a relatively high percentage for IOS device at 30–37 %.

4.3   Assess Requirements

The findings of the Collect Requirements stage are used to identify which stakeholders to connect with, in terms of relevant bus routes and businesses providing the rewards. The findings will also influence the development technology for the mobile application. From the responses, the mobile application can expect better support from students and youths in the age group 15–29 years, with low income level. Hence bus routes selected for the study are more relevant to this group. Brunei has only six private bus companies and each company is allowed to operate specific routes. Amongst the six companies, only one company (PHLS Sdn. Bhd.) agreed to participate on six of their routes. These routes cover some of the areas highlighted in the survey as relevant to this group, such as places of worship, shopping malls and youth centers. Similarly, the vendors supplying the rewards are selected based on the preferred choices of rewards identified from the survey. Hence, the six businesses participating in the scheme include apparel, restaurant franchises and companies promoting activities which appeal to youths such as a fitness center. Finally, the predominant mobile technology is identified as Android based (66 %) though the IOS based devices are relatively high at 30 %. Therefore the mobile application was developed on Android. IOS based devices will access the scheme via Web app.

Though the assessment is focused on the youths, comments from the over 30-years category should be noted, that an incentive scheme would have minimal success without taking into account factors such as high ownership of private vehicles, generous fuel subsidies and sufficient number of bus stops within walking distance for commuters.

4.4   Develop Prototype: R4R Mobile Application

Developing the prototype takes into consideration the needs of the main stakeholders and the potential that the R4R incentive scheme can evolve into a business platform to sustain the public transportation industry in Brunei.

Description of the R4R Ecosystem. R4R is designed to support an incentives-based ecosystem comprising several stakeholders: bus riders, bus companies, vendors providing the rewards and administrator. Figure 2 shows a high level use case diagram of the main actors i.e. stakeholders. The main role of the bus rider is to earn points and redeem points for rewards. Points are earned per bus ride, by referral and a once-off for registration. During the study period, riders are issued special R4R tickets with unique codes which can be scanned to earn points using the mobile application. After accumulation of sufficient points, riders select respective rewards to redeem. Once they claimed a reward, a unique code is issued. This code will enable the rider to use the reward at the respective vendor's business place by simply showing the code. The role of the vendors is to supply details of the rewards and to verify the rewards claim upon presentation of the code by the rider. For the bus companies, their role is to supply details of the bus routes and to issue the special R4R tickets to the riders. The administrator's role is to manage the platform for the R4R ecosystem. This includes managing all components relating to riders, vendors, bus companies, the rewards system, the code system and in addition, the R4R platform itself.

For the prototype, management of the ecosystem is highly dependent on the administrator. If the R4R ecosystem goes "live" and operates as a business entity, it is to be expected that some of the other actors (stakeholders) will assume better control of their respective functionality. For example, vendors should have autonomous control to initiate their own rewards and bus companies would directly update their route information.

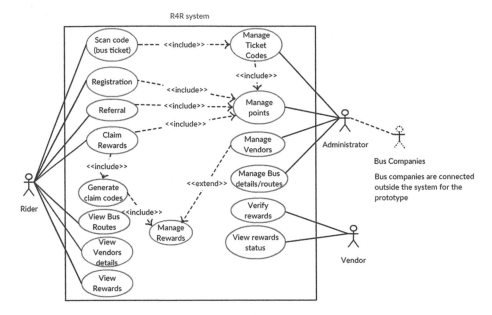

Fig. 2. Use case diagram for the main actors of the R4R system.

Architecture Overview of R4R. The architecture overview for the R4R application is based on a Backend as a Service (BaaS) approach to connect the front-end mobile app to the backend resources in the cloud using unified application programming interfaces (APIs) and software developer's kits (SDKs) (Fig. 3).

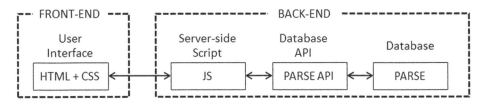

Fig. 3. Architecture overview of the R4R system.

The front-end R4R mobile app is designed using Intel XDK software [12]. It is created as a hybrid app of web and native technologies using HTML5 technologies, integrated with Cordova plugin which enable the app to access the device's native features such as camera and GPS. Pure JavaScript is used to communicate data between the client (app) and the server. Parse is used for the back-end comprising the Parse server and online database [13]. Parse has its own API to communicate with the database using JavaScript, Java or Objective-C. This is an advantage as it provides a unified interface between HTML5 and the back-end technology. Hence developers need not switch between different programming languages during development and can focus more on the front-end app development.

Features of the R4R Mobile App. The R4R mobile app prototype is focused on two actors, Riders and Vendors. Hence the app features two different login modes, Rider login and Vendor login. Riders have access to four main features: Rewards, Scan for points, Feedback, and Profile. Rewards feature lists available rewards that can be redeemed with the required number of points. Scan for Points feature allows riders to earn points by scanning the QR code on the special R4R bus ticket or by manually keying in the code found below the QR code. Figure 4 shows the process of how Riders earn points and redeem rewards. Feedback feature is a form whereby riders' feedback is captured live on the Parse database. The Profile feature is where Riders edit account details. In addition, Profile comes with additional views: "MyRewards", "MyTasks", "Sponsor Profiles" "Bus Routes" and "About". All these views basically list the information relevant to the Rider and inform various statuses of their points and rewards.

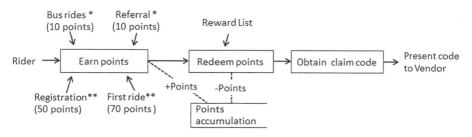

Fig. 4. Process to earn and redeem points for reward; *Points earned each time; **Points earned once-off

Vendors have access to two features, view the status of their rewards offered and use the Scan Code feature to scan codes from Riders claiming their rewards. Administrator oversees the R4R platform i.e. manages the technical operations of the platform as well as the business process(es). For example, a business process is capturing sponsor and reward details on the system, after negotiating and agreed upon on the terms with vendors. "Bus Companies" is currently an actor that does not interact direct with the R4R system. They feed details of the bus routes manually to the Administrator for input into the system.

R4R Mobile App User Interface (UI) Design. The overall UI is designed for familiarity in user experience. Hence the design is influenced by minimalistic art movement for simplicity and to reduce graphic noise. Button and input boxes are designed with rounded edges to soften the look. Icons are designed for familiarity to engage riders (users). Serif font is chosen as it is web-browser friendly. "Roboto Slab" font is selected for highlights as it gives off a semi-formal look and its thickness is ideal for legibility against the Serif font. Finally the main color scheme is "teal/green" to reflect the sustainability nature of the study and because "Green" colour has a feel of nature, balance and harmony. Figure 5 shows the UI design for the app and the theme is consistently applied throughout.

(1)   **(2)**   **(3)**   **(4)**   **(5)**

Fig. 5. User Interface (UI) for the R4R mobile app. (1) Login page (2) List of available rewards (3) Scan QR code on R4R bus ticket (4) Riders' feedback (5) Riders' profile page. Noted the picture is shown in shades of grey while the actual colour presentation is in shades of "teal/green". (Color figure online)

4.5   Conduct Pilot Study

A pilot study is conducted to test the prototype on the field and to observe claimed and actual travel behaviour during the trial period. The trial period is three weeks. The participants are one bus company with six routes (Routes 34–39), six vendors sponsoring the rewards and riders and bus Riders. The Riders also include a focus group of seventeen survey respondents randomly selected. The focus group will test and provide technical feedback of the R4R app and participate in the data validation where the main purpose is to observe the claimed and actual travel behaviour with the R4R incentive scheme during the trial period. Public riders were informed of the study and where to access the app through posters on the buses and ticket conductors. The Android version is downloadable from Google Play [14]. The web app version is downloadable via a URL link [15]. Riders who participate in the R4R pilot study are required to request for QR codes from the conductor. One QR code would be issued per ride. The codes are scanned via the mobile app to earn ten points. The point system was designed in such a way that riders would earn points not only by riding the bus but also by completing other tasks such as referrals of the mobile app. During the trial period, riders are rewarded generously to encourage them to take the first step towards sustainable travel behaviour as shown in Fig. 6.

	Register	First ride	Refer a friend	Total
Points	50	70	10	130

Fig. 6. R4R pilot study bonus points

Technical bugs with the application as reported by the focus group were resolved with relative ease within the study period, though non-technical issues were not resolved, such as some buses being unable to distribute the codes as they were not adequately briefed about the trial study.

5   Results

Survey results from the pre-initiation stage show some evidence that commuters will switch to taking bus transport with an incentive scheme, as claimed in 59 % of the respondents, who are mainly tertiary level students. The basis of this percentage justify developing the R4R mobile application to support the pilot study. The follow-up survey from the Collect Requirements stage is targeted for two age groups 15–29 and over 30. Results show that the highest percentage of feedback comes from the age group 20–24, of which 87 % are students (Table 1). This is expected as the incentive scheme would appeal more to those who are low income owners and are dependent on others for transport. This age group is also tech-savvy according to Socialbakers which stated a majority of Facebook users in Brunei are 18–24 (33 % population) [16]. Not surprisingly, the survey shows 96 % in this age group own a mobile smart device (Table 1). This is crucial as the R4R incentive scheme is developed for mobility.

Findings from both surveys influenced the development of the R4R system as a platform as well as the user interface design of the R4R mobile app. The platform has to support a mobile solution where commuters can do live updates to earn and redeem points and vendors are able to do live validation of claim codes. The mobile app is both Android and web based for broad coverage. The R4R platform and mobile app went live during the pilot study stage. The process flow of earning, redeeming and claiming points (Fig. 4) using the R4R system was successfully demonstrated. The results of the study period showed that R4R gained thirty two registered users and eleven riders successfully earn points by scanning or keying codes from bus tickets. The highest point earner from one rider is 140 points. A total of 10 reward vouchers were claimed by the riders.

In the pilot study stage, the focus group of 17 participants are sorted into two groups where 9 of 17 answered 'Yes' to taking the public bus with R4R incentive scheme, meanwhile 8 out of 17 answered 'No' to taking the bus with R4R incentive scheme. The number of participants in each group is decided in such a way to replicate similar claimed behaviour percentage (Table 1) from the survey conducted at the Collect Requirement stage. In the first week, none of the participants took the public bus. In the final week, only 4 out of 9 participants who claimed that they will take the public bus with R4R incentive scheme actually took the public bus, the other 5 did not. For the 8 participants who claimed that they will not take the public bus with R4R incentive scheme, 6 performed as claimed while 2 shifted their behaviour and took the bus. Based on this, 35 % of the participants took the public bus with R4R incentive scheme and 65 % did not. A discrepancy can be observed between the participants' claimed behaviour from the survey responses (51 %) and the actual behaviour (35 %) from the pilot study. This is as expected it takes time to change habits or behaviours [17].

6   Conclusions and Future Work

There are technical and business aspects to sustaining the R4R incentive scheme for public bus transport. The technical aspects involve the development of the front-end mobile app and the back-end cloud infrastructure. This led to the implementation of the

R4R mobile and web apps accessible on Google Play and URL respectively, while the R4R cloud is on a BaaS service provider (Parse). The business aspect of R4R includes integrating all the services of the vendors (rewards sponsors) and the bus companies with the back-end to serve the riders.

The R4R technical platform has demonstrated its workability to capture data especially ride information details and emulated the process of earning, redeeming and claiming rewards. This technical platform has potential to be evolved into a business to support not only bus transport but other forms of public transports. The business aspect of the R4R scheme would have to encompass the main stakeholder, riders, vendors, bus companies and the business itself. To be a viable business, the technical platform has to be further developed to include features to let vendors and transport companies do live updates and provide advertisement space. It would also need to go beyond the current basic gamification features to better engage riders. Furthermore, as data accumulate, there is potential to include analytics clients to explore data.

At the conclusion of the study, it is not possible to determine if the R4R scheme was able to increase bus ridership based on current conditions. This is due to several factors such as: the study period of three weeks being too short to get realistic data, insufficient bus operators are participating and therefore there are insufficient bus routes to simulate normal travel conditions, insufficient propaganda on the scheme hence attracting minimal riders. However, the focus group results (35 %) show some evidence that riders will actually take the bus with an incentive program in place. This is a good indication to further the research of the R4R scheme to better understand its viability as an incentive mechanism and a sustainable business to address the low ridership issue.

It should be noted that the CSPS whitepaper has discussed that the main issue which causes low ridership in bus transport is poor quality of the bus transport system in terms of conditions of buses and bus terminals, connectivity to places, overall journey speed and high ownership of private vehicle [2]. Collaborative efforts with several organisations such as the Land Transport Department, policy makers and bus operators are necessary to manage this issue. The R4R platform aspires to be the catalyst to initiate such organisations to start prioritising improvement of the public bus transport system to address the low ridership issue.

Acknowledgements.   The authors wish to thank PHLS Sdn. Bhd. for allowing the team to conduct the pilot study on its buses and the participating businesses (CBTL, Oh My Wings, The Apparel Corner BN, Genki Teppanyaki, Kaleidoscope and Food Panda Brunei) which generoussly provided the rewards for the incentive scheme.

References

1. Litman, T.: Are vehicle travel reduction targets justified? Victoria Transport Policy Institute (2013). http://www.vtpi.org/vmt_red.pdf. Accessed 7 June 2016
2. Center for Strategic and Policy Studies: Review to formulate a roadmap and draft national masterplan for a sustainable land transportation system for Brunei Darussalam, vol. 5, pp. 39–41. Brunei Darussalam (2014)

3. Radio Television Brunei, News Center.: Launching of Brunei Darussalam public bus service's new image (2013). http://goo.gl/0dVF6X. Accessed 8 June 2016
4. Bandial, Q.: Number of new vehicles increased by 19 % between 2011 and 2013. The Brunei Times (2015). http://goo.gl/RTAl7h. Accessed 10 June 2016
5. Djoen, S., Masaru, Y., Kunihiro, S., Hisashi, K.: Opportunities and strategies for increasing bus ridership in rural Japan: a case study of Hidaka City. Transp. Policy **24**, 320–329 (2012)
6. Land Transport Authority, Singapore: Travel smart measures for organisations (2016). http://goo.gl/JIS5hG. Accessed 9 June 2016
7. Land Transport Authority, Press Room, Singapore: More companies to "travel smart" (2014). http://goo.gl/qDPPyd. Accessed 9 June 2016
8. Lam, L.: Travel smart programme. Travel smart showcase. Singapore (2015)
9. Chew, L.: Shifting travel demand. Institute of Policy Studies, Singapore (2015). http://www.ipscommons.sg/shifting-travel-demand/. Accessed 12 June 2016
10. Merugu, D., Prabhakar, B., Rama, N.: An incentive mechanism for decongesting the roads: a pilot program in Bangalore. In: Proceedings of the 10th ACM Conference on Electronic Commerce, Workshop on the Economics of Networked Systems, 6–10 July 2009
11. Coombe, R.D.: Urban transport policy development: two case studies in the Middle East. Transport Rev. **5**(2), 165–188 (1985)
12. Intel Developer Zone: Intel® XDK (2016). https://software.intel.com/en-us/intel-xdk
13. Parse (2016). http://parse.com/
14. Noorazam, S.: R4R Ride4Rewards Beta. Google Play (2016). http://goo.gl/EgTKe0
15. Azammeh, H.: R4R Ride4Rewards Beta (2016). http://ride4rewards.comli.com/
16. Hayat, H.: Brunei tops in Facebook usage ranks No. 1 in Asia (2012). BruDirect Local News. http://goo.gl/nMHAj7. Accessed 29 Mar. 2016
17. Pluntke, C., Prabhakar, B.: INSINC: a platform for managing peak demand in public transit. JOURNEYS, pp. 31–39, September (2013)

Mobile *mBus* System Using Near Field Communication

Hexi Yeo, Phooi Yee Lau[⊠], and Sung-kwon Park

Universiti Tunku Abdul Rahman,
Kampar Campus, 1 Jalan Universiti, 31900 Kampar, Malaysia
yhx222@lutar.my, laupy@utar.edu.my,
sp2996@hanyang.ac.kr

Abstract. Near field communication (NFC) is a form of short range contactless and wireless communication between devices, a subset of RFID, with a much shorter communication range for security purposes. Any objects can become a passive device by tagging a NFC tag on them. In the past, RFID and smart card was a feasible choice for bus fare/token ticketing as their tag can cost as low as 10 cents/piece, being also the average price for a NFC tag. However, their tag readers are expensive, such as passive RFID and smart card reader, and not universal. Thus, this project proposes an economical and portable ticketing system, named *mBus*, designed to (1) simplify payments for bus rides and (2) low-cost fare/token tracking using NFC. The proposed *mBus* consist of (1) *bTag*, (2) *bReader*, and (3) *bData*. The *bTag* are able to (1) store information, and (2) communicate with an active reader device, *bReader*. User can purchase a *bTag*, being a NFC tag, and a bus pass. *bTag* contain crucial information such as user information and remaining token/fare. A mobile application, named as *bReader*, allows the bus driver to (1) read *bTag*, (2) act as a payment gateway, and (3) create new users. The system record user data in *bData*, being a remote SQL database using PHP web pages. The system is being designed as mobile application since mobile devices has huge market in the portable devices industry and recent mobile system has NFC chips installed in them, i.e. anyone who desires to use this bus fare/token ticketing system can find and obtain inexpensive NFC-embedded mobile device easily.

Keywords: Near field communication (NFC) · *mBus* · *bData* · *bTag* · *bReader*

1 Introduction

Near field communication (NFC) tags, lately, is being deployed as bus pass especially as close-range bus fare/token ticketing system, i.e. NFC reader is used to track and detect NFC tags, while NFC passive tags are being used as bus pass and store bus pass information. NFC, being a subset of RFID, shares the same characteristics, being (1) the low cost, (2) low power usage, and (3) no necessity of reading in direct line of sight characteristic of RFID. NFC provides data transmission rate up to 424 Kbit/and uses 13.56 MHz radio frequency to communicate. However, NFC have a much shorter effective range compared to the best effective range, being 10 cm (reported), i.e. normal effective range is 3 cm. The benefits of NFC communication having shorter range is

© Springer International Publishing AG 2017
S. Phon-Amnuaisuk et al. (eds.), *Computational Intelligence in Information Systems*,
Advances in Intelligent Systems and Computing 532, DOI 10.1007/978-3-319-48517-1_7

that NFC communication will have much lower vulnerability to noise and interference compared to RFID and other short range wireless communication technology such as Infrared and Bluetooth.

In 2013, Manuel et al. reveals that NFC-Powered Bridge between mobile application and NFC embedded system has various new security-related attack vectors to the system [1]. Some of these vectors could be found in (1) the mobile application itself, (2) the NFC communication itself, and (3) the communication to the target system. To solve these problems, concepts to secure the NFC communication are made such as encrypt the target system itself. However, this will cause additional battery usage for the calculation of the cryptographic algorithms. Nonetheless, this issue will be considered in the system. In 2006, Esko et al. discuss about the capability of the NFC tags to be emulated as a RFID tags in card interface mode [2]. This capability allows NFC tag to be tracked by both NFC tracker and RFID tracker. This work highlights one of the communication mode of NFC intended for peer-to-peer data communication which is NFCIP-1, defined under ECMA-340 standard. There are two variants of NFCIP-1, active mode and passive mode. By using NFCIP-1, it is possible to minimize the power usage and extend the lifespan of the target device or NFC tags which is beneficial to device that has restricted energy source. In 2009, Morak et al. proposed a method to register new patients with NFC-implemented system [3]. The moment the system touched a non-associated new user ID card, it implicitly register the ID card with a time stamp while the card data can be edited online later. This method will be referred when building up system to create new tag. In 2011, Finžgar et al. proposed a system which enables the use of phones for acquiring electronic public transport ticket [4]. However, in their proposed system, user are required to register as user, prior, and payment are divided into different subscribers, which could complicate the management and the price of tickets.

In this paper, a proof of concept using NFC as bus fare/token system is proposed. Overall, several advantages can be obtained, being (1) implementation of a easy-to-implement low-cost secure mobile ticketing system (encrypted) for public transport, (2) avoidance of bus operator related errors, such as fare/token calculation, (3) facilitate the monitoring of passenger and their preferences, and (4) distributed management for fare/token, compared with legacy-based centralized controlled system, such as fare/token purchase. The remainder of this paper includes: Sect. 2 that outlines the NFC characteristics; Sect. 3 that describes and discusses the proposed system; Sect. 4 discusses the *mBus* system and shows experimental results; and Sect. 5 concludes the paper.

2 NFC Characteristics in Related Works

Being a subset of RFID, NFC shares the low cost, low power usage and no necessity of reading in direct line of sight characteristic of RFID. NFC also provides a data transmission rate up to 424 Kbit/and uses 13.56 MHz radio frequency to communication. However, NFC have a much shorter effective range which the best effective range is 10 cm and normal effective range being 3 cm. The benefits of NFC communication having shorter range is that NFC communication will have much lower vulnerability to noise and interference compared to RFID and other short range wireless communication technology, such as Infrared and Bluetooth, and is desirable in this project.

For the close-range bus fare/token ticketing application, NFC reader is used to track and detect NFC tags, while NFC passive tags are used as bus pass and store information. The users are free to bring their own NFC tag as a bus pass. If the user don't have or can't get access to any NFC tag, the bus driver will have NFC tags to sell, being the NFC tag type 2. The NFC tag type 2 itself is cheap to manufacture thus easy obtainable, while the tag's high reusability reduce the overall cost. The 136 bytes memory in the tag is more than enough to store bus pass information required in this system.

3   Methodology

The target of this work is to deliver a bus fare/token ticketing application that utilizes NFC. The target audience of this work will be any bus companies that want to implement an emerging and economical *mBus* system. NFC will be used as the payment gateway, and to communicate and to modify NFC tag as bus pass. Mobile devices that are NFC enabled will be used as *bReader* for bus drivers. Passive NFC tag, named *bTag*, will be used as bus pass, which the tag contains crucial information to uniquely identify each tag stored in the *bData*. The bus fare/token ticketing system, at first, will verify bus drivers and check if the device is NFC-enabled. If the bus driver is verified and the device is NFC enabled, i.e. *bReader*, the system will shall scan *bTag*. The owner of the *bTag* can thereafter, (1) pay the bus fare/token, (2) check the balance bus fare/token, and (3) perform top-up. For a *bTag*, either (1) unrecognizable, or (2) new, the system will then enquire if the user wants to purge the tag and modify to become a new *bTag*, allowing the creation of new *bTag* on-the-go. When the *bTag* is categorized as unrecognizable, either (1) modified illegally or/and (corrupted) and/or (2) missing, the owner of the *bTag* can proceed to repair the tag, by checking the *bData*. If the owner of *bTag* loses the tag, the owner may also apply for a new *bTag*, and reclaim his remaining fare/token. *bData* store crucial information for the whole system, which include tables such as USER that stores users' information with their corresponding tag information, (1) HISTORY LOG that record the payment, top up or creation of the users, (2) BUS that stores bus related information and (3) MISSING TAG which has all the records of missing tag. Figure 1 shows the system flow and main component, with the following features.

- Function to detect *bTag* as either authorized bus pass or unrecognizable tag. If authorized as existing bus pass, allow owner of tag to proceed with payment, check for remaining balance or top up bus fare/token. If not, allows creation of new user.
- Function to detect if the authorized *bTag* has been illegally modified. If true, the *bTag* is categorized as unrecognizable.
- Function to store *bTag* activities. Only store *bTag* activity that affects the *bTag* balance or the *bTag* creation.

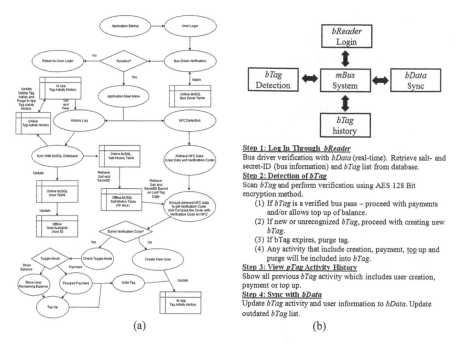

Step 1: Log In Through *bReader*
Bus driver verification with *bData* (real-time). Retrieve salt- and secret-ID (bus information) and *bTag* list from database.
Step 2: Detection of *bTag*
Scan *bTag* and perform verification using AES 128 Bit encryption method.
  (1) If *bTag* is a verified bus pass – proceed with payments and/or allows top up of balance.
  (2) If new or unrecognized *bTag*, proceed with creating new *bTag*.
  (3) If bTag expires, purge tag.
  (4) Any activity that include creation, payment, top up and purge will be included into *bTag*.
Step 3: View *pTag* Activity History
Show all previous *bTag* activity which includes user creation, payment or top up.
Step 4: Sync with *bData*
Update *bTag* activity and user information to *bData*. Update outdated *bTag* list.

(a)   (b)

Fig. 1. *mBus* System (a) system flow, and (b) main component and steps

- Function to pass the history of *bTag* activity to *bData* through Internet connection. The stored history in the *bTag* is then purged.
- Function to detect if the authorized *bTag* has expired, as the owner of the tag reported it to be missing and lost. If true, the *bTag* is purged.

bData, using mySQL database, is used to store all information for the system. Administrator is able to modify the *bData* through phpMyAdmin while the bus fare/token ticketing application may only indirectly access the database through PHPfiles. Being an open-source license-free database, more cost can be saved while security measure can be applied freely to make the database more secure. MySQL also can be hosted in Linux operating system, which can be installed in inexpensive low-end equipment which further reduces the cost of *mBus* system.

4 *mBus* System

mBus system consist of (1) *bReader*, (2) *bTag*, and (3) *bData*. *bReader* can be any devices that have NFC chip installed and able to connect to the Internet. This *bReader* should have .nfc package to provide access to NFC function and allows mobile application to read and to write NDEF message located at the NFC tag. *bTag* can be any type-2 rewritable NFC tag that has 136 bytes memory, able to communication with active NFC reader devices and can be programmable using NDEF. *bData* tracks the bus fare/token to the MySQL database for updating and retrieving required data. The following subsections discussed important elements of *mBus*.

4.1 System Database

XAMPP, being a free Apache distribution software, include installation of several web application is used in the project–see Fig. 2(a). With its control panel, the process of running the Apache webserver and MySQL database can be simplified. Apache webserver is used to host the MySQL online to be accessible through Internet while MySQL is used to store database information for the users–ERD diagram is shown in Fig. 3. Aptana Studio 3 is used to create PHP that connects the bus fare ticketing application to the MySQL Database for updating and retrieving required data–see Fig. 2(b).

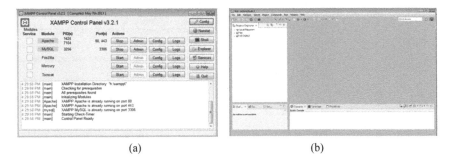

(a)   (b)

Fig. 2. *bData* system and application (a) XAMPP (b) Aptana Studio 3

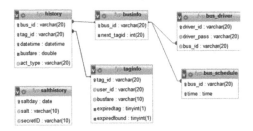

Fig. 3. ERD for *mBus*

4.2 Security–Verification Code

The verification code applied is encrypted using 128 AES encryption with TagID + UserID + Bus Fare + Last Tag Year + Last Tag Month + Last Tag Day + Salt (re-trieved from salt local storage based-on the date) + a specific secretID (retrieved from salt local storage based on date) since last creation, payment or top up. An encryption will be re-done on the retrieved data with the same salt and the same secretID. The verification code will be the same if the information in the tag is unaltered or uncorrupted. If match, the system will then check the toggle mode, while changing the Create New User button to Top Up selection.

- If the **toggle** mode is **Payment**, the bus fare will be deducted and the *bTag* will be rewritten with new bus fare, new current date and verification code from encryption with salt and secretID of the day. This action will be recorded into the *bTag* history local storage. The tag owner is revised to hold the *bTag* in place to avoid writing error.
- If the **toggle** mode is **Scan**, there will be no change while the *bTag* owner can check the remaining balance. If the verification code does not match, the *bTag* is categorized as unrecognized and the selection will revert back to **Create New User** if the selection is **Top Up**.

4.3   Analysis on *bTag*

In our first analysis, we analyses different *bReader*, being mobile devices, performance for the 25 mm diameter NTAG203 *bTag*. Analysis results show that the types of *bReader* do not greatly affect the *bTag* scan range, i.e. sharing almost similar scan range. It is assumed that mobile device manufacturers uses the same type of commercial NFC chip which emits the same power for detecting NFC tag, resulting in identical scan range for different *bReader*–see Fig. 4(a). In our second analysis, we analyses the performance of different *bTag* using **Huawei Ascend Mate 7**. Based on the analysis results, the type of *bTag* greatly affects the scan range–see Fig. 4(b). These *bTag* have different design, different size and different transparency, which may contribute to the scan range: (1) the more transparent the *bTag*, the longer the scan range, and (2) the larger the *bTag* the further the scan range–see Fig. 4(c). Besides that, it was found also that android devices also share similar spot when tagging *bTag*, which is shown in Fig. 4(d).

4.4   *mBus* GUI

Figure 5(a) shows the *bReader* login screen. When a bus driver clicks on the *mBus* icon on the *bReader*, the login screen is loaded which requires the driver to authenticate himself as a valid driver – see Fig. 5(a). The driver must enter correct driverID and correct corresponding password to use the application. When successful, the driver will proceed to the MAIN MENU. Otherwise the user is notified why they cannot log in – Fig. 5(b). The verified bus driver will be greeted with this initial login screen–see Fig. 5(c). Now the application can scan the *bTag* and perform verification. Verification is done using AES 128 bit encryption with *bTag* detail followed by the salt- and secretID, determined by last *bTag* date on the tag. If new generated encrypted text match encrypted text on the *bTag*, the *bTag* is verified and the MAIN MENU will change–see Fig. 5(d).

By clicking the Create New User button, the bus driver is taken into the add screen, where the bus driver can insert userID and bus fare to create a new user–see Fig. 6(a). For existing *bTag*, different mode available; (1) **Scan** - only show user ID and bus fare, (2) **Payment** - the *bReader* will require the tagging of *bTag* to confirm payment–see Fig. 6(b), and (3) **Top Up**–driver should key-in the amount of bus fare

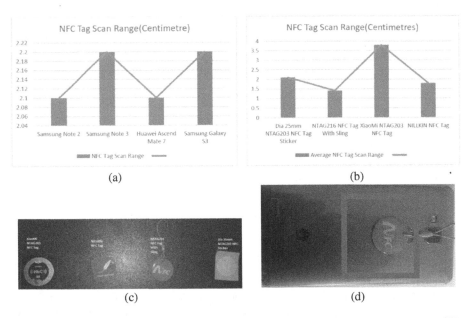

Fig. 4. Analysis of *mBus* system (a) *bReader* scan range (b) *bTag* scan range (c) *bTag* appearance (d) *bTag* scan spot

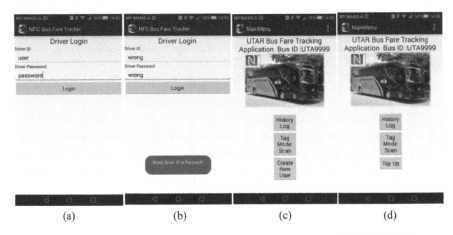

Fig. 5. *mBus* System (a) – (b) *bReader* login and (c) – (d) MAIN MENU

and the *bTag* owner have to confirmed the value by retagging, whereby owner will be notified when the top-up is successful showing available fare–see Fig. 6(c). If the tag_ID inside *bTag* is outdated, the *bTag* will be purged, as it is not a valid *bTag*. By clicking View History Log button, the driver will be taken into the *bData* screen, where s/he can see all previous tagged activities, including the creation of the new bus pass, payment, top up and purge tag–see Fig. 7(a). By clicking the Sync With

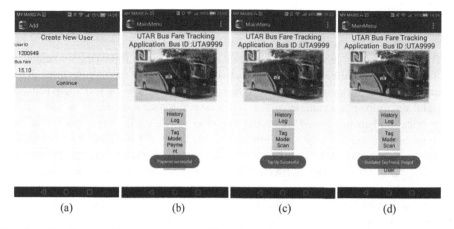

Fig. 6. *mBus* System (a) create new user (b) make payments (c) top-up (d) purging old *pTag*

Fig. 7. *bData* component (a) view log (b) loading *bTag* data into *bData*

Database button, the system will attempt to either inserting or updating *bData* into the history, thereafter, notify the driver that the system log will be cleared.

5   Discussion and Future Works

This work aim to prove that even with inexpensive or low requirement, an effective solution can still be produced. With the license-free attribute of NFC and MySQL, followed by the inexpensive price of mobile devices, it is possible to create a bus

Table 1. Comparison with other NFC applications

	Manuel [1]	Esko [2]	Luka [4]	Proposed Method
System completeness	No*	Yes	Yes	Yes
Encryption	Yes	No	No	Yes
Battery consideration	Yes	Yes	Yes	Yes
De-Centralized	No*	No	No	Yes

fare/token ticketing system. The *mBus* system developed was being tested, in the campus, to measure its performance, in real-time. Table 1 shows the comparison of the proposed system with some existing NFC applications.

There is considerable work yet to be done especially in the following two elements. The first, is due to the extreme short effective range of NFC communication, though sometimes this seems to appear as one of the weaknesses, can be of great help if used in the right situation. For example, if the wireless communication range is too far, the communication between the *bReader* and the *bTag* can be easily interrupted or sniffed, which required more security to be introduced, which could increase the workload of the whole system. As NFC is still a newly developed wireless communication technique, the NFC can still be improved to have a much better security during communication and be more robust to surrounding disturbance. It should be noted that while the communication between the *bReader and bTag* system is considered to be secured because of the short range, the communication between *bReader* and the *bData* may not be secured as the communication could still be tapped by third parties. Therefore, for future works, we would like to establish a secure channel between the *bReader* and the *bData*. Secondly, it was found that the more sub-system used to connect *bReader* to *bData*, the more problems will emerge. As an example, when we tried to use Connectify, being the software that could transform a computer into a wireless hotspot device, the connection attempts were mostly unsuccessful.

Acknowledgements. This work is supported by the UTAR Research Fund Project No. IPSR/ RMC/UTARRF/2014-C2/L03 "A Framework for V2 V/V2I Technologies for Smart Car System" from the Universiti Tunku Abdul Rahman, Malaysia.

References

1. Manuel, M., Norbert, D., Manuel, T. F., Christian, S., Reinhold, W., Holger, B., Josef, H.: PtNBridge–a power-aware and trustworthy near field communication bridge to embedded systems. In: Proceedings of the 2013 Euromicro Conference on Digital System Design (DSD 2013), Los Alamitos, CA, pp. 907–914 (2013)
2. Esko, S., Jouni, K., Juha, P., Arto, Y., Ilkka, K.: Application of near field communication for health monitoring in daily life. In: Proceedings of the Engineering in Medicine and Biology Society (EMBC 2006), New York City, USA, pp. 3246–3249 (2006)
3. Morak, J., Hayn, D., Kastner, P., Drobics, M., Schreier, G.: Near field communication technology as the key for data acquisition in clinical research. In: Proceedings of the First International Workshop on Near Field Communication (NFC 2009), Washington, USA, pp. 15–19 (2009)
4. Luka, F., Mira, T.: Use of NFC and QR code identification in an electronic ticket system for public transport. In: Proceedings of the International Conference on Software, Telecommunications and Computer Networks (SoftCOM 2011), Split, Croatia, pp. 1–6 (2011)

An Agent Model for Analysis of Trust Dynamics in Short-Term Human-Robot Interaction

Azizi Ab Aziz$^{(\boxtimes)}$, Wadhah A. Abdul Hussain, Faudziah Ahmad,
Nooraini Yusof, and Farzana Kabir Ahmad

Human-Centred Computing Research Lab, School of Computing,
College of Arts and Sciences, Universiti Utara Malaysia,
UUM, 06010 Sintok, Kedah, Malaysia
{aziziaziz,fudz,nooraini,farzana58}@uum.edu.my,
wadhah28edu@gmail.com

Abstract. Trust between human and robot is one of the crucial issues in robot-based therapy. It is highly important to provide a clearer and richer understanding and also to answer the questions why trust occurs in machines and how it can maintain successful interaction. In this paper, an agent based model for trust dynamics in short-term human-robot interaction is discussed and formally analysed. Three different cases were implemented to simulate various scenarios that explain the development of trust during short-term human-robot interaction; namely, (1) high level of trust, (2) moderate level of trust, and (3) low level of trust. Furthermore, simulation traces for fictional characters under different cases have pointed out realistic behaviours as existed in the literature. The developed model was verified by using mathematical (stability analysis) and automated verification (Temporal Trace Language).

Keywords: Trust · Human-Robot interaction · Robot-based therapy · Formal model

1 Introduction

Robots are poised to create a number of new roles in today's modern society. In the beginning of its development, robots are designed to serve important functions in factory automation, they are becoming in human daily functioning. This new function has prompted entirely new perspectives in human-robot interaction (HRI). It is fundamentally accepted principle by many researchers that HRI aims to study how humans interact with robots to accomplish specific tasks in human environments. This concept involves the design of robotic systems, interfaces, algorithms that make those robots capable to deliver effective interaction with humans. Therefore, exploring issues that are related to maintain the interaction between these new technologies and human beings are extensively researchable to get deeper insight to the phenomenon of trust in human robot interaction [1–3]. Thus, one of the grandest challenges for fruitful collaboration and communication between human and any other peer such as human, animal or machine is the existence of appropriate level of trust which is an important

© Springer International Publishing AG 2017
S. Phon-Amnuaisuk et al. (eds.), *Computational Intelligence in Information Systems*,
Advances in Intelligent Systems and Computing 532, DOI 10.1007/978-3-319-48517-1_8

characteristic needed to allow any human-robot interaction and relationship will achieve its designed objectives [4–8].

By modelling trust dynamics, it will provide a basic guidance to robot designers in designing a more trustworthy artefact that can increase the level of human trust from their design perspectives [7, 9]. The paper aims to discuss a formal analysis that can be modelled to generate trust's dynamics within short-term HRI and its possible application in robot-based cognitive therapy. The paper is organized as follows; Sect. 2 describes several grounding concepts of trust and HRI. Based on those concepts, a formal model is designed (Sect. 3). In Sect. 4, a number of simulation traces are presented to evaluate the proposed model by a mathematical analysis to determine stability of the model (Sect. 5), and followed by verification of the model as depicted in the literature using an automated temporal trace verification tool. Finally, Sect. 6 concludes the paper.

2 Trust and Human-Robot Interaction

Human robot interaction becomes an interesting area for the designers of robotic systems due to the ability of providing high potential of attracting and engaging user by applying significant factors for both implementation and appearance of robot conducive to effective interaction [2, 5, 7]. As a result, evaluating the capabilities of human and robots, and designing the technologies and training that produce desirable interactions are essential components of HRI. These components are inherently interdisciplinary in nature and requiring contributions from a number of fields such as psychology, cognitive science, mathematics, computer science and engineering. In a possible related scenario, a sociable robot will be able to assist a person to follow correct diet intake or to suggest some exercises in daily living. Despite of its simplicity, this scenario resembles an unstructured and natural way of human-robot interaction, which is one of the most complicated areas faced by HRI researchers [7, 9]. In addition, trust can be considered as one of the crucial factors, as this will increase the robot's acceptance in its role as an assistive technology in our daily routines [2, 5, 9].

2.1 Trust Dynamics

In the different perspectives, trust can be connected to the ability to persuade people in social interaction and collaboration. Thus, this could be one of the important keys that could directly influence individual's willingness to interact and cooperate with robot. One should aware that by having inappropriate level of trust, it could create a frustrating HRI experience [6, 8]. Normally, this negative influence will lead to another serious consequence such as failing to follow the advices/therapies recommended by the robot. For example, the effect of frustration will influence the neglect tolerance level; a condition when the robot's performance is declining but human attention is increasing as a result of performed tasks or as the rising level of task complexity [5, 7]. The fundamental construct of HRI trust is centrally related to the extent how robot performs its function properly. Believable behaviours, cues, physical appearances and level of automation are among important constructs to reflect the functionality of the robot [5, 10–12, 14, 15]. In robotic therapy (therapeutic), the HRI concept is important to establish support-provision concepts such as emotional, appraisal, and instrumental support [2, 3]. This type of robot aims to play significant role in helping individuals.

2.2 Long-Term and Short-Term Human Robot Interaction

The duration of HRI can be classified into two defined categories, namely; short-term interactions and long-term interactions. In *short-term HRI*, it occurs when many people interact with a robot for a limited time interval. For instance, the interaction occurs between hotel's receptionists, information kiosks, or with tour-guides. In the same vein, users spend restricted time (time constraint) with a robot. However, within this constraint, robots are expected to have a number of abilities to interact with people in novel modalities [16]. For example, a sociable museum tour-guide robot (mobile robot) was used as a case study to illustrate the concept of short-term interaction. This robot has been programmed with the function as an interactive digital tour-guide in museums with special function to help people and briefed to them important information for each displayed artefacts during the tour. The most critical task for this robot is to attract people to participate in a new tour and maintaining their attention. In the case of a tour-guide robot, this interaction could last for less than twenty minutes, and a person has to understand how to utilize his or her interaction with that robot within specified period of times [17]. In robot-based therapy for anxiety, a robot in a lab interacts in single sessions (up to 30 min) with individuals with the aim to reduce their anxiousness [18].

For *long-term HRI*, a human can interact with a robot over a prolonged period or even live together at the natural settings [19]. This type of interaction will allow a human to gradually acquire important knowledge of the robot (e.g., learning ability). In the domain of health management, the robotic weight loss coach (Autom) was developed to encourage behavioural change intervention and long-term interaction with robot among adults with overweight or obesity issues [20]. The proposed robotic system was compared to a touch screen computer (with an equivalence functionality) and to paper-based logbook for up to six weeks. The results have shown Autom's users used their system longer (< 52 days) compared to the touch-screen computer (< 37 days) and paper-based logbook users (< 30 days). However, the main challenge of the interaction will occur when the robot always displayed the same reaction of behaviour [21]. This cause a human would soon become bored with the robot and would discontinue his or her interaction with it.

3   Computational Modeling

This section covers the formalization of the trust dynamic. The proposed model combines ideas from research in theories in communication, trust, human-robot interaction, and dynamic modeling. These important constructs are embedded to simulate how people develop their trust with the robot. All of these constructs and its interactions are discussed in the following paragraphs in this section. Conceptually

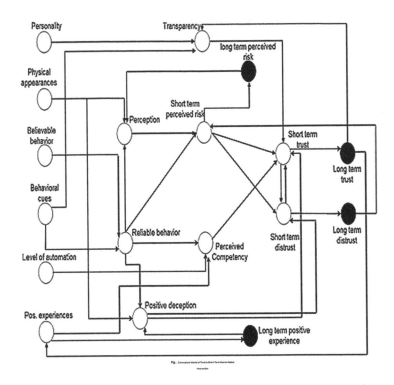

Fig. 1. Conceptual model of trust in short term human-robot interaction

there are nine main components are interacting to each other to simulate the dynamics of trust. These components are as *transparency, risk, trust, distrust, perception, behaviour, competency, deception,* and *experience* (as depicted in Fig. 1).

The model can be formalized once the conceptual and relationships of the model have been determined. All nodes are formulated to have values ranging from 0 (low) to 1 (high).

3.1   Instantaneous Relationships

The instantaneous relations are derived from a number of formulae, namely; perception (*Pc*), reliable behaviour (*Rb*), positive deception (*Pd*), competency (*Cy*), transparency (*Tr*), and positive experience (*Pe*). The short-term concepts (as a precursor to the long-term relations) can be considered as instantaneous relations. These are short-term perceived risk (*Sp*), distrust (*Sd*), and trust (*Sr*). These relationships were designed by the following formulae. Perception is calculated using the combination of long-term perceived risk (*Lp*), physical appearances (*Pa*), and reliable behaviour (*Rb*). The effect of reliable behaviour is measured by multiplying the value of believable behaviour (*Ba*) and behavioural cues (*Bc*).

$$Pc(t) = [\alpha_{pc} \cdot Pa(t) + (1 - \alpha_{pc}) \cdot Rb(t)] \cdot (1 - Lp(t)) \tag{1}$$

$$Rb(t) = Ba(t) \cdot Bc(t) \tag{2}$$

Later, the positive deception (Pd) is computed by regulating concepts in long-term positive experience (Le), physical appearances (Pa) and reliable behaviour (Rb).

$$Pd(t) = [\lambda_{pd} \cdot Rb(t) + (1 - \lambda_{pd}) \cdot Pa(t)] \cdot Le(t) \tag{3}$$

Competency is high when level of automation (La), positive experiences (Pe), and reliable behaviour (Rb) are high

$$Cy(t) = [\varphi_{cy} \cdot Rb(t) + (1 - \varphi_{cy}) \cdot (w_{c1} \cdot Rb(t) + w_{c2} \cdot Pe(t))] \cdot La(t), \sum w_{ci} = 1 \tag{4}$$

Short-term perceived risk (Sp) occurs when users foresee any potential threat in the relationship based on the behaviour of the robot. It has a positive correlation with long-term distrust (Ld), and contrary for both perception (Pc) and reliable behaviour (Rb). This concept will contribute towards the development of short-term distrust (Sd). It refers to the temporary non-confident in machine's behaviours and related to the relationship between short-term perceived risk, positive deception and short-term trust.

$$Sp(t) = Ld(t) \cdot [(1 - Pc(t)) \cdot (1 - Rb(t))] \tag{5}$$

$$Sd(t) = [\beta_{sd} \cdot Sp(t) + (1 - \beta_{sd}) \cdot (1 - Pd(t))] \cdot (1 - Sr(t)) \tag{6}$$

Contrary to this, short-term trust (Sr) refers to in the outcomes of a behaviour or relationship between human and robot through the interactions of positive deception (Pd), competency of advices given by robot (Cy), transparency (Tr), and short-term distrust (Sd). Another important concept, namely transparency plays an important role to understand the action or behaviour generated by robots. This concept combines all aspects in long-term trust (Lr), personality (Ps), and behavioural cues. Feeling about the interaction with the robot can be computed by evaluating norms in positive experience (Pe_{norm}) and long-term trust (Lr).

$$Sr(t) = [\beta_{sr} \cdot (\alpha_{sr} \cdot Tr(t) + (1 - \alpha_{sr}) \cdot Cy(t)] + (1 - \beta_{sr}) \cdot [Pd(t) \cdot (1 - Sd(t)) \cdot (1 - Sp(t))] \tag{7}$$

$$Tr(t) = [w_{r1} \cdot Lr(t) + w_{r2} \cdot Ps(t) + w_{r3} \cdot Bc(t)], \sum w_{ri} = 1 \tag{8}$$

$$Pe(t) = \beta_{Pe_{norm}} \cdot Pe_{norm} + (1 - \beta_{Pe_{norm}}) \cdot Lr(t) \cdot Pe_{norm} \tag{9}$$

Note that, α_{pc}, λ_{pd}, φ_{cy}, β_{sd}, β_{sr}, α_{sr} and $\beta_{Pe_{norm}}$ represent the proportional contribution factor, while, w_{ci} and w_{ri} are the weightage factor used for those respective relations. These proportional contribution and weightage factors have values ranging from 0 to 1.

3.2   Temporal Relationships

Short-term perceived risk contributes towards the development of long-term response (Lp), while the accumulated short-term trust and short-term distrust produce long-term trust (Lr) and long-term distrust (Ld) respectively. The formation of long-term positive experience (Le) is modelled using the accumulation of short-term positive experiences.

$$Lp(t+\delta t) = Lp(t) + \lambda_{Lp} \cdot Lp(t) \cdot [Sp(t) - Lp(t)] \cdot$$
$$(1 - Lp(t)) \cdot \delta t \tag{10}$$

$$Lr(t+\delta t) = Lr(t) + \varphi_{Lr} \cdot Lr(t) \cdot [Sr(t) - Lr(t)] \cdot$$
$$(1 - Lp(t)) \cdot \delta t \tag{11}$$

$$Ld(t+\delta t) = Ld(t) + \psi_{Ld} \cdot Ld(t) \cdot [Sd(t) - Ld(t)]$$
$$(1 - Ld(t)) \cdot \delta t \tag{12}$$

$$Le(t+\delta t) = Le(t) + \alpha_{Le} \cdot [Pe(t) - Le(t)] \cdot$$
$$(1 - Le(t)) \cdot \delta t \tag{13}$$

Note that the temporal contribution is measured in a time interval between t and $t + \delta t$. Furthermore, the temporal changes for all temporal formulas are computed by flexibility rates λ_{Lt}, φ_{Lr}, ψ_{Ld}, and α_{Le}. Based on those aforementioned formal specifications, a numerical prototype was developed as a platform to explore potential results and traces that explain the dynamics of the trust development during human-robot interaction.

4   Simulation

In this section, to illustrate the mechanism of the model, a number of experiments have been carried out to study effect of different variants and conditions on experimental setup (with all proportional contribution and weightage factors were assigned as 0.5 and 0.3 respectively) (Table 1).

Case #1 (High Level of Trust in HRI): This case aims to model positive humans' characteristics and good appearances robots (equipped with *believable cues*, *reliable behaviours*, and *high level of automation*) [9, 15, 21]. This is consistent for a person has good experiences in dealing with robot (as depicted in Fig. 2(a)). Typically, in human's natural settings, this condition can be related to human-expert interactions or perhaps with human-pet interactions [1, 10].

Table 1. Initial values for the simulation experiments

Factors	Case #1	Case #2	Case #3
Personality	0.9	0.5	0.1
Physical appearance	0.9	0.5	0.2
Believable behaviour	0.8	0.4	0.1
Believable cues	0.8	0.5	0.1
Level of automation	0.7	0.4	0.1
Positive experience	0.9	0.5	0.2

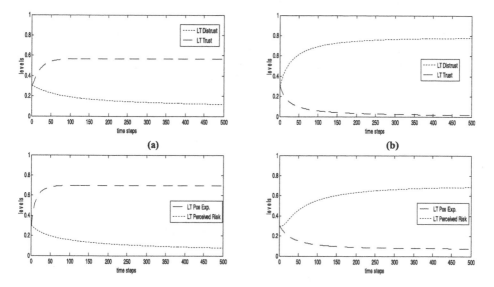

Fig. 2. Trust level for (a) Case #1 and (b) Case #3

Case #2 (Moderate Level of Trust in HRI): The given values were specified to determine possible conditions for moderate levels of trust. This condition exists when a person interacts occasionally with robots and expected fewer outcomes from the interaction. The same phenomena can be observed within our interactions with friendly strangers but expecting nothing from them [7, 10].

Case #3 (Low Level of Trust in HRI): This case represents the poor level of trust between human and robot (as depicted in Fig. 2(b)). All of those external factors have negative effects in initiating human-robot trust. Normally, humans are less willing to rely upon robot (negative traits) for achieving a particular task when robot's characteristics have less influences in terms of poor appearances, non-reliable performing, no social cues, and low level of automation [1, 6, 12, 14].

5 Evaluation

In this section, the temporal dynamics of the proposed model is verified using mathematical analysis and logical verification. Section 5.1 describes the stable (equilibrium) states of the model and Sect. 5.2 evaluates the dynamic properties of the model using an automated logical verification.

5.1 Mathematical Analysis

In this section the equilibria are analyzed that may occur under certain conditions. The equilibria describe situations in which a stable situation has been reached. Furthermore, the existence of reasonable equilibria is also a sign for the theoretical correctness of the formal specifications. As a first step, those formal specifications are replaced with values such that the differences between time point t and $t + \delta t$ are all 0 (in particular all temporal relationships). This leads to the following equations;

$$\lambda_{Lp} \cdot Lp \cdot [Sp{-}Lp] \cdot (1 - Lp) = 0$$

$$\varphi_{Lr} \cdot Lr \cdot [Sr{-}Lr] \cdot (1 - Lr) = 0$$

$$\psi_{Ld} \cdot Ld \cdot [Sd{-}Ld] \cdot (1 - Ld) = 0$$

$$\alpha_{Le} \cdot Le \cdot [Pe{-}Le] \cdot (1 - Le) = 0$$

Assuming the parameters λ_{Lp}, φ_{Lr}, ψ_{Ld}, α_{Le} nonzero, from (10) to (13), thus, the following equilibria cases can be distinguished for all possible simulation traces:

$$Lp = 0, \ Sp = Lp, \ Lp = 1$$
$$Lr = 0, \ Sr = Lr, \ Lr = 1$$
$$Ld = 0, \ Sd = Ld, \ Ld = 1$$
$$Le = 0, \ Pe = Le, \ Le = 1$$

These possible conditions will result up to $3^4 = 81$ possible equilibria. However, for some typical equilibrium cases, the formal analysis can be pursued further.

Case #1: *Lp* = 0
In this case, from (1), it shows that this equation is equivalent to

$$Pc = [\alpha_{Pc} \cdot Pa + (1 - \alpha_{Pc}) \cdot Rb)]$$

Assuming $\alpha_{Pc} = 0$, therefore $Pc = Rb$
Case #2: *Sp* = *Lp*
From Eq. (6), it follows that

$$Sd = [\beta_{Sd} \cdot Lp + (1 - \beta_{Sd}) \cdot (1 - Pd)] \cdot (1 - Sr)$$

However, the following subcases show how this case could be occurred;

$$[\beta_{Sd} \cdot Lp + (1 - \beta_{Sd}) \cdot (1 - Pd)] = 0 \text{ or } Sr = 1$$

Assuming $0 < \beta_{sd} < 1$, this is equivalent to:

$$Lp = -[(1 - \beta_{Sd}) \cdot (1 - Pd)]/\beta_{Sd}$$

By Eq. (7), it follows that;

$$Sr = [\beta_{Sr} \cdot (\alpha_{Sr} \cdot Tr + (1 - \alpha_{Sr}) \cdot Cy] + (1 - \beta_{Sr}) \cdot [Pd \cdot (1 - Sd) \cdot (1 - Lp)]$$

Moreover, it can be described as;

$$[\beta_{Sr} \cdot (\alpha_{Sr} \cdot Tr + (1 - \alpha_{Sr}) \cdot Cy] + (1 - \beta_{Sr}) \cdot [Pd \cdot (1 - Sd) \cdot (1 - Lp)] = 0$$

which later can be analyzed as;

$$\beta_{Sr} \cdot (\alpha_{Sr} \cdot Tr + (1 - \alpha_{Sr}) \cdot Cy = -(1 - \beta_{Sr}) \cdot [Pd \cdot (1 - Sd) \cdot (1 - Lp)]$$

Assuming $0 < \beta_{sr} < 1$, and $0 < \alpha < 1$, this is equivalent to:

$$Pd = [\beta_{sr} \cdot (\alpha_{Sr} \cdot Tr + (1 - \alpha_{Sr}) \cdot Cy] \cdot [(1 - Sd) \cdot (1 - Lp)]/(1 - \beta_{sr})$$

From Eq. (1), and assuming $\alpha_{pc} \neq 0$, this is equivalent to

$$Pc = [\alpha_{pc} \cdot Pa + (1 - \alpha_{pc}) \cdot Rb)] \cdot (1 - Sp))$$

Case #3: *Pe = Le*
For this case, using (4), it shows that

$$Cy = [\varphi_{cy} \cdot Rb + (1 - \varphi_{cy}) \cdot (w_{c1} \cdot Rb + w_{c2} \cdot Le)] \cdot La$$

Figure 3 visualizes the stability condition (equilibrium) for a number of selected relationships.

5.2   Logical Verification

In order to verify whether the model is capable to generate traces that adherence to related human-robot interaction literatures, a set of cases have been identified from related literatures. If these cases can be proved, it shows that the model produces traces that are coherent with the existing literature. For example, a case for boundary testing (*long-term trust* < 0 or *competency* > 1) can be explored and checked in order to avoid the occurrences of unexpected conditions. In addition, by executing a number of

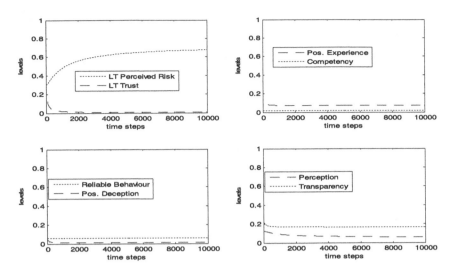

Fig. 3. Equilibria states for selected relationships

simulations traces and verifying these cases with respect to the resulting traces, we can easily detect any possible logical errors.

The Temporal Trace Language (TTL) is used as a basis to allow the verification process to take place. TTL is built on atoms referring to time stamps, states of the world, and simulation traces [22]. This relationship can be formalized as a *state (γ, t, output(R))| = p*. This representation means that state property p is true at the output of role R in the state of trace γ at time point t. For this purpose, special software has been developed for TTL, featuring both a property editor and a checking tool that enables formal verification of such properties against a set of simulated traces. A number of simulation traces including the ones described in Sect. 3 have been used as basis for the verification and were hold true (confirmed).

VP1: Physical Appearances will Improve Trust in Human Robot Interaction. If the robot has physical appearances that can be understood by human, it will increase the level of short-term trust during human-robot interaction [1, 3, 23].

VP1 ≡ ∀γ. TRACE, ∀t1, t2:TIME, ∀R1,R2,D1,D2:REAL
[state(γ,t1)|= has_value(physical_appearances, R1) &
state(γ,t2)|= has_value(physical_appearances, R2) &
state(γ,t1)|= has_value(long_term_trust, D1) &
state(γ,t2)|= has_value(long_term_trust, D2) &
t1 < t2 & R2 > R1] ⇒ Δ2 ε Δ1

VP2: High-risk Situation Reduces Trust. Humans normally intend to less trust in robot when the situation is risky which means less trust will be developed in human robot interaction [8, 12, 24].

VP2 ≡ ∀γ: TRACE, ∀t1, t2:TIME, ∀G1,G2,H1,H2,J1,J2:REAL
[state(γ,t1)|= has_value(level_automation, G1) &
state(γ,t2)|= has_value(level_automation, G2) &
state(γ,t1)|= has_value(believable_cues, H1) &
state(γ,t2)|= has_value(believable_cues, H2) &
state(γ,t1)|= has_value(long_term_trust, J1) &
state(γ,t2)|= has_value(long_term_trust, J2) &
t1 < t2 & G1 > G2 & H1 > H2] ⇒

VP3: Highly Competent Autonomous Robot Improves Trust. When intelligent systems such robots have high competency levels in performing a particular task, people intend to interact and develop trust in their interaction [14, 23].

VP3≡ ∀γ:TRACE, ∀t1, t2:TIME ∀M1, M2, D:REAL
[state(γ, t1)|= has_value(level_automation, M1) &
state(γ, t2)|= has_value(long_term_trust, M2) &
M1 ≥ 0.5 & t2= t1+D] ⇒ M2 ≥ 0.5

VP4: Individuals with Positive Personality Tend to Trust Autonomous Robots. Human with positive personality like openness, agreeableness, and extroversion are likely to trust the autonomous robots [3, 4, 10, 15].

VP4≡ ∀γ:TRACE, ∀t1, t2:TIME ∀Q1, Q2, D:REAL
[state(γ, t1)|= has_value(personality, Q1) &
state(γ, t2)|= has_value(long_term_trust, Q2) &
Q1 ≥ 0.8 & t2= t1+D] ⇒ Q2 ≥ 0.5

VP5: Verbal and Non-verbal Cues Play Important Roles to Regulate Trust. Normally, in human robot interaction, humans are developing a certain level trust with a sociable robot. Sociable cues (verbal and non-verbal) are playing major role in developing trust in HRI [3, 15, 25, 26].

VP5≡ ∀γ:TRACE, ∀t1, t2:TIME ∀G1,H1,J1,K1,D:REAL
[state(γ,t1)|= has_value(believable_cues, G1) &
state(γ,t1)|= has_value(believable_behaviours, H1) &
state(γ,t2)|= has_value(physical_appearances, J1) &
state(γ,t2)|= has_value(long_term_trust, K1) &
G1 ≥ 0.9 & H1 ≥ 0.9 & J1 ≥ 0.9 & t2= t1+ D] ⇒ K1 ≥ 0.5

6 Conclusion

The main idea addressed in this paper is to develop an agent based model that has basic abilities of analysing individuals' trust when interacting with social/assistive robots for specific tasks in cognitive therapy. This paper provides a first step to materialize that idea where a formal trust model has been designed and implemented to simulate different individuals' trust level related to their internal and external properties. This model is inspired by a large amount of research that has been conducted about the interpersonal trust related to machine/animal, and automation. To evaluate the model, a mathematical analysis has been conducted to prove the existence of equilibrium conditions as a basis to describe the convergence state of the model. Moreover, by using the Temporal Trace Language, the model has been verified to confirm the internal validity of the model. Therefore, the proposed model provides a foundation in designing a trust analysis module to improve robot-based therapy. In addition, future work of this project will be specifically focus how interactions, algorithms, and sensing properties can be further designed and developed in order to embed this model into related robot-based therapy system.

Acknowledgements. This project is based upon works partially supported by the Ministry of Higher Education Malaysia under RACE grant (Grant No. RACE/12633).

References

1. Hancock, P., Billings, D., Schaefer, K.: Can you trust your robot? Ergonomics Des. Q. Hum. Factors Appl. **19**, 24–29 (2011)
2. Fasola, J., Mataric, M.: A socially assistive robot exercise coach for the elderly. J. Hum. Robot Interact. **2**, 3–32 (2013)
3. Desai, M., Stubbs, K., Steinfeld, A., Yanco, H.: Creating trustworthy robots: lessons and inspirations from automated systems (2009)
4. Sanders, T.L., Wixon, T., Schafer, K.E., Chen, J. Y., Hancock, P.: The influence of modality and transparency on trust in human-robot interaction. In: 2014 IEEE International Inter-Disciplinary Conference on Cognitive Methods in Situation Awareness and Decision Support, pp. 156–159 (2014)
5. Taddeo, M.: Modelling trust in artificial agents, a first step toward the analysis of e-Trust. Minds Mach. **20**, 243–257 (2010)
6. Atkinson, D., Hancock, P., Hoffman, R.R., Lee, J.D., Rovira, E., Stokes, C., Wagner, A.R.: Trust in computers and robots: the uses and boundaries of the analogy to interpersonal trust. In: Proceedings of the Human Factors and Ergonomics Society Annual Meeting, pp. 303–307 (2012)
7. Grodzinsky, F., Miller, K., Wolf, M.: Responsibility for computing artefacts: the rules and issues of trust. ACM SIGCAS Comput. Soc. **42**, 15–25 (2012)
8. Robinette, P., Wagner, A.R., Howard, A.M.: Modeling human-robot trust in emergencies. In: 2014 AAAI Spring Symposium Series (2014)
9. Trang, S., Zander, S., Kolbe, L.M..: Dimensions of trust in the acceptance of inter-organizational information systems in networks: towards a socio-technical perspective (2014)
10. Billings, D.R., Schaefer, K.E., Chen, J.Y.C., Kocsis, V., Barrera, M., Cook, J., Ferrer, M., Hancock, P.A.: Human-animal trust as an analog for human-robot trust: a review of current evidence. Technical report ARL-TR-5949, DTIC Document (2012)
11. Coeckelbergh, M.: Humans animals, and robots: a phenomenological approach to human-robot relations. Int. J. Soc. Robot. **3**, 197–204 (2011)
12. Oleson, K.E., Billings, D., Kocsis, V., Chen, J.Y., Hancock, P.: Antecedents of trust in human-robot collaborations. In: Cognitive Methods in Situation Awareness and Decision Support (CogSIMA), pp. 175–178 (2011)
13. Simpson, J.A.: Foundations of interpersonal trust (2007)
14. Hancock, P.A., Billings, D.R., Schaefer, K.E., Chen, J.Y., De Visser, E.J., Parasuraman, R.: A meta-analysis of factors affecting trust in human-robot interaction. Hum. Factors J. Hum. Factors Erogon. Soc. **53**, 517–527 (2011)

15. Masthoff, J.: Computationally Modelling Trust: An Exploration. University of Aberdeen, Aberdeen (2007)
16. Dautenhahn, K.: Socially intelligent robots: dimensions of human-robot interaction. Philos. Trans. R. Soc. B: Biol. Sci. **362**, 679–704 (2007)
17. Thrun, S., Schulte, J., Rosenberg, C.: Interaction with mobile robots in public places. In: IEEE Intelligent Systems, pp. 7–11 (2000)
18. Aziz, A.A., Ahmad, F., Yusoff, N., Mohd-Yusof S.A., Ahmad, F.K.: Designing a robot assisted therapy for individuals with anxiety traits and states. In: The International Symposium on Agents, Multi-agent Systems and Robotics (ISAMSR 2015), pp. 80–87. IEEE Computer Society Press (2015)
19. Leite, I., Martinho, C., Paiva, A.: Social robots for long-term interaction: a survey. Int. J. Soc. Robot. **5**, 291–308 (2013)
20. Kidd, C., Breazeal, C.: Robots at home: understanding long-term human-robot interaction. In: Proceedings of the 2008 IEEE/RSJ International Conference on Intelligent Robots and Systems (2008)
21. Rodić, A., Jovanović, M.: How to make robots feel and social as humans. In: The 6th International Conference on Advanced Cognitive Technologies and Applications (Cognitive 2014), pp. 133–139 (2014)
22. Bosse, T., Jonker, C.M., van der Meij, L., Sharpanskykh, A., Treur, J.: Specification and verification of dynamics in agent models. Int. J. Coop. Inf. Syst. **18**, 167 (2009)
23. Bainbridge, W., Hart, J., Kim, E.S., Scassellati, B.: The effect of presence on human-robot interaction, RO-MAN 2008. In: The 17th IEEE International Symposium on Robot and Human Interactive Communication, pp. 701–706 (2008)
24. Robinette, P., Wagner, A.R., Howard, A.M.: Building and maintaining trust between humans and guidance robots in an emergency. In: AAAI Spring Symposium: Trust and Autonomous Systems, pp. 78–83 (2013)
25. Tapus, A.: Towards personality-based assistance in human-machine interaction. In: The 23rd IEEE International Symposium on Robot and Human Interactive Communication, RO-MAN, pp. 1018–1023 (2014)
26. Yagoda, R.E., Gillan, D.J.: You want me to trust a robot? The development of a human-robot interaction trust scale. Int. J. Soc. Robot. **4**, 235–248 (2012)

An Ambient Agent Model for a Reading Companion Robot

Hayder M.A. Ghanimi[✉], Azizi Ab Aziz, and Faudziah Ahmad

Human-Centred Computing Research Lab,
School of Computing College of Arts and Sciences,
Universiti Utara Malaysia, 06010 Sintok, Kedah, Malaysia
hayder.alghanami@gmail.com,
{aziziaziz, fudz}@uum.edu.my

Abstract. This paper presents the development of an ambient agent model (software agent) as an initial step to develop a reading companion robot to sup-port reading performance. This ambient agent model provides detailed knowledge (human functioning) about reader's dynamics states. Based on this human functioning knowledge, a robot will be able to reason about reader's conditions and provides an appropriate support. Several simulation traces have been generated to illustrate the functioning of the proposed model. Furthermore, the model was verified using an automated trace analysis and the results have shown that the ambient agent model satisfies a number of related properties as presented in related literatures.

Keywords: Human-ambient agent · Temporal dynamics · Reading performance · Companion robot

1 Introduction

Reading is an important task to all individuals to function in today's society as it is a fundamental way to develop minds. Likewise, reading considered as an indispensable part to learn new concepts, knowledge, and even to build someone's character. How-ever, despite of the remarkable advantages of reading, it is always associated with a number of hurdles, particularly, if it caters for solving complicated tasks (e.g., solving math) and this may deter effective learning experiences. For examples, readers are always subjected to cognitive overload that has great potentials to cause several negative implications such as mental exhaustion, frustration, boredom, and thereby disengagement definitely occurs [1, 2].

Furthermore, for a meaningful and seamless reading task, there is a need to develop a companion robot, represented as a table lamp, which monitors and assesses readers' conditions while performing serious reading tasks as determined earlier [3]. This paper presents the precursor to develop a reading companion robot where an ambient intelligent agent model is developed. Through integrating this ambient agent model into a robot, this robot will be able to understand its environment and readers' states, and thereby an appropriate support will be given [4]. The ambient agent model was designed using a set of dynamic properties that show how readers experience negative

© Springer International Publishing AG 2017
S. Phon-Amnuaisuk et al. (eds.), *Computational Intelligence in Information Systems*,
Advances in Intelligent Systems and Computing 532, DOI 10.1007/978-3-319-48517-1_9

effects, take observations as inputs, and belief-desire-intention (BDI) concept to determine its internal functions and actions.

This paper is organized as follows. Section 2 introduces the importance of companion robots and explains some of the prospective design elements of a reading companion robot. Later, the ambient agent model is described in Sect. 3. In Sect. 4, the properties of the ambient agent model were formalized and specified. Section 5 shows the simulation results. After that, the ambient model is verified in Sect. 6. Finally, Sect. 7 concludes the paper.

2 Companion Robots

In more recently, companion robotic technology has considerably yielded a remarkable achievements in terms of aiding humans and improving their well-beings. For example, there are several successful stories of developing companion robots that support people in social manners using verbal and non-verbal communications. Examples of such robots are, (1) Autom is for loss weight management [5], (2) Jibo is to help family members in a very human way [6], and (3) CAKNA is to help individuals with anxiety traits and states [7]. In addition, the key success factor of these robots is contributed to the physical embodiments and its reasoning abilities.

The embodiment of such robots plays an essential role to increase its success where many studies have shown that physically embodied robots are more preferred over virtual one due to its physical presence [8, 9]. To design a reading companion robot that offers intelligent support to readers while performing a difficult reading task as aforementioned in Section one, several components are needed as explained in Fig. 1. This paper is to discuss an ambient agent model (software agent) related to reading behaviours. The proposed model will be used as one of the main components in developing a reading companion robot. A detailed discussion on other components is beyond the scope of this paper.

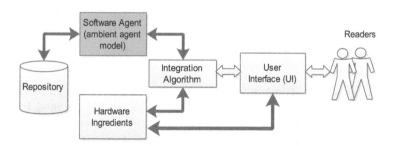

Fig. 1. The components of a reading companion robot

3   The Ambient Agent Model

In recent years, to develop ambient intelligent agents, it became very important to incorporate human-functioning models (or domain models) into ambient agent applications. By incorporating these models, the ambient agent will be able to reason about the human's states and its processes [10, 11]. In this paper, a dynamical model of cognitive load and reading performance is integrated as the overall functioning process of ambient agent model. The ambient agent model was made using four different com-ponents. These components are; (1) Domain Model, (2) Belief Base, (3) Analysis Model, and (4) Support Model. Moreover, a Belief-Desire-Intention (BDI) structure was used as the foundation of the ambient agent model's properties [12]. Figure 2 shows the complete integration of these models.

The solid arrows indicate the information exchange between process, and the dot-ted arrows represent the integration process of the domain model within the ambient agent model. Accordingly, two different ways to integrate the domain models within the agent model have been used (as in Fig. 2);

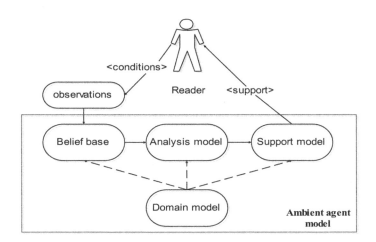

Fig. 2. The ambient agent model

- *Analysis component*: To perform analysis of the human's states and processes by (model-based) reasoning based on observations and the do-main model.
- *Support component*: To generate support actions for the human by (model-based) reasoning based on observations and the domain model.

The following sub-sections will discuss more details of the four different models that form the ambient agent model.

3.1   A Dynamical Domain Model of Cognitive Load and Reading Performance

To understand readers' mental states and environment processes, a domain model of cognitive load and reading performance has been developed earlier as seen in [13]. The model was conceptually made based on several factors of reader's cognitive load and its effects on reading task. These factors are interconnected to each other based on their relationships as they categorized into instantaneous and temporal relationships. Once the dynamical relationships were determined, the model was mathematically formalized and simulated. The results of the simulation have exhibited realistic patterns that reflect the impact of cognitive load on reading performance. In addition, the model was designed in a way to be tailored with several other sub-models. Detailed discussions of the model development and its simulation results can be found in [13].

3.2   Belief Base

The main function of the belief base is to generate initial beliefs (basic and derived beliefs) based on certain observations of the ambient agent [14]. In this work, robot's observations on reading task properties, environment and readers' conditions will generate basic beliefs, while derived beliefs will be generated based on derivation using the domain model. The observations of the robot can be achieved using several means, for example integrating existed sensors to measure sound and temperature levels or a set of questions that reflect each observation the robot aims to observe. Figure 3 shows observations, basic beliefs, and derived beliefs that form the belief base model. Note that, for a better resolution all the diagrams in this paper can be found at (http://goo.gl/uB3Iwf).

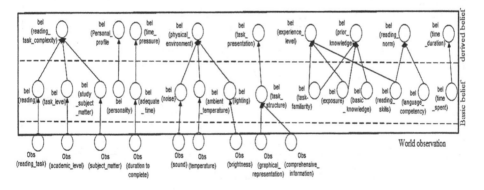

Fig. 3. Observations, basic beliefs, and derived beliefs in belief base model

One of the advantages to have such concept is it allows future extensions of the model. In addition, another model can make use this set of related beliefs without having to generate a new one.

3.3   Analysis Model

The main goal is to analyze the dynamics of reader's state and processes. In this case, the ambient agent must be equipped with the domain model and therefore a reasoning process will be performed to assess some concepts that cannot be measured directly only by observations. Thus, the ambient agent will use both of the domain model and belief base model to assess the reader's states. Related to this, four assessments were made to determine the reader's states, namely; (1) *assessment of cognitive load*, (2) *assessment of exhaustion*, (3) *assessment of persistence*, and (4) *assessment of reading performance* (as depicted in Fig. 4 with dark nodes). As a result, the ambient agent will be capable to monitor the related conditions of the reader. For example, if the assessment for both reading performance and persistence are low, then the model will consider the reader is not performing well due to the low level in persistence. Therefore, the ambient agent will trigger its desire to increaser reader's persistence level. Later, this will trigger the support model. Figure 4 depicts the analysis of the domain model.

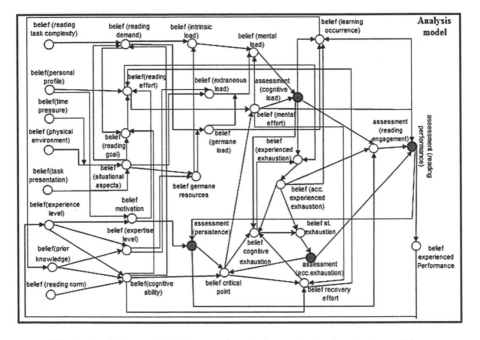

Fig. 4. Analysis model for human functioning analysis of reader's conditions

The next step was to formalize these concepts in logical atoms. Once these concepts have been identified, the temporal relations and properties for these concepts within the ambient agent can be formally described.

3.4 Support Model

If readers are learning new information that taxing their resources, then after a certain period of time they will experience some negative effects that deter their learning experience [15]. Therefore, it is necessary to help readers in a such condition. To support readers, the ambient agent model can use the result from the analysis model to generate support actions. In addition, the implementation of BDI in the support model will provide an action selection process, dedicated to decide which action should be chosen. On the conceptual level, the implementation of the BDI mechanism in the support model is illustrated in Fig. 5.

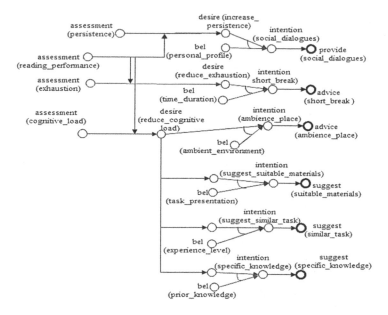

Fig. 5. Action selection process in the support model

In this model, desire to reduce cognitive load and exhaustion (or to increase persistent) is added as a desire, while intention process is to support the decision making process and improve reading performance. The activated desire derives an intention to perform some actions to support readers. The relationship between these constructs (e.g., R1 $\twoheadrightarrow$ R2) can be described as;

∀N:CONDITION, ∀X:ROBOT, ∀T:TASK
R1: desire(X, increase (N)) $\twoheadrightarrow$ intention(X, provide(T))
R2: intention(X, provide(T)) $\twoheadrightarrow$ performed(X, provide(T))

4   Ontology and Specifications

To specify properties on dynamic relationship, the ontology of the model was de-signed using Predicate Calculus. For example, any robot ability to observe the complexity level of subject matter can be expressed as:

observed(X:ROBOT, subject_matter (L:LEVEL)).

Ontology for Robot's Observation: Observation on reader's condition can be executed through a set of questions or related sensors for each input for the belief base model. In this case, the robot will request or detect some inputs on reading task, academic level of the task, subject matter, time to complete the task, temperature, bright-ness sound level, and the level of information representation associated to the task that needs to be solved.

```
observed(X: ROBOT, reading_task(S:STATUS))
observed(X: ROBOT, academic_level(L:LEVEL))
observed(X: ROBOT, subject_matter(C:COMPLEXITY))
observed(X: ROBOT, duration_to_complete(D:DURATION))
observed(X: ROBOT, sound(L:LEVEL))
observed(X: ROBOT, temperature(T:TEMP_LEVEL))
observed(X: ROBOT, brightness(I:INTENSITY))
observed(X: ROBOT, graphical_presentation(S:STATUS))
observed(X: ROBOT, comprehensive_information(L:LEVEL))
```

Ontology for Belief Base: The ontologies of *Basic beliefs* that are generated after several observations as follows:

```
belief (X: ROBOT, reading(S:STATUS))
belief (X: ROBOT, task_level(L:LEVEL))
belief (X: ROBOT, study_subject_matter(C:COMPLEXITY))
belief (X: ROBOT, personality(T:TYPE))
belief (X: ROBOT, adequate_time(D:DURATION))
belief (X: ROBOT, noise(L:LEVEL))
belief (X: ROBOT, ambient_temperature(S:STATUS))
belief (X: ROBOT, lighting (I:INTENSITY))
belief (X:ROBOT, task_structure(L:LEVEL))
belief(X:ROBOT,  task_familiarity(L:LEVEL))
belief (X: ROBOT, exposure(L:LEVEL))
belief (X: ROBOT, basic_knowledge(L:LEVEL))
belief (X: ROBOT, reading_skills(L:LEVEL))
belief (X: ROBOT, reading_competency(L:LEVEL))
belief (X: ROBOT, time_spent(D:DURATION))
```

Then, *Derived beliefs* as follows:
```
belief(X: ROBOT, reading_task_complexity(C:COMPLEXITY))
belief(X: ROBOT, personal_profile (T:TYPE))
belief(X: ROBOT, time_presure(D:DURATION))
belief(X: ROBOT, ambient environment (S:STATUS))
belief(X: ROBOT, task_presentation(L:LEVEL))
belief(X: ROBOT, experience_level(L:LEVEL))
belief(X: ROBOT, pror_knowledge(L:LEVEL))
belief(X: ROBOT, reading_norm(L:LEVEL))
belief(X: ROBOT, time_duration(D:DURATION))
```

Ontology for Analysis Model: Four assessments were made to evaluate the condition of the reader and a support action is triggered based on the evaluation results.

```
assessment(X:ROBOT, persistence(L:LEVEL))
assessment(X:ROBOT, cognitive_load(L:LEVEL))
assessment(X:ROBOT, exhaustion(L:LEVEL))
assessment (X: ROBOT, reading_performance(L:LEVEL))
evaluation(X:ROBOT, persistence(L:LEVEL))
evaluation(X:ROBOT, cognitive_load(L:LEVEL))
evaluation(X:ROBOT, exhaustion(L:LEVEL))
```

Ontology for Support Model: A set of actions will be performed by the robot using belief and analysis models with respect to the specified desire. Moreover, the BDI approach regulates action selection process.

```
belief(X:ROBOT, persistence(L:LEVEL))
belief(X: ROBOT, cognitive_load(L:LEVEL)) be-
lief(X: ROBOT, exhaustion(L:LEVEL))
desire(X:ROBOT, increase (N:CONDITION)) de-
sire(X: ROBOT, reduce (N:CONDITION))
intention(X:ROBOT, provide(T:TASK))
intention(X:ROBOT, advice(T:TASK))
intention(X:ROBOT, suggest(T:TASK))
performed(X:ROBOT, provide(T:TASK))
performed(X:ROBOT, advice(T:TASK))
performed(X:ROBOT, suggest(T:TASK))
```

The formalization of some properties makes use of sorts. These sorts are shown in Table 1.

Table 1. Sorts used.

Sort	Elements
STATUS	{yes, no}
LEVEL	{low, medium, high}
TYPE	{positive, negative}
COMPLEXITY	{easy, moderate, difficult}
INTENSITY	{too_bright, adequate, too_dim)
DURATION	{short, moderate, long}
TASK	{find_an_ambience_ place, specific_knowledge, short_break, suitable_materials, similar_task, social_dialogues}
TEMP_LEVEL	{too_warm, neutral, too_cool}

The ontologies mentioned above were used to generate temporal rules specifications which are used to provide a set of knowledge for the robot to reason with. To utilize the specifications, a forward reasoning method for belief generation is used. Besides, to specify simulation model, a temporal specification language (LEADSTO) has been used. In this section, the domain model is used, which consists of two inter-acting dynamical models, one to determine the human's cognitive load and performance, and one to determine the support that will be offered by a robot. The approach used to specify the domain model is based on the hybrid dynamical modelling language LEADSTO [16]. In this language, direct temporal dependencies between two state properties in successive states are modelled by executable dynamic properties.

Consider the format of $\alpha \twoheadrightarrow_{e,f,g,h} \beta$, where α and β are state properties in form of a conjunction of atoms (conjunction of literals) or negations of atoms, and e,f,g,h represents non-negative real numbers, then it can be interpreted as follows:

If state property α holds for a certain time interval with duration g, after some delay (between e and f), state property β will hold a certain time interval of length h.

In addition, this representation also holds a temporal traceγ, denoted by $\gamma \mid = \alpha \twoheadrightarrow$

$\forall t1[\forall t1[t1 - g \le t < t1 \Rightarrow \alpha$ holds in γ at time t]
$\Rightarrow \exists d [e \le d \le f \& \forall t' [t1 + d \le t' < t1 + d + h]$
$\Rightarrow \beta$ holds in γ at time t']

A more detailed discussion about this language can be found in [16]. Following are some examples of the generated rules specifications in exhaustion condition.

DB1: Derived Belief on Time Duration. When the robot believes that the reader already spent long time on the task, then the robot believes that the time duration a reader consumed without a pause is long.

belief(robot, time_spent(long)) $\twoheadrightarrow$ belief (robot, time_duration(long)).

EEX: Evaluation on Exhaustion Condition. When a robot assesses that reader is exhausted and it is no longer performing well, then the robot evaluates the condition as a high level of exhaustion.

assessment(robot, exhaustion(high)) ^
assessment(robot, reading_performance (low)) $\twoheadrightarrow$
evaluation(robot, stage(exhaustion, high))

INT: Intention to Advice for Short Break. When the robot desires to reduce the level of exhaustion through advising for a short break and the robot believes that the time a reader already consumed is long, then the robot will have an intention to advice reader for getting a short break.

desire(robot, reduce(exhaustion)) ^ belief(robot, time_duration(long)) $\twoheadrightarrow$ intention(robot, advice(short_break))

ACT: Action to Advice for Short Break. When the robot intends to advice the reader to get a short break, then the robot will advise the reader for a short break

intention(robot, advice(short_break)) $\twoheadrightarrow$
performed(robot, advice(short_break))

5   Simulation Results

Based on the models presented earlier, a number of simulations have been conducted using LEADSTO [16]. For this paper, two simulation experiments were presented. In the figures below, the timeline (as depicted in both figures) is presented on the horizontal axis, the state properties are on the vertical axis and a dark blue box indicates that a state property is true. To perform a reading task simulation, 500 time steps are used (i.e., four hours of reading were approximately simulated).

Simulation #1: Demanding Task with Insufficient Reader's Resources. In this simulation, the robot observes several conditions concerning reading task, such as; difficult subject meant for a higher academic level, distraction environment due to high level of sound, temperature, and brightness. Likewise, reading task is not presented with comprehensive and graphical information. A reader also has no enough knowledge and experience on the reading task. As a result, the robot will be able to assess reader's condition through the time and an appropriate action will be per-formed if all beliefs hold true as well. The unwanted conditions in the case are high exhaustion, high cognitive load, low persistence, and low reading performance. The result of the simulation is explained in Fig. 6 (a).

Simulation #2: Not Demanding Task with Insufficient Reader's Resources. In this simulation, the robot observed that reading task has no impact on reader conditions where it was not difficult, meant for the right academic level, and presented with

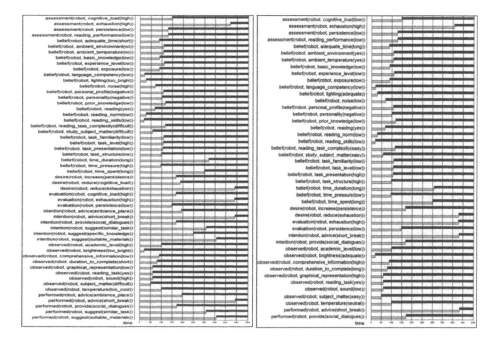

Fig. 6.   Simulation results (a) Demanding task (b) Non-demanding task

graphical and comprehensive information. The environment was not distraction as well. In addition, the robot believes that the reader is not skilled enough to perform the task. In this case, the robot will be able to assess three unwanted conditions through the time which are low persistence, high exhaustion, and low reading performance. With the time, the robot is able to tackle all the unwanted conditions as appropriate actions will be performed to each condition. The result of the simulation is explained in Fig. 6(b).

6 Automated Verification

This section aims to verify relevant dynamic properties of the cases considered in the ambient agent model. To do so, several properties were identified from the related literatures and an automated verification using Temporal Trace Language (TTL) is performed. This language allows formal specification and analysis of dynamic proper-ties; it is either a qualitative or a quantitative representation [16]. TTL is designed on atoms, to represent the states, traces, and time properties. This relationship can be presented as a state(γ, t, $output(R)$) | = p, means that state property p is true at the output of role R in the state of trace γ at time point t. It is also comparable to the *Holds*-predicate in the Situation Calculus. Based on that concept, dynamic properties can be formulated using a sorted First-Order Predicate Logic (FOPL) approach.

VP1: Provide social dialogues when the reader is not persistent to continue reading as its reading performance getting low [17].

$\forall\gamma$:TRACE, t:TIME
[state (γ,t) |= belief(robot, persistent(low)) $\wedge$
state (γ,t) |= assessment(robot ,reading_peformance(low)]
$\Rightarrow \exists t'$:TIME > t:TIME [state(γ,t') |=performed (robot, provide(social_dialogue)]

VP2: Advice for a short break session when the reader encounters high mental exhaustion [18].

$\forall\gamma$: TRACE, t:TIME
[state(γ, t) |= evaluation(robot, exhaustion(high))
$\Rightarrow \exists t'$:TIME > t:TIME [state(γ, t') |=performed (robot, advice(short_break))]

VP3: Suggestion to find an ambience place when reader's encounters a high level of cognitive load and his/her reading performance is not good enough [19].

$\forall\gamma$: TRACE, t1,t2:TIME
[state(γ,t1) |= belief(robot, ambient_environment(no)) $\wedge$
state(γ,t2) |= evaluation(robot, cognitive_load(high))
$\Rightarrow \exists t3$:TIME > t1:TIME [state(γ,t3) |=performed (robot, advice(ambience_place))]

VP4: Suggest specific knowledge related to the subject matter (e.g. providing hints) when the reader encounters high level of cognitive load that deters its reading performance.

∀γ: TRACE,t:TIME
[state(γ,t)|= belief(robot, prior_knowledge(low)) ∧
state(γ,t)|= evaluation(robot, cognitive_load(high))
⇒ ∃t':TIME > t:TIME [state(γ, t')|=performed (robot, suggest(specific_knowledge))]

7 Conclusion

In this paper, an ambient agent model to monitor reader's conditions such as persistence, cognitive load, exhaustion, and reading performance is introduced. By compiling knowledge from the domain model into the agent model, the agent is able to reason about the state of the reader. Thus, it is capable to analyze the level of cognitive load and reading performance based on several observable features and beliefs. The model has been specified using a formal modelling approach, which enables a qualitative specification. These formal temporal properties will allow the ambient agent to reason about specific conditions related to reader's states. Furthermore, several simulations results pointed out that the proposed ambient agent model is able to evaluate a number of conditions and base on its assessments a set of actions were given to sup-port the reader. For the next step, a thorough evaluation process will be considered. Moreover, an integration mechanism to incorporate the ambient agent model into a companion robot will be developed.

Acknowledgement. The project is partially funded by UUM Postgraduate Research Scholarship programmes.

References

1. Kalyuga, S.: Cognitive load theory: implications for affective computing. In: proceeding of the Twenty-Fourth International Florida Artificial Intelligence Research society Conference, pp. 105–110, Amsterdam (2011)
2. Schnotz, W., Fries, S., Horz, H.: Motivational aspects of cognitive load theory. In: Contemporary Motivation Research: From Global to Local Perspectives, pp. 69–96 (2009)
3. Mohammed, H., Aziz, A.A., Ahmad, R.: Exploring the need of an assistive robot tosupport reading process: a pilot study. In: International Symposium on Agents, Multi-Agent Systems, and Robotics (ISAMSR), pp. 35–40 (2015)
4. Bosse, T., Both, F., Gerritsen, C., Hoogendoorn, M., Treur, J.: Methods for model-based reasoning within agent-based Ambient Intelligence applications. Knowl. Based Syst. **27**, 190–210 (2012)
5. Kidd, C.D., Breazeal, C.: Robots at home: understanding long-term human-robot interaction. In: International Conference on Intelligent Robots and Systems, pp. 3230–3235 (2008)

6. Rane, P., Mhatre, V., Kurup, L.: Study of a home robot: JIBO. Int. J. Eng. Res. Technol. **3** (10), 94–110 (2014)
7. Aziz, A.A., Ahmad, F., Yusof, N., Ahmad, F.K., Yusof, S.A.M.: Designing a robot-assisted therapy for individuals with anxiety traits and states. In: International Symposium on Agents, Multi-Agent Systems and Robotics, pp. 98–103 (2015)
8. Jung, Y., Lee, K.M.: Effects of physical embodiment on social presence of social robots. In: Procceding of PRESENCE, pp. 80–87 (2004)
9. Berland, M., Wilensky, U.: Comparing virtual and physical robotics environments for supporting complex systems and computational thinking. J. Sci. Educ. Technol. **24**, 1–20 (2015)
10. Aziz, A.A., Klein, M.C.: Incorporating an ambient agent to support people with a cognitive vulnerability. In: Zacarias, M., Oliveira, J.V. (eds.) Human-Computer Interaction. SCI, vol. 396, pp. 169–192. Springer, Heidelberg (2012)
11. Bosse, T., Callaghan, V., Lukowicz, P.: On computational modeling of human-oriented knowledge in ambient intelligence. J. Ambient Intell. Smart Environ. **2**(3), 3–4 (2010)
12. Kinny, D., Georgeff, M., Rao, A.: Amethodology and modelling technique for systems of BDI agents. In: Van de Velde, W. (ed.) MAAMAW 1996. LNCS, vol. 1038. Springer, Heidelberg (1996)
13. Ghanimi, H.M.A., Aziz, A.A., Ahmad, F.: An agent-based modeling for a reader's cognitive load and performance. Adv. Sci. Lett. (2016, to appear)
14. Aziz, A.A., Klein, M.C.A., Treur, J.: An integrative ambient agent model for unipolar depression relapse prevention. J. Ambient Intell. Smart Environ. **2**(1), 5–20 (2010)
15. Kalyuga, S.: Cognitive load theory: implications for affective computing. In: proceeding of the Twenty-Fourth International Florida Artificial Intelligence Research society Conference (2011)
16. Bosse, T., Jonker, C.M., Van Der Meij, L., Treur, J.: A language and environment for analysis of dynamics by simulation. Int. J. Artif. Intell. Tools **16**(3), 435–464 (2007)
17. Mumm, J., Mutlu, B.: Designing motivational agents: the role of praise, social comparison, and embodiment in computer feedback. Comput. Hum. Behav. **27**(5), 1643–1650 (2011)
18. Gulz, A., Silvervarg Flycht-Eriksson, A., Sjödén, B.: Design for off-task interaction–rethinking pedagogy in technology enhanced learning. In: The 10th IEEE International Conference on Advanced Learning Technologies, pp. 204–206 (2010)
19. Choi, H.H., van Merriënboer, J.J.G., Paas, F.: Effects of the physical environment on cognitive load and learning: towards a new model of cognitive load. Educ. Psychol. Rev. **26** (2), 225–244 (2014)

Student Acceptance and Attitude Towards Using 3D Virtual Learning Spaces

Hamadiatul Hayati Abdul Hamid[(⊠)], Zulhilmi Ahmad Sherjawi,
Saiful Omar, and Somnuk Phon-Amnuaisuk

Universiti Teknologi Brunei, Gadong, Brunei Darussalam
hamadiatul.hamid@moe.gov.bn, zulhilmi.as@gmail.com,
{saiful.omar,somnuk.phonamnuaisuk}@utb.edu.bn

Abstract. The aim of this paper is to investigate the factors influencing student's acceptance and attitude towards using 3D virtual learning spaces for education. Extended Technology Acceptance Model has been utilized as its hypothetical premise by incorporating self-efficacy and perceived enjoyment as new external variables. The model is tested through a survey administered to 85 students who took an interest in using the 3D virtual learning spaces. We conducted a regression analysis to examine the potential influence of independent variables on the acceptance and attitude towards using 3D virtual learning spaces. Our result showed that attitude towards using was the significant influence on behavior intention to use. In addition, the reconciliation of self-efficacy and perceived enjoyment are also significant antecedents to perceived ease of use and perceived usefulness. This study confirms that self-efficacy, perceived enjoyment, perceived ease of use, and perceived usefulness are important variables of acceptance and attitude towards using 3D virtual learning spaces.

Keywords: 3D virtual learning spaces · Virtual worlds · Education · TAM · Self-efficacy · Enjoyment · Universiti Teknologi Brunei · Brunei Darussalam

1 Introduction

Learning spaces are places, physical or virtual, where learning happens. Information and Communication Technologies play an important role in provides education facilities creating and facilitating 21st century learning environments such as the digital world (i.e. digital learning objects, digital learning resources, digital game-based learning and etc.), virtual worlds (i.e. virtual world learning spaces, virtual world learning games, virtual world language learning and etc.), and the physical environment (campus, lecture room, library, computer lab and etc.) accommodate to support students and educators to achieve the 21st century knowledge and skills, to empower and enable students to be deep rooted learners and dynamic members in the public eye. A virtual learning environment is a powerful and purposeful tool for teaching and learning.

Three-dimensional (3D) virtual learning spaces, offer a stimulating and new environment to improve learning for students to generate interactive learning experience. 3D virtual worlds ordinarily give three fundamental components: the deception of

© Springer International Publishing AG 2017
S. Phon-Amnuaisuk et al. (eds.), *Computational Intelligence in Information Systems*,
Advances in Intelligent Systems and Computing 532, DOI 10.1007/978-3-319-48517-1_10

3D space; avatar that serves as the visual realistic of 3D representations of users; and intelligent 3D chat environment for users to interconnect with one another.

User acceptance in the proposed study refers to investigate the factors influencing student's acceptance and attitude towards using Universiti Teknologi Brunei (UTB) 3D Virtual Learning Spaces (3DVLSs) as learning tool which make use of an 3D immersive environment such as that presented by virtual worlds for learning, analyzing the facilitation of technology use, amongst others, factors affecting the Perceived Ease Of Use (PEOU), and Perceived Usefulness (PU). The conceptual framework guiding this study was the Technology Acceptance Model (TAM). UTB 3DVLSs combines aspects of game-based learning and simulations inside the virtual worlds of *OpenSim*, designed to resemble the Universiti Teknologi Brunei phase 3 building which typically include a lecture hall, meeting rooms, and different learning spaces for presentations, and group discussion as to develop the real world and flexible learning spaces.

The next section provides with a concise overview of the literature review on virtual worlds and virtual learning spaces, technology acceptance model, and technology acceptance model and virtual worlds. Section three presents the proposed research model and hypotheses, examines six variables. Section four discussed the research methodology outlines the data collection processes. Section five describes data analysis and results. The final section delivers a conclusion and discussion about student's acceptance and attitude towards using 3D virtual learning spaces for education.

2   Literature Review

2.1   Virtual Worlds and Virtual Learning Spaces

Virtual worlds have existed since the early 1980 s, there is no for the most part acknowledged meaning of the virtual world, however they do require that the world is constant; the world must keep on existing even after a user leaves the world, and user rolled out improvements to the world ought to protect [1, 2]. In a virtual world, the user makes an "avatar", a character that signifies a user in a recreated 3D space. Avatars can travel through the virtual world, and collaborate synchronously with other user and with objects in the world. Regularly, most depend upon text-based chat tools, despite the fact that a couple manage of audio chat, to communicate with another user. 3D virtual worlds are relatively new; however, preliminary research shows that they support various types of educational resourcefulness, such as studies examine whether 3D virtual world can be used for: learning science, used in the biology course and forensic science [3], teaching 3D virtual world in the chemistry virtual classroom [4], using 3D virtual classroom simulation for teachers' continuing professional development [5], and virtual classrooms for an online and a blended course [6].

Virtual learning spaces are designed for teaching, learning, communication, program delivery, knowledge development, and content access [7]. According to [8], the two case studies which prove that Virtual Space can be a new framework for learning are the framework of English as Second Language Academic class and 3D simulation of archeological excavation site, which demonstrate general advantages of 3D virtual environment technology which can be utilized as learning apparatus by giving a virtual learning space that (i) assists activities that are not possible in physical locations; (ii) gives users access to facilities not accessible physically; (iii) allows people in different locations to interact, and (iv) offers a assortment of observation and measurement tools for execution assessment and improvement.

The Polytechnic University of Bucharest, [4] created a virtual learning space as a 3D virtual chemistry class using EON Creator software package which offers new opportunities for teaching in immersive and creative spaces. From the experiments of the chemical virtual environment, the students can travel through a 3D space and navigate through all the 3D objects, interaction in a virtual world increases student commitment in an online class and furthermore students' feeling of group with the class. They concluded that virtual world can be compelling and creative tool for a better and more advanced educational environment.

2.2   The Technology Acceptance Model (TAM)

The research studies focused of user acceptance of new innovation of utilizing virtual world as a learning domain by applied TAM. One of the outstanding models identified with technology acceptance and use is the TAM, initially presented by Davis [9]. TAM was observed to be much less difficult, simpler to utilize, and greater powerful mannequin of determinants of user acceptance of Information System and Technology, which were found to palatable foresee an individual's expectations. TAM is initially an extension of Theory of Reasoned Action (TRA). It is based on the TRA developed by Martin Fishbein and Icek Ajzen [10], a broadly concentrated on model from social psychological science which concerned with determinants of intentionally proposed individual's voluntary behaviors.

TAM sets that PEOU is liable to impact PU, where the increase of PEOU leads to improve the performance of an activity. Subsequently, PEOU impacts PU. Both PEOU and PU predict Attitude Towards Using (ATU), the evaluation of the user desirability of using the information system and technology. ATU alongside PU impacts individual user's Behavioral Intention to Use (BIU) of the information system and technology.

Past studies have suggested TAM is a useful framework for considerate user acceptance of new information system and technology. Since 2006, many attempts to adapt the TAM in various fields were found in literature such as: [11, 12] heavily applied TAM to understand user acceptance in e-learning. On the other side, the adaption of TAM to the m-learning is found to gain increasing importance lately, as can be found in published research of [13, 14]. Then again, TAM additionally utilized as a part of blended learning [15], virtual learning environment [16, 17], the virtual world [18–21].

2.3   The Technology Acceptance Model (TAM) and Virtual Worlds

A few studies have inspected TAM as a model how people accept, adopt and utilize the virtual world, specifically at the higher education level. [18] construct and moderating variables analyzed to evaluate user acceptance and adoption of virtual worlds using *Second Life (SL)* based on TAM. Results demonstrated that community factors for instance communication, collaboration, and cooperation unequivocally impact the PU of Virtual Worlds and it demonstrates that probability to intermingle in a 3D environment in combination with Voice over IP plays a key theatrical role in user acceptance and technology adoption of Virtual Worlds.

TAM and Media Richness Theory have been utilized as a part of the study to looking at the media richness of *SL* and effect on the user acceptance which conducted by [19]. Found that the relationships among media richness and PEOU and PU lead to the actual usage and acceptance of the *SL*, validating that these three constructs are practical and useful in understanding the virtual experience in the context of use of education in virtual worlds. The beneficial effect of media richness on user acceptance of *SL* has all the earmarks of being a huge finding of the study.

In another study by [20] were surveys undergraduate business students' intention to accept and use the virtual world *SL* for education. This study is based on TAM and extended variables that add to the acceptance or abandonment of new Information Technology. The outcomes recommend that PEOU influences user's intention to adopt *SL* through PU and the "fun factor" is also specified which describes that the computer playfulness and computer self-efficacy is significantly related to acceptance and use of virtual worlds.

3   Research Model and Hypotheses

Base on TAM and extended TAM theories, the research model analyzes 6 variables: Self-Efficacy (SE), Perceived Enjoyment (PE), Perceived Usefulness (PU), Perceived Ease of Use (PEOU), Attitude towards Using (ATU), and Behavioral Intention to Use (BIU) to determine students' acceptance and attitude towards using 3DVLSs. The original TAM incorporated additional variables to produce extended TAM. SE and PE are the new or extended variables added to the TAM model to fit this study. Figure 1 portrays the research model utilized in this study.

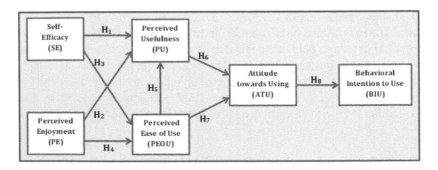

Fig. 1. The research model

There are eight hypotheses in this study. The hypotheses were formed according to the extended TAM that proposed as below:

Hypothesis 1: SE of 3DVLSs will have a significant influence on PU.

Hypothesis 3: SE of 3DVLSs will have a significant influence on PEOU.

SE refers to "the degree to which a person perceived his or her ability to perform a system specific computer related task" [22]. The four measurement items utilized and adjusted from [20, 23] are: I could complete the task using the 3DVLSs if there is no one around to tell me what to do as I go (SE1), I could complete the task using the 3DVLSs if I had never used a package like it before (SE2), I could complete the task using the 3DVLSs if I had only the software manuals for reference (SE3), I could complete the task using the 3DVLSs if I had seen someone else using it before trying it myself (SE4).

Hypothesis 2: PE of 3DVLSs will have a significant influence on PU.

Hypothesis 4: PE of 3DVLSs will have a significant influence on PEOU.

PE refers to "the degree to which performing an activity is perceived as providing pleasure or joy in its own right, aside from performance consequences" [23]. PE attempted to measure the degree to which users perceived their interaction with 3DVLSs to be enjoyable. Four measurement items used and revised from [17, 21, 23] are: Using the 3DVLSs makes learning more enjoyable (PE1), Using the 3DVLSs was pleasant (PE2), Using the 3DVLSs was an interesting experience (PE3), Overall, I found learning using the 3DVLSs is fun (PE4).

Hypothesis 5: PEOU of 3DVLSs will have a significant influence on PU.

Hypothesis 7: PEOU of 3DVLSs will have a significant influence on ATU.

PEOU refers to "the degree to which a person believes that using a particular system would be free of effort" [22]. Seven measurement items were applied and adapted from [9, 15, 16, 18–20, 23] are: Learning to operate 3D virtual learning spaces would be easy for me (PEOU1), I found it easy to get 3D virtual learning spaces to do what I want to do (PEOU2), My interaction with 3D virtual learning spaces was clear and understandable (PEOU3), 3D virtual learning spaces are flexible to interact with (PEOU4), It's easy for me to become skillful at using 3D virtual learning spaces (PEOU5), It's easy to play the role of the avatar (PEOU6), Overall, I believe 3D virtual learning spaces easy to use (PEOU7).

Hypothesis 6: PU of 3DVLSs will have a significant influence on ATU.

PU refers to "the degree to which a person believes that using a particular system would enhance his or her job performance" [22]. Eight measurement items were applied and adapted from [9, 12, 15, 16, 18–20, 23] are: Using 3DVLSs enables me to accomplish my tasks in the coursework more quickly(PU1), Using 3DVLSs would improve my coursework performance (PU2), Using 3DVLSs would increase my productivity in my course work (PU3), Using 3DVLSs would enhance my effectiveness in learning (PU4), Using 3DVLSs improves my communication with my colleagues (PU5), Using 3DVLSs improves my collaboration with my colleagues (PU6),

Using 3DVLSs improves my cooperation with my colleagues (PU7), Overall, I found 3DVLSs useful in my course work (PU8).

Hypothesis 8: ATU of 3DVLSs will have a significant influence on BIU.

ATU refers to "the degree to which a person associated positive feelings with target system" [22]. Four measurement items were applied and adapted from [12, 18] are: 3DVLSs make learning more interesting (ATU1), 3DVLSs motivate learning (ATU2), I have a positive attitude toward using 3D virtual learning spaces (ATU3), Overall, I believe using 3DVLSs would be a good idea (ATU4).

BIU denotes "the degree to which a person has formulated conscious plans to perform or not to perform some specified future behaviors" [22]. Four measurement items were applied and adapted from [12, 15, 19, 20, 23] are: Assuming I had access to 3DVLSs, I intend to use it (BIU1), Given that I had access to 3DVLSs, I predict that I would use it (BIU2), I am going to positively utilize 3DVLSs (BIU3), I intent to use 3DVLSs frequently in future (BIU4).

4   Research Methodology

4.1   Data Collection and Procedure

Data were collected from a random sampling of around 85 students toward the end of the academic year (April and May 2016). The surveys were given to selected postgraduate and undergraduate students studying full time which enrolled in Computing, Business, and Engineering Courses.

The course conducted using UTB 3DVLSs platform to enhance technology use, acceptance, and experience. During the testing session the participating students are guided to explore the UTB 3DVLSs divided into four sessions; demonstration, hands-on, task-oriented, and questionnaires session. In the demonstration session (Fig. 2), all participants were briefly explained about the nature of the UTB 3DVLSs. Then, in brief hands-on they simply followed the instruction and played around with the UTB 3DVLSs. In task-oriented interaction session (Fig. 3), a list of tasks was given where the respondents were required to complete. After that, the respondents were asked to answer questionnaires that also included in the UTB 3DVLSs.

Fig. 2. Demonstration session

Fig. 3. Task-oriented interaction session

4.2   Instrumentation

The survey instrument was developed and designed in two parts. Part 1 contained personal information and part 2 captured the questionnaire, using a seven-point Likert-style scale ranging from "strongly disagree" (1) to "strongly agree" (7) which consists of 31 items for measuring the SE, PE, PEOU, PU, ATU and BIU. Questionnaire items were broadly utilized as a part of previous research identified with TAM; be that as it may, the questions were changed to fit the particular setting of the current research.

5   Results

5.1   Background Profile

The background data of participating students has been outlined in Table 1. Table 1 defines the characteristics of students. The majority are males within the age group of 21 to 30 years (78%).

The data was collected from the 85 students and analyzed through SPSS. At that point, the basic model was surveyed, giving results for hypothesis testing. The extended TAM hypotheses were affirmed by conducting the validity and reliability tests, presenting descriptive statistics of the research variables, then carry out the regression analyses and correlation analyses.

5.2   Validity and Reliability

In order to ensure the quality of the survey data, the validity and reliability tests were employed, two metrics are used: average variance extracted and composite reliability.

A Cronbach Alpha was conducted to determine the reliability for each construct variable. The overall Cronbach Alpha for all items is 0.96. It changes for corresponding construct variables somewhere around 0.77 and 0.90. [24] give the accompanying dependable guidelines: ">0 .9 – Excellent, >0.8 – Good, >0.7 – Acceptable, >0.6 – Questionable, >0.5 – Poor, and <0.5 – Unacceptable" (p. 231). As showed in Table 2, the estimation of all our variables exceed the minimum value of 0.70, propose that 0.5 to be a valid estimation of component, above 0.5 demonstrates a good convergent [25], which was significantly above the 0.70 level, indicating the reliability of the instrument.

Additionally, a validity test was performed using factor analysis. All constructs in the model satisfactorily pass the test; the variance is greater than 50%. Factor analysis shows the total no. of variance is greater than 50%, which is 73.5%. Thus, it was determined that this instrument had achieved acceptable levels of validity.

Construct validity of more than 0.50 and composite reliability of 0.70 or above are deemed acceptable. Therefore, its show that strong empirical support for the validity and reliability of the construct variables used in the research model. The study demonstrated that extended TAM is valid, reliable, and fit to assess the acceptance of the users of 3DVLSs processes. The accompanying Table 2 outlines the construct variables and number of basic measurement items, Cronbach Alpha and the Mean value and Variance.

Table 1. Profile of the respondents

Item	Variable	Description	Frequency	Percentage
1	Gender	Male	45	53%
		Female	40	47%
2	Age	20 or Under	4	5%
		21 to 30	66	78%
		31 or above	15	18%
3	Education level	Undergraduate (Degree)	71	84%
		Postgraduate (Masters)	14	16%
4	Faculty	Computing	48	56%
		Engineering	26	31%
		Business	11	13%
5	Internet experience	Less than 5 years	7	8%
		6 to 10 years	44	52%
		More than 11 years	34	40%
6	3D virtual experience	None	7	8%
		Less than 3 years	46	54%
		4 to 7 years	23	27%
		More than 8 years	9	11%
7	To what extent using 3D Virtual	Not at all	7	8%
		Very little	31	36%
		Average	29	34%
		More than average	5	6%
		Very much	13	15%
8	Speed of current internet	Very slow	6	7%
		Slow	28	33%
		Acceptable	33	39%
		Good	17	20%
		Excellent	1	1%

Table 2. Descriptive statistics of the constructs

Variable	No. of measurement items	Cronbach alpha	Mean	Variance
SE	4 items	0.77	5.08	0.80
PE	4 items	0.83	5.93	0.72
PEOU	7 items	0.84	5.26	0.68
PU	8 items	0.92	5.10	0.87
ATU	4 items	0.89	5.85	0.86
BIU	4 items	0.90	5.62	0.88
Overall	31 items	0.96		

5.3   Hypothesis Testing: Regression Analysis

To test the hypotheses, we conducted linear regression analysis in which the BIU was set as dependent variable. ATU explains 73% of the variance in BIU ($R^2 = 0.730$). PEOU and PU combined to explain 46% of the variance in ATU ($R^2 = 0.460$). SE and PE combined to explain 50% of the variance in PEOU ($R^2 = 0.504$). SE and PE combined to explain 47% of the variance in PU ($R^2 = 0.471$). The model (73%) has a strong prediction value and good parsimony as shown in Fig. 4 below.

All of the beta coefficients are significant at p<0.01 level. ATU is a strong antecedent to BIU with a beta coefficient of 0.88 supports H8. PEOU has a higher beta coefficient to ATU (0.54) than PU to ATU (0.32), supports H7 and H6. PEOU also has a high beta coefficient to PU (0.82), supports H5. PE has a high beta coefficient to PEOU (0.45) than SE to PEOU (0.20), supports H4 and H3. Additionally, SE has a higher beta coefficient

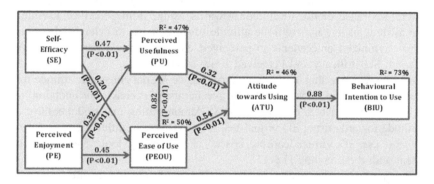

Fig. 4. The structural model results

Table 3. The summary of hypothesized results

H#	Path		R^2	Path	t-value	Results of
	From	To		coefficient		hypotheses
H1	SE →	PU		0.471	4.385	Supported (P<0.01)
H2	PE →	PU		0.318	2.662	Supported (P<0.01)
H3	SE →	PEOU		0.202	3.925	Supported (P<0.01)
H4	PE →	PEOU		0.451	5.462	Supported (P<0.01)
H5	PEOU →	PU		0.823	7.672	Supported (P<0.01)
H6	PU →	ATU		0.320	3.042	Supported (P<0.01)
H7	PEOU →	ATU		0.538	3.999	Supported (P<0.01)
H8	ATU →	BIU		0.876	14.997	Supported (P<0.01)
	PU		47%			
	PEOU		50%			
	ATU		46%			
	BIU		73%			

to PU (0.47) than PE to PU (0.32) found that SE and PE have a significant positive influence on PU of 3D virtual learning spaces, supporting H1 and H2.

Correlation analyses showed that all the construct variable; SE, PE, PEOU, PU, and ATU had a high correlation among one another (p<0.01) as shown in Table 3. All eight hypotheses are supported. All the construct variables are statistically significant at 1% level of significance as the p value corresponding to all the construct variables are less than 0.01. Hence H1, H2, H3, H4, H5, H6, H7, and H8 are accepted.

6   Discussion and Conclusion

This study attempts to investigate the factors influencing student's acceptance and attitude towards using 3D virtual learning spaces for education. With survey data from 85 students, the research model with 6 variables were proposed and investigated. Overall the model clarified 73% of the variance in behavioral intention to use. The results supported the causal path from perceived ease of use to perceived usefulness, from perceived ease of use to attitude towards using, from perceived usefulness to attitude towards using, and from the attitude towards using to behavioral intention to use. Two significant antecedents to perceived ease of use and perceived usefulness were found: Self-Efficacy and Perceived Enjoyment.

The study revealed that Technology Acceptance Model construct: attitude towards using (ATU) was the most significant predictors of behavioral intention to use (BIU) with significant beta coefficient (0.88) demonstrating that students have a positive attitude towards using 3D virtual learning spaces, can influence their behavioral intention to use 3D virtual learning spaces as a learning tool. These findings are consistent with the literature [12–15].

Then again, perceived ease of use was influential in making its effect on perceived usefulness and attitude towards using, consistent with [12, 15]. The positive influence of perceived ease of use suggests that students discovered 3D virtual learning spaces simple to utilize, so they thought that it was more useful and they ought to have a positive attitude towards using the 3D virtual learning spaces. Perceived usefulness has reliably been significant antecedent in anticipating user's attitude towards using, were found similar to those of [12–15]. When students agreed that 3D virtual learning spaces represent useful environment for learning i.e. improve and increase productivity in coursework and enhance effectiveness in learning, this means that perceived usefulness has a positive influence on students attitude towards using of 3D virtual learning spaces.

The literature review led us to develop an extended Technology Acceptance Model incorporated two new constructs applicable to the understanding of student's acceptance and attitude towards using 3D virtual learning spaces for education: Self-Efficacy and Perceived Enjoyment. An appropriate consideration of these two constructs can increase student's perceived ease of use and perceived usefulness, thereby forming a more positive attitude and intention to use the 3D virtual learning spaces. The presence of self-efficacy and perceived enjoyment are as expected, play a significant role in determining student's behavior, by means increase student's ability and interactivity, where students feel more confident and enjoyable of accepting and using towards the 3D virtual learning spaces.

Student's self-efficacy in virtual world learning can play a major role, ability to accomplish actions effect on usage and a high degree of effort effect easy to use, findings revealed that self-efficacy is a significant predictor for both perceived ease of use and perceived usefulness [11, 12]. Enjoyment also an important factor of virtual world's usage, considered that technology for fun denotes the extent to believe participating in 3D virtual learning spaces is enjoyable and high impact on how easy the student perceived the 3D virtual learning spaces can be used, findings revealed that perceived enjoyment is a significant predictor for both perceived ease of use and perceived usefulness [17, 21]. Our findings specify that "user's confidence" and the 'enjoyment factor" in technology use are significantly related to acceptance and attitude towards using the 3D virtual learning spaces for education.

References

1. Nurulhidayati, H.M.S., Mohd Saiful, O., Authien, W.: Virtual campus: a case study of development using open sources software. In: Proceedings of 2016 International Conference on Artificial Intelligence & Manufacturing Engineering (ICAIME 2016), pp. 1–9 (2016)
2. Bell, M.W, Castronoya, E., Wagner, G.G: Surveying the virtual world - a large scale survey in second life using the virtual data collection interface (VCDI). Research Notes series. German Council for Social and Economic Data (RatSWD) Research Note No. 40, pp. 1–47 (2009)
3. Carolyn, L., Mary A.C.: Student perceptions of learning science in a virtual world. In: Proceedings of the 24th Annual Conference on Distance Teaching & Learning, pp. 1–5 (2008)
4. Shudayfat, E., Florica, M., Alin, D.B.M.: A 3D virtual learning environment for teaching chemistry in high school. In: Proceedings of the 23rd International DAAAM Symposium, vol. 23(1), pp. 423–428 (2012)
5. Pavlos, K., Demetrios G.S.: Implementing a 3D virtual classroom simulation for teachers' continuing professional development. In: Proceedings of the 18th International Conference on Computers in Education, pp. 36–44 (2010)
6. Michele, A.P., Florence, M.: Using virtual classrooms: student perceptions of features and characteristics in an online and a blended course. MERLOT J. Online Learn. Teach. 6(1), 135–147 (2010)
7. Anita, Z.B.: Personalizing virtual learning spaces: a participatory approach. In: Proceeding of World Conference on E-learning in Corporate, Government, Healthcare, and Higher Education 2011, pp. 1790–1795 (2011)
8. Ali, A., Peggy, H., Shawn, G., Nuket, N.: Virtual space as a learning environment: two case studies. In: Proceedings of the 11th International Education Technology Conference'de Sunulmus Bildiri Istanbul, Turkey, vol. 1, pp. 1278–1285 (2011)
9. Fred, D.D.: A technology acceptance model for empirically testing new end-user information systems: theory and results. Doctoral dissertation, Sloan School of Management, Massachusetts Institute of Technology, Cambridge (1986)
10. Martin, F., Icek, A.: Belief, Attitude, Intention, and Behavior: An Introduction to Theory and Research. Addison-Wesley, Reading (1975)

11. Chorng, S.O., Jung, Y.L.: Gender differences in perceptions and relationships among dominants of e-learning acceptance. Comput. Hum. Behav. **22**, 816–829 (2006)
12. Sung, Y.P.: An analysis of the technology acceptance model in understanding university students' behavioral intention to use e-learning. Educ. Technol. Soc. **12**(3), 150–162 (2009)
13. Sung, Y.P., Min, W.N., Seung, B.C.: University students' behavioral intention to use mobile learning: evaluating the technology acceptance model. Br. J. Educ. Technol. **43**(4), 592–605 (2012)
14. Afzaal, H.S., Noah, A.R., Rudy, R., Armanadurni, A.R.: A preliminary study of students' attitude on m-learning: an application of technology acceptance model. Int. J. Inf. Educ. Technol. **5**(8), 609–614 (2015)
15. Nikolaos, T., Stelios, D., Maria, P.: Assessing the acceptance of a blended learning university course. Educ. Technol. Soc. **14**(2), 224–235 (2011)
16. Erik, M.V.R., Jeroen, J.L.S.: The acceptance and use of a virtual learning environment in China. Comput. Educ. **50**(3), 838–852 (2008)
17. Reinder, V.: An empirical investigation of pupils' acceptance of a virtual learning environment (2007)
18. Marc, F., Christoph, L.: User acceptance of virtual worlds. J. Electron. Commer. Res. **9**(3), 231–242 (2008)
19. Nauman, S.,Yun, Y., Suku, S.: Media richness and user acceptance of second life. In: Proceedings Ascilite Melbourne, pp. 851–860 (2008)
20. Jia, S., Lauren, B.E.: Intentions to use virtual worlds for education. J. Inf. Syst. Educ. **20**(2), 225–234 (2009)
21. Saniye, T.T., Veysi, İ.: Acceptance of virtual worlds as learning space. Innov. Educ. Teach. Int. **52**(3), 254–264 (2015)
22. Fred, D.D.: Perceived usefulness, perceived ease of use, and user acceptance of information technology. MIS Q. **13**(3), 319–340 (1989)
23. Viswanath, V.: Determinants of perceived ease of use: integrating control, intrinsic motivation, and emotion into the technology acceptance model. Inf. Syst. Res. **11**(4), 342–365 (2000)
24. Darren, G., Paul, M.: SPSS for Windows Step by Step: A Simple Guide and Reference, 4 (11.0 Update) edn, pp. 1–63. Allyn & Bacon, Inc., Boston (2003)
25. Jum. C.N.: Psychometric Theory. 3rd ed. The University of Michigan: McGraw-Hill, (1967)

Data Mining and Its Applications

Class Noise Detection Using Classification Filtering Algorithms

Zahra Nematzadeh[1], Roliana Ibrahim[1(✉)], and Ali Selamat[1,2]

[1] Faculty of Computing, Universiti Teknologi Malaysia,
81310 Johor Bahru, Johor, Malaysia
zahra_nematzadeh@yahoo.com, {roliana, aselamat}@utm.my
[2] UTM-IRDA Digital Media Centre of Excellence,
Universiti Teknologi Malaysia, 81310 Johor Bahru, Johor, Malaysia

Abstract. One of the significant problems in classification is class noise which has numerous potential consequences such as reducing the overall accuracy and increasing the complexity of the induced model. Subsequently, finding and eliminating misclassified instances are known as important phases in machine learning and data mining. The predictions of classifiers can be applied to detect noisy instances, inconsistent data and errors, what is called classification filtering. It creates a new set of dataset to develop a reliable and precise classification model. In this paper we analyze the effect of class noise on six supervised learning algorithms. To evaluate the performance of the classification filtering algorithms, several experiments were conducted on six real datasets. Finally, the noisy instances are removed and relabeled and the performance was then measured using evaluation criteria. The findings of this study show that classification filtering have a potential capability to detect class noise.

Keywords: Class noise · Classification filtering · Machine learning · Noise detection

1 Introduction

The errors and data inconsistencies, which reduce the quality of real data, are commonly referred as noise. Noisy data categorized into two types, which are class noise and attribute noise. Class noise is referred to instances with wrong labels, and attribute noise is referred to instances with unusual attribute values [1]. The main issues of class noise are diminishing the accuracy of predictions and increasing the complexity of induced model and number of required training data [2]. It also affect the quality of information extracted from the data and decisions made using noisy data [3]. Although, removing the noisy data is almost challenging, but it plays a significant role in machine learning to have reliable model and high performance [4]. Therefore, identifying noisy data and eliminating or emending them became well-known field in data mining research [2]. Classification filtering is the filtering approach which is presented by Brodley & Friedl [5]. It uses the predictions of classifiers as a tool to identify mislabeled samples [6, 7]. The assumption of the classification filtering is to trust the classifiers which are able to predict the label of instance correctly and the wrong

S. Phon-Amnuaisuk et al. (eds.), *Computational Intelligence in Information Systems*,
Advances in Intelligent Systems and Computing 532, DOI 10.1007/978-3-319-48517-1_11

classified samples are known as noise [5]. While it is necessary to identify which instances are noisy and should be reduced in real world datasets, this information is not available for the datasets. This paper focuses on the study of the machine-learning algorithms as classification filter to detect class noise. The proposed model is observed for noise detection without adding artificial noise. The paper is organized as follows. Section 2 presents related works on noise detection. Section 3 describes the methodology of the proposed model. Datasets and performance measures are presented in Sect. 4. Section 5 explains results and discussions. Finally, Sect. 6 concludes this paper.

2 Related Works

Classification Filtering is known as the predictions of classifiers for noise identification [7]. There are some studies, which have been done based on classification filtering. Thongkam et al. [13] applied SVM on training set to detect and eliminate all samples which misclassified by the SVM. It was an effective technique for detecting misclassified outliers. Also, Jeatrakul et al. [14] applied same approach using neural networks which enhances the confidence of cleaning noisy training instances. Sluban et al. [15] developed new class noise detection algorithms including the high agreement random forest filter which detect the noisy samples with high precision. Likewise, the proposed approach of [13] was extended by Miranda, Garcia [16] using four classifiers with diverse machine learning algorithms and then combined by voting technique for noise identification. Segata et al. [17] proposed Fast Local Kernel Machine Noise Reduction method which is the fastest and achieves the highest improvements in NN accuracy. Moreover, local Support Vector Machines noise reduction technique is proposed by [18] which is considerably outperform in comparison with other analyzed techniques for real dataset while it was not good in artificial datasets. These methods highlight that presence of good classifiers in classification filtering is important and existence of class noise produces poor classifiers [19]. Different from previous studies which only evaluated the impact of removing technique on noise detection and classification performance, this study attempts to examine the effect of both removing and relabeling techniques using classification filtering algorithms.

3 Methodology

This section presents the methodology employed in this study which identifies noisy instances. Based on Fig. 1, this model comprises three main phases, which are data preparation, noise detection, and noise classification.

3.1 Phase1: Data Preparation

To evaluate the noise detection techniques quantitatively it is essential to know which objects are noisy. Because this information was not accessible for the datasets, class

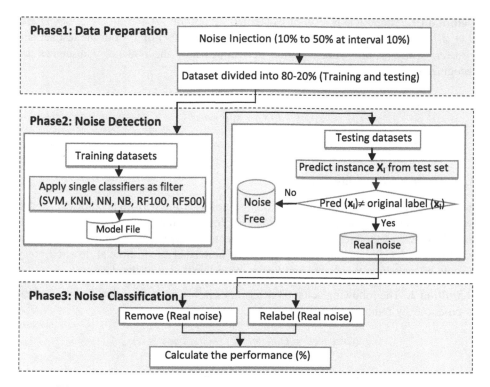

Fig. 1. The overall architecture of the proposed classification filtering model

noise was randomly injected by replacing labels of 10 %, 20 %, 30 %, 40 % and 50 % objects of each dataset. In this phase the data sets were randomly and without replacement divided into 80 % and 20 % train and test set respectively. The experiments were conducted based on 20 runs for each dataset and each noise level to attain average evaluation criteria.

3.2 Phase2: Noise Detection

In this part, we employed six classifiers to detect noisy instances in validation set which are Support Vector Machine (SVM) [8], Random Forests (RF100) [9], Random Forests (RF500) [9], Naïve Bayes (NB) [10], Neural Network (NN) [11] and K-Nearest Neighbor (KNN-k = 10) [12] respectively. Let "X" as noisy samples with "n" samples $X = \{x_1, x_2, \ldots, x_n\}$ and "x_i" is the feature $x_i = (x_{i1}, x_{i2} \ldots x_{im})$ where m is the number of features in X. The data sample "X" includes two class labels while $X = (Y, Z)^t$ where $Y = \{y_l = x_i | (x_i) = 1\}$ where $l = 1, 2, \ldots, h$ and h the number of samples that their labels are $+1$ and $Z = \{z_r = x_i | L(x_i) = -1\}$ where $r = 1, \ldots, s$ and s is the number of samples that their labels are -1 and $n = s + h$ and t is the transpose. $L(x_i)$ represents the class label for each sample $L(x_i) = \{label(x_i) | Label(x_i) = 1 \text{ or } -1\}$.

Definition 1. Suppose φ is a set of k classifier algorithms ($\varphi = \{\varphi_1, \varphi_2, \ldots, \varphi_k\}$) and $B = \varphi(test, train) = \{b_i\}_{i=1}^n$ where b_i is the predicted label of x_i from the test set and $|test\ set| = n$. The following equation determines how the real noisy instances are recognized.

$$b_i = \begin{cases} L(x_i) & x_i \text{ is noise free} \\ -L(x_i) & x_i \text{ is noise} \end{cases} \tag{1}$$

3.3   Phase3: Noise Classification

In order to consider the effect of the noisy samples on the performance of the model, two methods are used to deal with noisy samples, which are "removal" and "relabeling". Removal method removes all the real detected noisy samples and produces a new lessened dataset. Relabeling method assigns a new label to all the real detected noisy objects by switching their label and keeps the original size of the dataset.

Definition 2. The following equation determines how a new set of data set (noise-free) is produced by removing technique.

$$Noise\ free = Dataset\ (X) \ - \ real\ noise(x_i) \tag{2}$$

Definition 3. The relabeling technique produces noise free set as follow: An instance (x_i) from noisy set with label of $L(x_i)$ or $L'(x_i)$, $L(x_i)$ transformed into $L'(x_i)$, and vice versa.

4   Experimental Studies

In this section, the experimental datasets and the performance evaluation criteria used in this study are considered.

4.1   Datasets

In this study, six machine learning datasets from the UCI repository [20] have been used. Table 1 lists the datasets used in this research with the number of classes (#Class), number of features (#Feature) and number of examples (#Ex).

Table 1. Distribution of datasets [20]

Dataset	#Ex	#Features	#class
Pima	768	8	2
Wisconsin	683	9	2
Liver	345	7	2
Parkinson	197	23	2
Heart (statlog)	270	13	2
Ionosphere	351	34	2

4.2 Performance Measure

Since it is necessary to know which samples are noisy for quantitative evaluation of noise detection methods [1], this study randomly injected and controlled the noisy instances. We can then evaluate the effectiveness of the proposed model using precision and recall [1].

$$Precision = \frac{\text{number of true noisy instances detected}}{\text{number of all instances identified as noisy}} \quad (3)$$

$$Recall = \frac{\text{number of true noisy instances detected}}{\text{number of all noisy instances in the dataset}} \quad (4)$$

Also, combining of precision and recall, which is called F-measure, is computed to have a preferred precision-recall tradeoff. F-measure weights precision twice as much as recall [1]. According to previous works, precision should be preferred in noise detection such that the noisy instances recognized are real noise. So, the value of β in f-measure was 0.5 to give more importance to precision than to recall [21]. The F-measure equation is stated following:

$$F_\beta = \left(1 + \beta^2\right) \times \frac{\text{Precision} \times \text{Recall}}{\left(\beta^2 \times \text{Precision}\right) + \text{Recall}} \quad (5)$$

Moreover, accuracy formula is used to calculate the performance of the proposed technique in classification. Classification accuracy is calculated using confusion matrix [22]. In following formula, TN (True Negative) referred as correctly rejected samples, TP (True Positive) referred as correctly identified samples, FP (False Positive) referred as incorrectly identified samples and FN (False Negative) means incorrectly rejected samples.

$$Accuarcy = \frac{TP + TN}{TP + TN + FP + FN} \quad (6)$$

5 Results and Discussions

In order to evaluate the proposed model in terms of noise detection, the results of precision, recall and f-measure are presented and discussed. The results of noise classification performance in terms of accuracy is presented to evaluate the noise classification techniques as well.

5.1 Noise Detection Evaluation Results in Terms of Precision

As it is mentioned previously, a high value of precision means, the proposed model can correctly detect those samples as noises which really are noise. Based on the

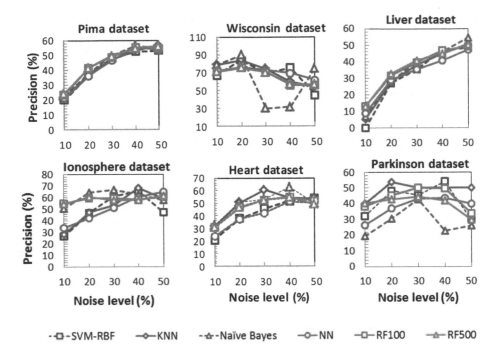

Fig. 2. Comparing the precision results obtained using six classification filtering algorithms on six UCI datasets

experimental results presented by Fig. 2, the best classification filtering algorithm is KNN for Pima dataset with 56.639 % at noise level of 50 %, Naïve Bayes for Liver dataset with 54.401 % at noise level of 50 %, Naïve Bayes for Wisconsin dataset with 89.560 % at noise level of 20 %, KNN for Ionosphere dataset with 68.328 % at noise level of 40 %, SVM-RBF for Parkinson dataset with 54.166 % at noise level of 40 % and KNN for heart dataset with 60.398 % at noise level of 30 % in terms of precision respectively. The results of precision show that by increasing the noise level, the precision is almost improved and the classifiers able to detect more noisy instances. These algorithms can correctly detect real noisy.

5.2   Noise Detection Evaluation Results in Terms of Recall

Figure 3 shows the recall of each dataset using six classification filtering algorithms in various noise levels. A high recall rate shows that many of the noisy data introduced were recognized, while a low recall rate shows that the major of the noisy cases were ignored [1]. Based on the experimental results, the best classification filtering algorithm is RF500 for Pima dataset with 75.490 % at noise level of 20 %, KNN for Liver dataset with 62.318 % at noise level of 40 %, NB for Wisconsin dataset with 94.485 % at noise level of 20 %, NB for Ionosphere dataset with 91.428 % at noise level of 10 %, KNN and NN for Parkinson dataset with 89.473 % at noise level of 10 % and NB for

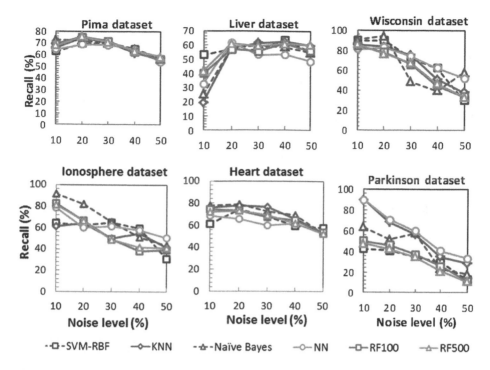

Fig. 3. Comparing the recall results obtained using six classification filtering algorithms on six UCI datasets

heart dataset with 78.703 % at noise level of 20 % in terms of recall respectively. These algorithms can detect many of the noisy data injected to data correctly.

5.3   Noise Detection Evaluation Results in Terms of F-Measure

In this section, the F-measure of each dataset, which are obtained using six classification filtering algorithms in various noise levels, are analyzed separately and illustrated in Fig. 4. Based on the experimental results, the best classification filtering algorithm is NB for Pima dataset with 57.861 % at noise level of 40 %, NB for liver dataset with 54.448 % at noise level of 50 %, NB for Wisconsin with 90.502 % at noise level of 20 %, NB for Ionosphere dataset at noise level of 20 % with value of 67.455 %, KNN achieved highest F-measure with 55.918 % at noise level of 20 %, Naïve Bayes obtained highest F-measure at noise level of 40 % with value of 63.784 % for Heart dataset.

5.4   Noise Classification Results in Terms of Accuracy

This section discusses noise classification results in terms of accuracy. In order to achieve the accuracy of the datasets, SVM classifier with RBF kernel was used. As

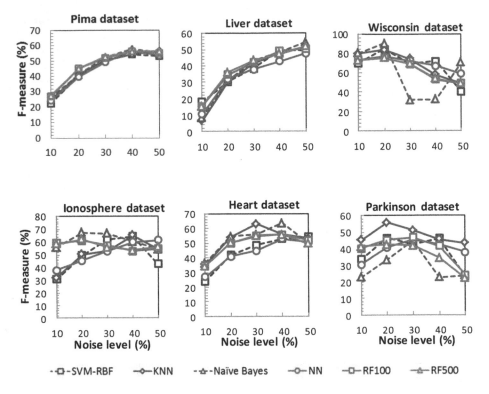

Fig. 4. Comparing the F-measure results obtained using six classification filtering algorithms on six UCI datasets

shown in Table 2, the best accuracy related to Pima dataset obtained by Naive Bayes using removal technique, which was 94.735 % at 30 % noise level, while the higher accuracy of relabeling technique belongs to Naïve Bayes with 92.909 % at 30 % noise level. In addition, the best accuracy obtained by KNN in Liver dataset using removal technique with 88.100 % at noise level 10 %, whereas the relabeling technique obtained accuracy with 82.049 % using SVM at noise level 10 %. Moreover, the best accuracy of Wisconsin dataset obtained by SVM after removing noisy instances with 98.671 % at 10 % noise level while the higher accuracy after relabeling the detected noisy instances by SVM was 98.462 % at noise level 10 %. Furthermore, the highest accuracy in Parkinson dataset achieved with 95.148 % at noise level 10 % using SVM after removing noisy instances while the higher accuracy using relabeling achieved by SVM with 93.482 % at noise level 10 %. Likewise, the best accuracy of heart dataset obtained with 90.579 % at noise level 20 % after removing noise, which detected by KNN, while the best accuracy after relabeling noisy instances achieved by Naïve Bayes with 88.797 % at noise level 10 %. The best accuracy of Ionosphere dataset also obtained with 87.743 % at noise level 50 % after removing noise using RF500, while the higher accuracy after relabeling noise detected by SVM was 92.416 % at noise

Table 2. The best classification technique for six various datasets in terms of accuracy

Dataset	Noise level	Classification filtering algorithm	Accuracy	Noise classification technique
Pima	30 %	NB	94.735	Removal
Liver	10 %	KNN	88.100	Removal
Wisconsin	10 %	SVM-RBF	98.671	Removal
Heart	10 %	KNN	90.579	Removal
Parkinson	10 %	SVM-RBF	95.148	Removal
Ionosphere	10 %	SVM-RBF	94.801	Removal

level 10 %. Table 2 summarized the best noise classification technique for various machine learning dataset in different noise levels.

6 Conclusion

This paper focuses on the study of the six machine-learning algorithms as classification filter to detect class noise. It attempts to study and evaluate the performance of the supervised learning algorithms with the existence of certain number of mislabeled instances in the data using the precision, recall, F0.5 and accuracy measures. Although, the results of this study show an improvement in evaluation metrics values by increasing the noise injection levels, the value of the evaluation metrics especially precision is not too high in some datasets, which highlights the need for improvement in noise detection using classification filtering in future investigations. Finally, the real noisy samples are removed or relabeled and then the accuracy was measured. However, the results prove that classification filtering have a potential capability to detect noisy instances.

References

1. Sluban, B., Lavrač, N.: Relating ensemble diversity and performance: a study in class noise detection. Neurocomputing **160**, 120–131 (2015)
2. Frénay, B., Verleysen, M.: Classification in the presence of label noise: a survey. IEEE Trans. Neural Netw. Learn. Syst. **25**, 845–869 (2014)
3. Zhu, X., Wu, X.: Class noise vs. attribute noise: a quantitative study of their impacts, pp. 177–210 (2004)
4. Lowongtrakool, C., Hiransakolwong, N.: Noise filtering in unsupervised clustering using computation intelligence. Int. J. Math. Anal. **6**(59), 2911–2920 (2012)
5. Brodley, C.E., Friedl, M.A.: Identifying mislabeled training data. J. Artif. Intell. Res. **11**(1), 131–167 (1999)
6. Gamberger, D., Lavrac, N., Groselj, C.: Experiments with noise filtering in a medical domain. In: ICML, pp. 143–51. Citeseer (1999)

7. Khoshgoftaar, T.M., Rebours, P.: Generating multiple noise elimination filters with the ensemble-partitioning filter. In: Proceedings of the 2004 IEEE International Conference on Information Reuse and Integration, IRI 2004. IEEE (2004)
8. Vapnik, V.N., Vapnik, V.: Statistical Learning Theory. Wiley, New York (1998)
9. Breiman, L.: Random forests. Mach. Learn. **45**, 5–32 (2001)
10. Yuan, L.: An improved Naive Bayes text classification algorithm in Chinese information processing. In: Proceedings of the Third International Symposium on Computer Science and Computational Technology (ISCSCT 2010) (2010)
11. Folorunsho, O.: Comparative study of different data mining techniques performance in knowledge discovery from medical database. Int. J. **3**(3) (2013)
12. Cover, T., Hart, P.: Nearest neighbor pattern classification. IEEE Trans. Inf. Theory **13**, 21–27 (1967)
13. Thongkam, J., Xu, G., Zhang, Y., Huang, F.: Toward breast cancer survivability prediction models through improving training space. Expert Syst. Appl. **36**, 12200–12209 (2009)
14. Jeatrakul, P., Wong, K.W., Fung, C.C.: Classification of imbalanced data by combining the complementary neural network and SMOTE algorithm. In: Wong, K.W., Mendis, B.U., Bouzerdoum, A. (eds.) ICONIP 2010, Part II. LNCS, vol. 6444, pp. 152–159. Springer, Heidelberg (2010)
15. Sluban, B., Gamberger, D., Lavra, N.: Advances in class noise detection. Front. Artif. Intell. Appl. **215**, 1105–1106 (2010)
16. Miranda, A.L., Garcia, L.P.F., Carvalho, A.C., Lorena, A.C.: Use of classification algorithms in noise detection and elimination. In: Corchado, E., Wu, X., Oja, E., Herrero, Á., Baruque, B. (eds.) HAIS 2009. LNCS, vol. 5572, pp. 417–424. Springer, Heidelberg (2009)
17. Segata, N., Blanzieri, E., Cunningham, P.: A scalable noise reduction technique for large case-based systems. In: McGinty, L., Wilson, D.C. (eds.) ICCBR 2009. LNCS, vol. 5650, pp. 328–342. Springer, Heidelberg (2009)
18. Segata, N., Blanzieri, E., Delany, S.J., Cunningham, P.: Noise reduction for instance-based learning with a local maximal margin approach. J. Intell. Inf. Syst. **35**(2), 301–331 (2010)
19. Angelova, A., Abu-Mostafa, Y., Perona, P.: Pruning training sets for learning of object categories. In: IEEE Computer Society Conference on Computer Vision and Pattern Recognition, CVPR 2005, pp. 494–501. IEEE (2005)
20. Frank, A., Asuncion, A.: UCI machine learning repository. University of California, School of Information and Computer Science, Irvine, CA (2010). http://archive.ics.uci.edu/ml
21. Garcia, L.P.F., Lorena, A.C., Carvalho, A.C.: A study on class noise detection and elimination. In: 2012 Brazilian Symposium on Neural Networks (SBRN), pp. 13–18. IEEE (2012)
22. Nematzadeh, Z., Ibrahim, R., Selamat, A.: A method for class noise detection based on k-means and SVM algorithms. In: Fujita, H., Guizzi, G. (eds.) SoMeT 2015. CCIS, vol. 532, pp. 308–318. Springer, Heidelberg (2015)

A Novel Robust R-Squared Measure and Its Applications in Linear Regression

Sougata Deb$^{(\boxtimes)}$

Analytics Professional and Independent Researcher, Singapore, Singapore
deb.sougata@gmail.com

Abstract. R^2, despite being a widely used goodness-of-fit measure for linear regression shows erratic behavior in presence of data contamination. Several alternate measures have been proposed that show some improvement under specific conditions. However, no single universal measure exists as such that can be used to assess and compare performance of linear regression models without being concerned about composition of data. This paper proposes a new **robust** R^2 measure that is found to work better than existing measures across scenarios. Performance superiority has been demonstrated using extensive simulation results and three real publicly available datasets. Proposed methodology also shows significant improvement in outlier detection and comparable performance to other established methods for robust linear regression.

1 Introduction

Linear regression is one of the most extensively used statistical models due to its easy interpretation and wide applicability. With this ubiquitous usage, it is imperative to have a consistent and scalable goodness-of-fit (**GoF**) measure for developing, comparing and selecting the best-suited linear regression model in all cases. Traditional measures are mostly error variance based (R^2, AIC, BIC, $RMSE$) or statistical test based (χ^2, F, LR). Some less popular robust measures use either truncated error variance (median-R^2, LTS [7]) or rank information (Wilcoxon dispersion [8]). These measures perform reasonably in cases of no or low contamination but start showing erratic results as contamination increases.

However, contamination is natural in real life data due to various data capture, entry, interpretation or other random issues. To overcome this, many robust linear regression techniques were proposed over past few decades. Needless to say that the problems of robust regression and contamination detection are closely related [6]. Effectiveness of a robust regression technique lies in identifying contaminated observations and subsequently adjusting for their impact in the final model. While robust estimates are largely viewed as more stable and apt, it often becomes challenging to establish this superiority quantitatively in absence of a universal GoF measure. This paper proposes a new R^2-based measure (**RoR^2**) that shows the required robustness, scalability and precision. Consistency of performance and intuitive interpretation similar to R^2 can make it the desired measure for evaluating any linear regression model.

© Springer International Publishing AG 2017
S. Phon-Amnuaisuk et al. (eds.), *Computational Intelligence in Information Systems*,
Advances in Intelligent Systems and Computing 532, DOI 10.1007/978-3-319-48517-1_12

Section 2 provides a brief overview of data contamination, existing GoF measures and how they behave in presence of contamination. Section 3 introduces the RoR^2 measure and shows how its calculation leads to outlier detection and robust regression estimates. Section 4 elaborates on the empirical results derived from the simulated, as well as real datasets. Finally Sect. 5 summarizes the findings and future research scope to build on this topic further.

2   Existing Measures and Improvement Scope

2.1   Outliers and Leverage Points

Data contamination can be of two fundamental types - outliers and leverage points [7]. Outliers are observations where dependent variable ($\mathbf{Y}$) values lie far from the regular zone. A regular zone is where most of its values are located. Similarly, leverage point is an outlier in $\mathbf{X}$-space. Leverage points can be categorized as *good* or *bad*. *Good* leverage points do not deviate from the underlying pattern despite being contaminated while *bad* leverage points do. In this paper, the phrase leverage point is always used to define the *bad* leverage points.

This paper further classifies contamination into random and biased. *Random* denotes presence of both positive and negative contamination *i.e.* arbitrarily large and small values. *Biased* implies presence of only one type for all contaminated observations. Hence *biased* contamination is a stronger variation because some contamination effects under *random* may cancel each other out.

2.2   Contamination vs. Traditional Measures

To demonstrate limitations of SSE-based measures, a simple linear regression construct has been used to generate 100 observations.

$$y_i = 30 + 2x_i + \epsilon_i \qquad (x_i = i = 1, 2, \ldots, 100) \qquad \epsilon_i \sim N(0, 2)$$

R^2, AIC and BIC are calculated on $\hat{y}_i$s coming from OLS and actual regression estimates. Graphs in Fig. 1 show how these measures get affected even with

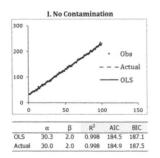
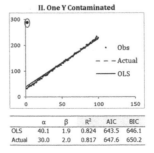
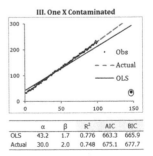

I. No Contamination						II. One Y Contaminated						III. One X Contaminated					
	α	β	R^2	AIC	BIC		α	β	R^2	AIC	BIC		α	β	R^2	AIC	BIC
OLS	30.3	2.0	0.998	184.5	187.1	OLS	40.1	1.9	0.824	643.5	646.1	OLS	43.2	1.7	0.776	663.3	665.9
Actual	30.0	2.0	0.998	184.9	187.5	Actual	30.0	2.0	0.817	647.6	650.2	Actual	30.0	2.0	0.748	675.1	677.7

Fig. 1. Illustrating limitations of traditional measures

a single contaminated observation and highlights the glaring concerns with OLS estimation due to its extreme contamination sensitivity. Following section discusses some alternate robust regression techniques and related GoF measures that can be used to address these concerns.

2.3 Robust Regression and Related GoF Measures

Several techniques including Quantile regression, Huber's M, Rousseeuw and Leroy's Least Trimmed Square (LTS), Yohai's MM estimation are used for robust linear regression on contaminated data. Objective functions for each of these techniques can serve as a potential GoF measure [4], *e.g.* median-R^2 [1] that replaces sum by median of squares in the R^2 equation or LTS. Both are, however, based on a fixed subset ($\approx 50\%$) of observations and ignore model performance on remaining observations. Furthermore, LTS has scalability issues. Usage of F-statistic and D_R (Wilcoxon dispersion) are also recommended [8] in such scenarios. All these measures indeed select the actual estimates over OLS (Fig. 2) for the same scenarios illustrated in Fig. 1. RoR^2 combines the best features of these measures into a single methodology by restricting performance assessment only to the set of *well-behaved* or regular observations. Instead of using a fixed cut-off, the algorithm finds out regular observations dynamically based on data characteristics.

	I. No Contamination				II. One Y Contaminated				III. One X Contaminated			
	med R²	LTS	f-stat	D_R		med R²	LTS	f-stat	D_R		med R²	LTS	f-stat	D_R
OLS	0.999	57	52420	7071	OLS	0.992	206	458	25361	OLS	0.969	907	339	38263
Actual	0.998	56	52453	7088	Actual	0.998	60	514	19404	Actual	0.998	60	423	21266

Fig. 2. Comparing performance of alternate measures

3 Proposed Methodology

Calculation follows a two-step process. Set of regular observations (**R**) is identified in step I. **R** is then used to calculate RoR^2 for any regression estimator at hand. RoR^2 computation process is detailed in the next section. This forms the core of both steps I and II as detailed in Sect. 3.2.

3.1 RoR² Computation Process

Consider the linear model

$$Y = \beta_0 + \beta_1 X_1 + \beta_2 X_2 + \cdots + \beta_p X_p + \epsilon$$

Let $\{y_1, y_2, \ldots, y_n\}$ be a set of n actual Y values. Let $\hat{\beta}_\mathbf{T}$ be an estimate of $\beta = (\beta_0, \beta_1, \ldots, \beta_p)^T$ derived using estimator T and $\{\hat{y}_1, \hat{y}_2, \ldots, \hat{y}_n\}$ be the predicted Ys. Let $\{e_1^2, e_2^2, \ldots, e_n^2\}$ be the set of squared residuals such that

$$e_i = y_i - \hat{y}_i \qquad (i = 1, 2, \ldots, n)$$

An ordered set $\{e_{(1)}^2, e_{(2)}^2, \ldots, e_{(n)}^2\}$ is then formed such that

$$e_{(1)}^2 \le e_{(2)}^2 \le \cdots \le e_{(n)}^2$$

Let $\{y_{(1)}, y_{(2)}, \ldots, y_{(n)}\}$ be the corresponding set of actuals. Using the first m elements of these two ordered sets, a *progressive* R^2 is calculated as

$$pR_{(m)}^2 = 1 - \frac{\sum\limits_{i=1}^{m} e_{(i)}^2}{\sum\limits_{i=1}^{m} (y_{(i)} - \bar{y})^2} \qquad (m = 1, 2, \ldots, n) \qquad (1)$$

$$\text{where} \qquad \bar{y} = \frac{1}{n} \sum_{i=1}^{n} y_{(i)} = \frac{1}{n} \sum_{i=1}^{n} y_i$$

$pR_{(m)}^2$ reflects prediction performance of T in the ordered subset of size m. A sample proportion factor is then used to arrive at a *scaled-progressive* R^2

$$spR_{(m)}^2 = \sqrt{\frac{m}{n}} \cdot pR_{(m)}^2 \qquad (m = 1, 2, \ldots, n) \qquad (2)$$

It readily follows that

$$pR_{(n)}^2 = spR_{(n)}^2 = R^2 = 1 - \frac{SSE}{SST}$$

RoR^2 is defined as

$$RoR^2 = max_m(spR_{(m)}^2) \qquad (3)$$

Furthermore, $c = \text{argmax}(spR_{(m)}^2)$ helps form a partition on two subsets- **W** (first c observations) and **E** from the ordered sample data. Sections 4.2 and 4.5 results will show that **E** indeed contains the contaminated observations. This RoR^2 computation process (**RCP**) forms the core of the main algorithm, as described in Fig. 3.

3.2   Main Algorithm

As already highlighted, computation of RoR^2 involves two steps. First is to identify the set of regular observations (**R**) and the second is to calculate RoR^2 for any estimator based on **R**. It should be noted that the first step is independent of the actual regression estimator being evaluated. Rather, effective identification of **R** is reliant on predictions from a representative linear regression estimator. There can be alternate ways to decide about this most representative estimator based on a given dataset. OLS estimation offers a significantly lower $O(np^2)$ time complexity while robust techniques generally have $O(n^2)$ or higher complexities [3]. However, OLS, as demonstrated earlier, is the most contamination sensitive technique. To overcome this, a local partitioning of data, in **X**-space into k subsets, is done before running OLS. Partitioning is based on Euclidean distance from

$$\mathbf{x}_{med} = (median(X_1), median(X_2), \ldots, median(X_p))$$

This partitioning scheme imparts two distinct benefits

1. It limits the impact of contaminated observations to specific partitions instead of exposing them to the entire dataset.
2. It also improves runtime as only n/k observations are used at a time (k=10 used in this paper, but a higher value can be chosen based on n/p ratio).

In the next stage, these initial OLS estimates for each partition are updated iteratively by using **RCP**. Subset **E** of contaminated observations are excluded from estimation in each subsequent iteration and the final estimate for a partition is reached when no further observation gets excluded through **RCP**.

To avoid selection of non-representative estimates, $|E_j| > 0.5|P_j|$ check is used. This helps discard estimates coming from biased or localized models due to high contamination within a partition. Once all partitions are evaluated and resulting estimates identified, next step is to discard estimates that are significantly different from the majority of estimates. This works as an additional check to eliminate estimates coming from contaminated partitions. This decision is taken by calculating *similarity* among the normalized estimates. Cosine similarity is used for this purpose which ensures that only the $k^*(\leq k)$ most similar estimates are aggregated to form $\hat{\beta}_{RoR}$.

$$\hat{\beta}_{RoR} = (\frac{1}{k^*}\sum_{j=1}^{k^*}\hat{\beta}_{j0}, \frac{1}{k^*}\sum_{j=1}^{k^*}\hat{\beta}_{j1}, \ldots, \frac{1}{k^*}\sum_{j=1}^{k^*}\hat{\beta}_{jp})$$

$\hat{\beta}_{RoR}$ is then used on the selected data partitions to get the set of regular observations using **RCP**. Additionally, $\hat{\beta}_{RoR}$ is used on the discarded partitions to swap-in additional observations which were discarded earlier as part of the contaminated partitions. The condition

$$e_j^2 \leq max(e_i^2 : \forall i \in W_{final})$$

ensures only well-behaved or regular observations get selected in W_{add}. Combined set of observations coming from both these datasets, results in **R**, the desired set of regular observations. Once **R** is identified, $\hat{y}_i$s (using any $\hat{\beta}_T$) for observations in **R** are run through **RCP** to obtain RoR^2 value for T. RoR^2 coming from **R** is not directly scalable across datasets. Applying a sample proportion multiplier resolves this and final reported value becomes

$$RoR_f^2 = \frac{|\mathbf{R}|}{n}.RoR^2$$

This final measure RoR_f^2 can be used to compare performances of different linear regression estimators within the same or across different datasets. Following section elaborates on its performance using simulated and real datasets.

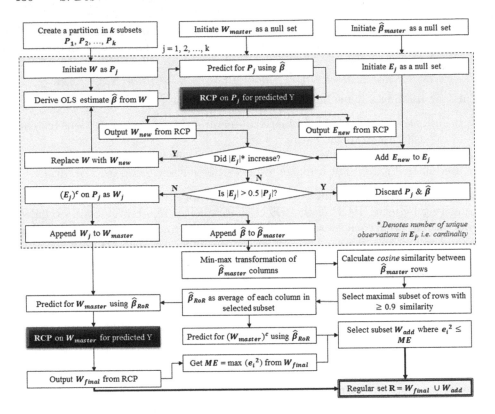

Fig. 3. Main algorithm to find set of regular observations **R**

4   Empirical Results

4.1   Simulation Construct

Base Model. Following linear model forms the base of this simulation study. Each sample simulates values of X_1-X_4 and $\mathcal{E}$ which are subsequently combined using (4) to get the Y values.

$$Y = 20 + X_1 - 2X_2 - 5X_3 + 10X_4 + \mathcal{E} \qquad (4)$$

$$X_1 \sim \Gamma(4) \qquad X_2 \sim Poi(10) \qquad X_3 \sim N(0, 10) \qquad X_4 \sim t(6) \qquad \mathcal{E} \sim N(0, 10)$$

Contamination Scheme. 0–60 % observations are rigged randomly (using $C_i \sim N(0, 1)$) for either or all of the three variables - Y, X_2 and X_4 within each sample. Contamination in a variable within a sample can be of one of five types as shown in Table 1. In 50 % cases, contamination is done in values closest to its median and randomly in other 50 %. This elaborate scheme ensures coverage across contamination scenarios, thus making the conclusions robust and holistic.

Table 1. Data contamination details

Type	None	Positive bias	Negative bias	Random-1	Random-2
New =	Original	Orig + $|C_i|^*200$	Orig − $|C_i|^*200$	C_i^*200	Orig + C_i^*200

Sample Size and Replication. Sample size is varied randomly from 1,000 to 30,000 in multiples of 500. 35,000 such samples are simulated using SAS® software (SAS Institute, Cary NC) for this study.

Regression Estimators. Eight linear regression estimators are analyzed using SAS procedures REG, QUANTREG and ROBUSTREG on the same datasets.

1. Ordinary Least Square (OLS)
2. Quantile Regression using finite smoothing (QNT)
3. Huber's M-estimator with Huber weight (HUB)
4. Huber's M-estimator with Tukey's bisquare weight (BIS)
5. Rousseeuw and Leroy's Least Trimmed Square estimator (LTS)
6. Yohai's MM-estimator with LTS initialization (MML)
7. Rousseeuw and Yohai's S-estimator (SES)
8. RoR^2 regression estimator, which is an OLS estimator used on the identified set of regular observations (ROR)

Identifying *Predicted Best* Estimators. *Predicted best* is the one with the best GoF value. For R^2, median-R^2, RoR^2 and F-statistic, estimator with maximum value within a dataset is chosen as *predicted best*. If GoF values for other estimators are extremely close to maximum, all such estimators are also categorized as *predicted best* to account for minor fluctuations due to the random error component. 0.0001 is used as closeness cut-off for the three R^2 measures and 99.99 % of maximum for F-statistic. Similarly, estimators with minimum LTS and D_R are chosen as *predicted best*. Other estimators within 0.1 (LTS) and 0.0001 (D_R) of minimum are also selected as *predicted best*.

Identifying *Actual Best* Estimators. *Actual best* estimators are defined by *closeness* of $\hat{\beta}_\mathbf{T}$ with the actual parameter vector, $\beta = (20, 1, -2, -5, 10)^T$. When all $\hat{\beta}_{Ti}$s obtained using estimator T are close to the actual β_is,

$$\hat{\beta}_{Ti} \in [\beta_i - 3^*\sigma_i, \beta_i + 3^*\sigma_i] \qquad (i = 0, 1, \ldots, 4)$$

T is categorized as *actual best*. σ_is are calculated using only the $\hat{\beta}_i$s coming from samples with no contamination. If no estimator satisfies the above criteria for a sample, the estimator with minimum $||\hat{\beta}_T - \beta||$ is chosen as *actual best*.

Performance of GoF Measures. Rejection of an *actual best* estimator by a GoF measure is termed as Type I error while selection of an estimator that is not *actual best* is termed as Type II error. Performances of GoF measures are compared based on these two types of errors.

4.2   Model Selection Performance

Table 2 shows aggregated performance based on all 35,000 simulated datasets. Additionally, graphs in Fig. 4 show how RoR^2 retains its strong and consistent performance across levels and types of contamination. This consistent and superior performance can help modelers assess and select the right models across contamination scenarios with a higher confidence.

Table 2. Model selection performance of GoF measures

Measure	Type-I	Type-II	Measure	Type-I	Type-II
R^2	27.0 %	28.8 %	F-statistic	44.0 %	7.1 %
D_R	29.1 %	22.2 %	median-R^2	23.3 %	5.6 %
LTS	4.4 %	10.3 %	RoR^2	**1.4 %**	**3.1 %**

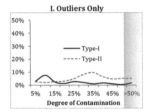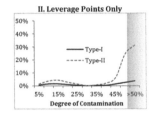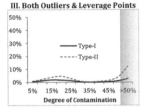

Fig. 4. Model selection performance by contamination type and level

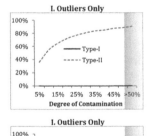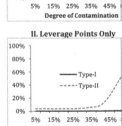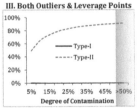

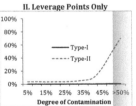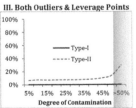

Fig. 5. Cook's D (top) vs. RoR^2 (below) performance by contamination type and level

4.3   Contamination Detection Performance

This section compares contamination detection performance using similar Type I and Type II errors. A non-contaminated observation in $\mathbf{R}^c$ is defined as Type I error while a contaminated observation retained in $\mathbf{R}$ is categorized as Type II error for RoR^2 methodology. Graphs in Fig. 5 compare performance of RoR^2 based contamination detection against the established Cook's D measure [2] using $(4/(n-p-1))$ as cut-off [10] for influential points. It demonstrates how the proposed methodology significantly outperforms Cook's D across scenarios.

4.4   Regression Estimator Performance

While Sect. 4.2 results establish relevance of RoR^2 in selecting the *actual best* models, this section further drills down to exactly which estimator becomes

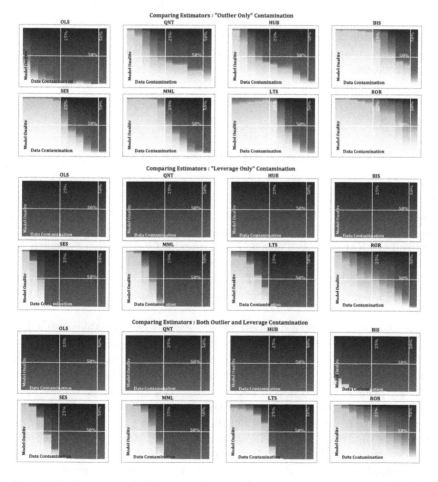

Fig. 6. Performance of different estimators by contamination type and level

actual best in different contamination scenarios. It casts significant light on the applicability of different estimators in different situations. This is of particular importance in cases when modeler has prior information or belief about the type and degree of contamination present in data.

Graphs in Fig. 6 show performance of the eight estimators by levels and types of contamination. Each graph shows mix of Superior/Average/Poor parameter estimates at different levels of contamination. Superior are the estimates where all $\hat{\beta}_i$s lie within $\pm 3\sigma_i$ limits of the actual β_is, same as the *actual best* definition. Average estimates are where all $\hat{\beta}_i$s lie within $\pm 6\sigma_i$ limits and Poor otherwise. The darker the shades of grey, the worse the estimates are. It readily stands out that estimators are more sensitive to leverage points than outliers. Furthermore, it is worth highlighting that

- OLS, QNT, HUB, BIS go awry as soon as leverage points are introduced.
- SES, MML and LTS show some resistance but as leverage contamination crosses 15, 20 and 25 % respectively, proportion of Poor estimates increases sharply to 100 %.
- ROR estimator outperforms all others at very high contamination levels. Performance is comparable to the other robust estimators at low and moderate contamination.
- ROR estimator performance is primarily dependent on level of contamination and not so much on contamination type, *i.e.* outliers/leverage points or biased/random.

4.5   Performance Assessment Based on Real Datasets

Having observed the encouraging performance on simulated datasets, the same methodology is now put to test on real datasets. Three publicly available datasets are sourced from the UC Irvine Machine Learning Repository [11] for this purpose. These are the Combined Cycle Power Plant (CCPP; 9,568 obs, 4 IVs and 1 DV; source: Kaya, H. & Tüfekci, P.), Concrete Compressive Strength (CCS; 1,030 obs, 8 IVs and 1 DV; original owner and donor: Prof. Yeh, I-Cheng) and Boston Housing (BH; 506 obs, 13 IVs and 1 DV; source: StatLib library maintained at Carnegie Mellon University, creators: Harrison, D. & Rubinfeld, D.L.) datasets.

These three are of very different types, *e.g.* all the variables in CCPP are measured and captured by sophisticated sensors. As a result, this dataset is not expected to contain outliers or leverage points as such. CCS dataset has a mix of machine captured and manually input variables. Prior researches on this dataset [5] primarily used neural network models for robust prediction, hinting at the presence of some contamination. BH dataset, on the other hand, is known for contamination where several transductive and semi-supervised algorithms were applied in the past. All the variables in BH dataset are captured and aggregated manually. This provides a good mix of real-life situations to test the proposed methodology. Also it is important to note that there is no concrete documentation on the exact set or number of contaminated observations in any

Dataset	CCPP				CCS				BH			
Decision	Total	Excluded	**Regular** (Included)		Total	Excluded	**Regular** (Included)		Total	Excluded	**Regular** (Included)	
Obs.	9,568	53	9,515		1,030	173	857		506	166	340	
Model	RoR^2	r	r	RMSE	RoR^2	r	r	RMSE	RoR^2	r	r	RMSE
OLS	**0.931**	0.496	**0.967**	**4.314**	0.630	0.347	0.850	8.502	0.563	0.778	0.910	0.128
QNT	**0.931**	0.492	**0.967**	4.326	0.666	0.266	0.874	7.793	0.596	0.745	0.932	0.111
HUB	**0.931**	0.494	**0.967**	4.315	0.645	0.319	0.860	8.214	0.589	0.760	0.928	0.114
BIS	**0.931**	0.494	**0.967**	4.315	0.699	0.175	0.906	6.751	0.598	0.746	0.934	0.109
SES	0.930	0.492	**0.967**	4.335	**0.707**	0.155	**0.913**	6.522	**0.600**	0.722	**0.937**	**0.108**
MML	**0.931**	0.493	**0.967**	4.322	**0.707**	0.155	**0.913**	**6.500**	0.595	0.710	0.934	0.110
LTS	0.927	0.489	**0.967**	4.474	0.705	0.152	0.911	6.599	0.596	0.704	0.935	0.113
ROR	**0.931**	0.495	**0.967**	**4.314**	0.686	0.133	0.895	7.211	0.595	0.500	0.921	0.121

Fig. 7. Performance assessment based on three real datasets

of these datasets. Hence the goal is to assess the outcomes qualitatively to form a logical view on performance.

The same set of eight estimators is used on each dataset. To avoid any overfitting bias, a 3-fold cross validation is performed to record the test performance for each estimator. Simultaneously, the datasets are passed through the proposed algorithm to arrive at the subset of regular observations. Effectiveness of contamination detection is assessed by comparing the correlation coefficient (**r**) between the actual and predicted values, separately for the regular and excluded observations. For each dataset and estimator, Fig. 7 results show a significant difference in **r** between the two sets. This confirms that the excluded observations are indeed different from the regular ones.

Next, estimator performances for the regular observations are evaluated based on both **r** and root mean square error (RMSE). It shows that the estimators with the highest RoR^2 values indeed have the highest **r** and lowest RMSE for each dataset. Finally ROR estimator is found to perform at a similar level to that of the other established robust estimators. The results, however, do not indicate any superiority of ROR estimator over others apart from the OLS estimator.

5   Conclusions and Next Steps

Empirical results provide the necessary confidence on the proposed methodology. It clearly demonstrates how RoR^2-based model selection and contamination detection outperform the existing and established methods, while parameter estimation for linear regression produces comparable results. This methodology can easily be integrated and executed in standard Statistical packages as it leverages only the most fundamental techniques, $e.g.$ OLS estimation, Euclidean distance.

A direct extension of this study will be to apply and evaluate this methodology using more real life datasets from different domains. Additionally, future research should explore (1) generalizing the methodology for mixed linear models and splines, (2) customization feasibility of the same methodology in context of non-linear regression and (3) studying behavior of RoR^2 in cases of excess or limited information, $i.e.$ presence of additional or less independent variables than what is used in the underlying model.

References

1. Peter, J.R.: Least median of squares regression. J. Am. Stat. Assoc. **79**(388), 871–880 (1984)
2. James, P.S.: Outliers and influential data points in regression analysis. Psychol. Bull. **95**(2), 334–344 (1984)
3. John, M.S., William, L.S.: Algorithms and complexity for least median of squares regression. Discrete Appl. Math. **14**(1), 93–100 (1986)
4. Santiago, V.: On the behaviour of residual plots in robust regression. Stat. Econometrics Ser. **04**, 93–104 (1993). Working Paper
5. I-Cheng, Y.: Modeling of strength of high performance concrete using artificial neural networks. Cem. Concr. Res. **28**(12), 1797–1808 (1998)
6. Victoria, J.H., Jim, A.: A survey of outlier detection methodologies. Artif. Intell. Rev. **22**(2), 85–126 (2004)
7. Peter, J.R., Annick, M.L.: Robust Regression and Outlier Detection, vol. 589. Wiley, New York (2005)
8. Jeff, T.T., Joseph, W.M.: Rank-based analysis of linear models using R. J. Stat. Softw. **14**(7), 1–26 (2005)
9. Heysem, K., Pınar, T., Fikret, S.G.: Local and global learning methods for predicting power of a combined gas and steam turbine. In: Proceedings of the International Conference on Emerging Trends in Computer and Electronics Engineering (ICETCEE 2012), Dubai, UAE, pp. 13–18 (2012)
10. Khaleelur Rahman, S.M.A., Mohamed Sathik, M., Senthamarai Kannan, K.: Multiple linear regression models in outlier detection. Int. J. Res. Comput. Sci. **2**(2), 23–28 (2012)
11. Moshe, L.: UCI Machine Learning Repository. Irvine, CA: University of California, School of Information and Computer Science (2013). http://archive.ics.uci.edu/ml
12. Pınar, T.: Prediction of full load electrical power output of a base load operated combined cycle power plant using machine learning methods. Int. J. Electr. Power Energy Syst. **60**, 126–140 (2014). ISSN 0142–0615

An Improvement to StockProF: Profiling Clustered Stocks with Class Association Rule Mining

Kok-Chin Khor[(⊠)] and Keng-Hoong Ng

Faculty of Computing and Informatics, Multimedia University,
Jalan Multimedia, 63100 Cyberjaya, Selangor, Malaysia
{kckhor, khng}@mmu.edu.my

Abstract. Using StockProF developed in our previous work, we are able to identify outliers from a pool of stocks and form clusters with the remaining stocks based on their financial performance. The financial performance is measured using financial ratios obtained directly or derived from financial reports. The resulted clusters are then profiled manually using mean and 5-number summary calculated from the financial ratios. However, this is time consuming and a disadvantage to novice investors who are lacking of skills in interpreting financial ratios. In this study, we utilized class association rule mining to overcome the problems. Class association rule mining was used to form rules by finding financial ratios that were strongly associated with a particular cluster. The resulted rules were more intuitive to investors as compared with our previous work. Thus, the profiling process became easier. The evaluation results also showed that profiling stocks using class association rules helps investors in making better investment decisions.

Keywords: Profiling stocks · Class association rule mining · StockProF

1 Introduction

Stock markets are meant for trading stocks - capitals raised by companies via the share issuance. The markets provide marketability and liquidity not only to listed companies but also contribute directly to the economic growth of many countries [1]. Therefore, a stock market is sometimes viewed as the economic barometer of a country.

Many investors keen on stock markets because the markets provide the convenience of investing their cash into shares and convert the shares back to cash. Besides, companies listed on stock markets follow strict regulations set by the governments for their operations [2]. A certain level of protection is therefore ensured when investors trade their shares in the stock markets.

Information about listed companies is generally freely available to the public. In Malaysia, the companies listed on Bursa Malaysia have to provide financial reports quarterly to the public. Investing stock markets is sometimes difficult for novice investors because there are a lot of financial data and information to digest before understanding the intrinsic value of a company.

© Springer International Publishing AG 2017
S. Phon-Amnuaisuk et al. (eds.), *Computational Intelligence in Information Systems*,
Advances in Intelligent Systems and Computing 532, DOI 10.1007/978-3-319-48517-1_13

Understanding the intrinsic value of a company needs *fundamental analysis* [3]. Through fundamental analysis, investors can assess a company using its *qualitative* and *quantitative data*. Qualitative data involve quality of key management, strategies of marketing, branding, etc. Quantitative data, on the other hand, refer to financial ratios that can be obtained or derived directly from financial reports. Using financial report, a company listed on stock markets can be evaluated in different aspects: (1) its financial position via balance sheet, (2) its earnings and profitability via income statement, and (3) its utilisation of cash via cash flow statement. Popular financial ratios to assess a company such as total asset turnover, cash ratio, debt ratio, etc. can be derived from these statements.

Analysing these financial ratios to profile stocks and subsequently to build a stock portfolio is not an easy task. In our previous work, we developed a stock profiling framework, StockProF, to help investors in profiling stocks rapidly. However, there is still room for improvement.

2   An Overview of StockProF

StockProF aims to simplify the stock profiling process and build stock portfolios in a rapid manner [4]. The process flow of StockProF is as illustrated in Fig. 1. There are two main data mining techniques used in this framework: Local Outlier Factor (LOF) and Expectation Maximization (EM). Using LOF, the framework firstly identifies outliers from a pool of stocks. The outliers are either outperforming or poor performing stocks. The framework is then continued by using EM to cluster the remaining stocks. Stocks with the same financial characteristics will be gathered in the same cluster. With the resulted outliers and clusters, investors profile them using mean and 5-number summary.

The profiling outcome helps investors in making investment decisions based on their investment strategies. In our previous work, we managed to identify not only the outperforming and aggressive stocks, but also defensive stocks as well as average stocks. If investors wish to gain maximum profits, then they can consider investing in outperforming stocks or aggressive stocks. On the other hand, they can consider defensive stocks if a market/industry is showing signs of slowing down. A defensive stock does not reinvest much on its operation. Rather, it retains profits from its operation and pays to the shareholders in the form of dividends.

No doubt StockProF has provided a convenient way to investors in building stock portfolios rapidly, particularly in identifying outliers and forming clusters using data mining techniques. However, an improvement to StockProF is needed. The stock profiling step (refer to Fig. 1), especially on the resulted clusters, is done by manually interpreting the mean and 5-number summary of each and every cluster. Such interpretation requires knowledge on financial ratios and it is time consuming.

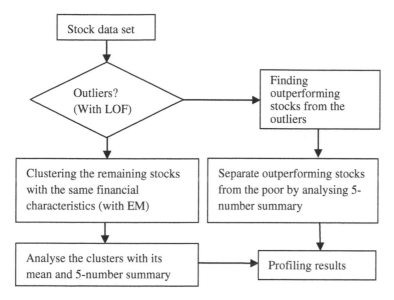

Fig. 1. Using StockProF, investors are able to profiles outlier stocks and stocks in resulted clusters.

3   Methodology

3.1   Preparation of the Stock Data Set

This study aimed at the 42 plantation stocks listed on Bursa Malaysia, as shown in Table 1. Their raw financial data of year 2014 were collected using DataStream, a financial database of Thomson Reuters, and complemented by financial reports posted on the Bursa Malaysia website if there are any missing data. Since companies listed on stock markets are different in sizes, capitals, etc., comparison thus become hard. To compare and evaluate companies in the same industry, the raw data were then converted into six financial ratios (refer to Table 2), as predetermined in our previous study [4]. These financial ratios are commonly used in financial markets for evaluating stocks [5]. However, the ranges of values for the financial ratios were very much different from each other. Therefore, the financial ratios were scaled [0,1] with min-max normalization to ease clustering in the later stage.

We explain these financial ratios in brief. *Total asset turnover* evaluates how effective a company is in generating sales based on its assets; the higher the ratio, the more effective a company is in utilising its assets. *Cash ratio* is used to evaluate the liquidity of a company; a higher ratio than others means that the company is capable of repaying its debts using its internal funds. *Debt ratio* shows the level of how a company leverages its own finance; the lower the ratio, the lower the loan default risk of the company. *Return on equity (ROE)* evaluates how good a company is in generating profits using its shareholder's funds; a high ratio indicates the high competence level of the company's management team. *Dividend yield (DY)* tells how much cash return can be earned by a shareholder in a financial year; a high dividend distributed to

Table 1. The plantation stocks listed on Bursa Malaysia

Name	Code	Name	Code	Name	Code
AASIA	7054	IJMPLNT	2216	PINEPAC	1902
BKAWAN	1899	INCKEN	2607	PLS	9695
BLDPLNT	5069	INNO	6262	RSAWIT	5113
BPLANT	5254	IOICORP	1961	RVIEW	2542
CEPAT	8982	KLK	2445	SBAGAN	2569
CHINTEK	1929	KLUANG	2453	SHCHAN	4316
DUTALND	3948	KMLOONG	5027	SOP	5126
FAREAST	5029	KRETAM	1996	SWKPLNT	5135
FGV	5222	KULIM	2003	TDM	2045
GENP	2291	KWANTAS	6572	THPLANT	5112
GLBHD	7382	MALPAC	4963	TMAKMUR	5251
GOPENG	2135	MHC	5026	TSH	9059
HARNLEN	7501	NPC	5047	UMCCA	2593
HSPLANT	5138	NSOP	2038	UTDPLT	2089

Table 2. The six financial ratios that were used to compare companies in the same industry.

Financial ratios	Formula
Total Asset Turnover	Total Sales/Total Assets
Cash Ratio	Cash/Current Liabilities
Debt Ratio	Total Liablities/Total Assets
ROE	Net Income/Shareholder's Equity
DY	Dividend per Share/Price per Share
PE	Price per Share/Earning per Share

shareholders means that the company has a healthy cash flow and it is able to generate well profits from its operation. Price earning (PE) ratio evaluates whether a company is under-valued, fully valued or over-valued as compared with its peers. Investors normally favour companies with low PE in a particular industry. There could be occasions that PE value is negative when earning per share is negative (refer to Table 2). In this case, reporting PE is not meaningful and therefore we set its negative value to zero [6]. An advantage of setting such value is that it would not skew any statistics calculated based on PE.

The data preparation yielded a stock data set for StockProF (refer to Fig. 1). Running the data set yielded a few outliers and clusters. We intended to profile the resulted clusters using class association rules (CARs). The outliers were not profiled because it contained both outperforming and poor performing stocks. At this stage of StockProF, a manual analysis on the outliers had to be done to separate the outperforming stocks from the poor.

3.2   Mining Class Association Rules

Analysing manually the mean and 5-number summary of the resulted clusters and profiling them is time consuming. To ease this step, we replaced the manual analysis method with CARs proposed by [7]. In CARs, association rule mining (Apriori) [8] was adapted to integrate with classification rule mining.

Apriori is a popular algorithm for mining frequent item sets for Boolean association rules – rules that check the presence or absence of an item. Items which occurs frequently in a data set are called frequent item sets. Association rules can be generated based on frequent item sets if they are associated with other [9]. The quality of association rules is measured using *support* and *confidence*. Let $\{X, Y\}$ denotes an item set. The support shows the relation of the number of events containing $\{X, Y\}$ to all events in a data set. On the other hand, the confidence shows the relation of the number of events containing both item sets $\{X\}$ and $\{Y\}$ to the number of events containing item set $\{X\}$. In addition to the support and confidence, any subset of the frequent item set, $\{X\}$ and $\{Y\}$, must be frequent item sets as well.

With association rule mining, the target is not determined ahead. Using CARs, a subset of rules is restricted to a target. In this study, the target was the clusters resulted from StockProF. Initially, the values of these financial ratios (of all stocks) were discretized using equal-width partitioning into three ordinal values, namely, *low*, *medium* and *high*. Then, CARs was applied to every cluster to generate class association rules.

$$Financial\,Ratio = \{low, medium, high\} = > Cluster = \{C1, C2\}$$

The meta-rule above was used to specify the constraint of class association rules. The purpose of the meta-rule was to obtain only rules that were suitable for profiling a resulted cluster. In addition, only rules that showed strong associations between financial ratios and a particular cluster were selected. To generate strong rules, both support and confidence must be high. The confidence value predetermined in this study was 0.90. To generate rules efficiently, the upper bound minimum support was 1.0 and the support was reduced iteratively up to point that the rules were found by not less than the predetermined confidence.

Profiling the stocks in every cluster has now become easier than before. The generated rules are also more intuitive than the 5-number summary.

4   Results and Discussion

As shown in Table 3, StockProF yielded eight outliers and two clusters. Cluster 1 and Cluster 2 (denoted as C_1 and C_2) had 24 and 10 stocks, respectively. Tables 4 and 5 provide the mean and 5-number summary for C_1 and C_2. The statistics gave us an insight of the data dispersion of the clusters; this allowed us to understand the characteristics of the clusters and profile them. The profiling results can be used by investors to build their stock portfolios based on the investment strategies. To ease investors and to save their time, CARs was used instead of the statistics. We then profiled the clusters using the generated class association rules.

Table 3. The outliers and the members of two clusters, resulted from StockProF.

Outliers	C_1			C_2
AASIA	BKAWAN	KLK	TDM	CHINTEK
GOPENG	BLDPLNT	KMLOONG	THPLANT	FAREAST
HARNLEN	BPLANT	KULIM	TMAKMUR	HSPLANT
IOICORP	CEPAT	KWANTAS	TSH	INCKEN
KLUANG	DUTALND	MHC		KRETAM
PINEPAC	FGV	NPC		MALPAC
RSAWIT	GENP	PLS		NSOP
SBAGAN	GLBHD	SHCHAN		RVIEW
	IJMPLNT	SOP		UMCCA
	INNO	SWKPLNT		UTDPLT

Table 4. The mean and 5-number summary of C_1.

Financial Ratios	Mean	Min	Q1	Median	Q3	Max
Total Asset Turnover	0.424686	0.052819	0.150559	0.303202	0.718241	1
Cash Ratio	0.007776	0.000129	0.001304	0.004981	0.009786	0.037501
Debt Ratio	0.534256	0.209956	0.416136	0.545666	0.685486	0.834199
ROE	0.466836	0.189688	0.419568	0.459	0.546043	0.626903
DY	0.182183	0	0.054165	0.186073	0.289107	0.481206
PE	0.069083	0	0.045419	0.076648	0.089675	0.136939

Table 5. The mean and 5-number summary of C_2.

Financial Ratios	Mean	Min	Q1	Median	Q3	Max
Total Asset Turnover	0.179353	0	0.059555	0.167802	0.303711	0.401879
Cash Ratio	0.125851	0.00414	0.054383	0.085944	0.159042	0.442159
Debt Ratio	0.131589	0	0.051993	0.121463	0.173673	0.348057
ROE	0.423002	0.306317	0.332382	0.41914	0.486159	0.59551
DY	0.246046	0	0.109429	0.219464	0.373226	0.6458
PE	0.073854	0	0	0.062246	0.09971	0.275229

4.1   Profiling the Clusters

We profiled C_1 using the generated class association rules, as shown in Table 6. Rules in Table 6 are all exact rules as the confidence values are 1. In general, stocks in C_1 had a below average financial performance. C_1 had a low cash ratio, which meant low in cash in relation to its short-term debts. The liability bore by C_1 was medium among the peers. This had probably resulted low DY for investors even though the profitability of C_1 were average among the peers (medium ROE). Stocks in C_1 were considered under-valued with their low PE. Total asset turnover is not discussed here as it was not strongly associated to C_1.

Table 6. The association rules of C_1 with confidence value 1 and minimum support 0.55.

No.	Association Rules
1	Cash ratio = low ==> Cluster = C_1
2	Debt ratio = medium ==> Cluster = C_1
3	ROE = medium ==> Cluster = C_1
4	DY = low ==> Cluster = C_1
5	PE = low ==> Cluster = C_1

C_2 were generally similar to C_1 in many aspects: low cash ratio, medium ROE, low DY and low PE (Table 7). In contrary, C_2 had a lower debt ratio than C_1. In addition, the rule related to total asset turnover was found as this financial ratio was strongly associated with C_2. To conclude, stocks in C_2 generally performed slightly better than C_1.

Table 7. The association rules of C_2 with confidence value 1 and minimum support 0.65.

No.	Association Rules
1	Total asset turnover = low ==> Cluster = C_2
2	Cash ratio = low ==> Cluster = C_2
3	Debt ratio = low ==> Cluster = C_2
4	ROE = medium ==> Cluster = C_2
5	DY = low ==> Cluster = C_2
6	PE = low ==> Cluster = C_2

4.2 Building Stock Portfolios

Building stock portfolios were the next step after profiling the clusters. Since the focus of this work was to apply CARs on the resulted clusters, therefore we will not discuss which stocks investors should choose from the outliers to build their stock portfolios. The technique of how to choose outperforming stocks from outliers was discussed in our previous work [4].

Based on the profiling results in Sect. 4.1, investors should choose stocks from C_2 rather than C_1. We will assess the performance of both clusters in average capital return in the next section.

4.3 Average Capital Performance

The average capital performance for both clusters was evaluated using moving average (MA) of stock prices in half year and 1-year basis (refer to Eq. 1).

$$Average\ capital\ return\ =\ (MA\ year\ 2015 - MV\ year\ 2014)\ /\ MA\ year\ 2014 \qquad (1)$$

MA is the average price of a stock over a specific number of days [10]. For instance, a 10-day MA is the summation of 10 days closing prices of a stock divided by

10. In our previous work, we used stock prices on the particular dates (30th June and 31st Dec for half year and 1-year evaluation) for evaluating the capital performance of the stocks in the resulted clusters. However, there could be speculative activities on these particular dates to cause price fluctuation and therefore affect the evaluation results. Even though a group of stocks was evaluated with the purpose of mitigating the price fluctuation, we feel that it was not enough. With MA, the effect of price fluctuations can be further mitigated.

It was not so promising for the palm oil industry of Malaysia in the year 2015. The palm oil price was USD 700 per metric tonne at the beginning of the year 2015. It was then dropped to the lowest price, USD 480 on 26th August 2015 (a decline of 31.4 %) and subsequently bounced back slightly to USD 560 at the end of the year 2015 (a decline of 20 %). The price fall was due to the weak demand from major markets and high supplies from the producers like Malaysia and Indonesia (both account for 85 % of the global supplies) [11]. We would like to see how plantation stocks defend themselves in such situation.

Table 8. The average capital performance of C_1 and C_2 based on the moving average of the stock prices.

C_1	1/2-year (%)	1-year (%)	C_2	1/2-year (%)	1-year (%)
BKAWAN	−5.79	−7.36	CHINTEK	−3.73	−8.34
BLDPLNT	−5.40	−2.40	FAREAST	6.86	3.51
BPLANT	−9.38	−9.67	HSPLANT	−3.07	−7.45
CEPAT	−16.51	−19.44	INCKEN	−12.31	−14.67
DUTALND	−1.86	−8.68	KRETAM	−20.60	−22.50
FGV	−46.17	−52.92	MALPAC	−5.04	−6.82
GENP	−4.88	−5.06	NSOP	−12.50	−18.00
GLBHD	13.02	27.31	RVIEW	−5.87	−9.57
IJMPLNT	0.45	−1.42	UMCCA	−9.04	−12.10
INNO	−16.85	−17.29	UTDPLT	−1.63	−0.38
KLK	−3.84	−4.12			
KMLOONG	−1.77	−1.69			
KULIM	−16.48	−11.93			
KWANTAS	−11.42	−19.53			
MHC	−11.22	−13.88			
NPC	3.54	2.23			
PLS	−22.72	−26.79			
SHCHAN	−11.11	−17.61			
SOP	−18.41	−23.97			
SWKPLNT	−13.79	−19.51			
TDM	−16.68	−24.09			
THPLANT	−15.10	−22.23			
TMAKMUR	−24.13	−25.77			
TSH	2.56	−3.04			
Average	−10.58	−12.87		−6.69	−9.63

As shown in Table 8, C_2 performed slightly better than C_1 in overall. The average capital losses for C_2 in half year and 1-year evaluations were -6.69 % and -9.63 %, respectively; the losses were slightly less than C_1. C_1 suffered capital losses -10.58 % and -12.87 % averagely in half year and 1-year evaluations.

5 Conclusion

In this study, we improved StockProF by providing a convenience to especially novice investors in profiling the resulted clusters. Instead of interpreting the mean and the 5-number summary of the clusters, class association rules which are more intuitive to them were provided. When evaluating the average capital performance of the clusters, we used 1-year MA of the stock prices instead of the stock prices on the particular dates. This is to reduce the effect of price fluctuations caused by speculative activities. The evaluation results showed that the rules generated are able to help investors in identifying the correct cluster of stocks for their investment portfolios.

Nevertheless, further improvement to StockProF is still possible. Although the scope of searching the right stocks has been greatly narrowed using StockProF, but the number of stocks in a cluster could still be large for investors with limited funds, particularly individual investors. They still need to further narrow the search themselves to pick the number of stocks they can afford. Our future research will focus on the possibility of narrowing further the search for the right stocks within the clusters (or outliers) automatically.

References

Mayo, H.B.: Investments: An Introduction. Cengage Learning, Boston (2016)

Rules of Bursa Malaysia Securities | Bursa Malaysia Market. http://www.bursamalaysia.com/market/regulation/rules/bursa-malaysia-rules/securities/rules-of-bursa-malaysia-securities

Schlichting, T.: Fundamental Analysis, Behavioral Finance and Technical Analysis on the Stock Market. GRIN Verlag, Munich (2013)

Ng, K.H., Khor, K.C.: StockProF: a stock profiling framework using data mining approaches. Inf. Syst. E-Bus. Manage. 1–20 (2016). doi:10.1007/s10257-016-0313-z

Pilbeam, K.: Finance and Financial Markets. Palgrave Macmillan, London (2010)

Damodaran, A.: Investment Valuation: Tools and Techniques for Determining the Value of Any Asset. Wiley, Hoboken (2012)

Liu, B., Hsu, W., Ma, Y.: Integrating classification and association rule mining. In: Fourth International Conference on Knowledge Discovery and Data Mining, pp. 80–86 (1998)

Agrawal, R., Srikant, R.: Fast algorithms for mining association rules. In: Proceedings of the 1994 20th International Conference on Very Large Data Bases, New York, NY, USA, pp. 1–12. ACM (1996)

Han, J., Pei, J., Kamber, M.: Data Mining: Concepts and Techniques. Elsevier, Amsterdam (2011)

Khan, A., Zuberi, V.: Stock Investing for Everyone: Tools for Investing Like the Pros. Wiley, New York (1999)

Craymer, L.: Palm Oil's Slide Looks Set to Continue (2015). http://www.wsj.com/articles/palm-oils-slide-looks-set-to-continue-1441087183

Empirical Study of Sampling Methods for Classification in Imbalanced Clinical Datasets

Asem Kasem[1], A. Ammar Ghaibeh[2(✉)], and Hiroki Moriguchi[2]

[1] School of Computing and Informatics, Universiti Teknologi Brunei,
Gadong, Brunei Darussalam
asem.kasem@utb.edu.bn
[2] Medical Informatics Department, The University of Tokushima,
Tokushima, Japan
ammargh@acm.org, h_moriguchi@ap6.mopera.ne.jp

Abstract. Many clinical data suffer from data imbalance in which we have large number of instances of one class and small number of instances of the other. This problem affects most machine learning algorithms especially decision trees. In this study, we investigated different undersampling and oversampling algorithms applied to multiple imbalanced clinical datasets. We evaluated the performance of decision tree classifiers built for each combination of dataset and sampling method. We reported our experiment results and found that the considered oversampling methods generally outperform undersampling ones using AUC performance measure.

Keywords: Clinical data mining · Imbalanced data · Undersampling · Oversampling · Decision tree C4.5

1 Introduction

There has been an increase of the use of data mining techniques in the different areas of medicine (bioinformatics, medical imaging, clinical informatics, and public health informatics) in the last decade. This is due to the impact data mining had on other domains, such as banking, marketing, and e-commerce, which gave high hopes for similar achievements in medicine, by extracting untapped knowledge contained in available medical data as well.

The aim of clinical data mining is to search for useful patterns and information within patients' data, and develop prediction models that can support clinical decision making [1, 2]. Data mining can be used to build predictive models in prognosis, diagnosis and treatment planning. Even when the data is collected for purposes other than directly diagnosing a disease or predicting treatment outcome, useful medical information can still be retrieved. Nakamura et al. [3] used data mining to predict the

---

First and second authors have contributed equally to produce this paper.

S. Phon-Amnuaisuk et al. (eds.), *Computational Intelligence in Information Systems*,
Advances in Intelligent Systems and Computing 532, DOI 10.1007/978-3-319-48517-1_14

development of pressure ulcer in hospitals from patients' data that were originally collected for the purpose of calculating nursing costs. Decision making in the medical field is rather more sensitive than many other fields because of its direct relation to life and death consequences, and the well-being of patients. Therefore, a decision should be made with strong belief that is supported by thorough evaluation and clear explanation. This makes clinical data mining distinctive than other data mining uses in various ways. For example, it is widely common in clinical data mining to use white-box classifiers such as rule-based learners or decision trees because the resulting model is represented in a readable format. This enables the physicians to interpret the model output based on their medical knowledge, and increases their confidence when making their final decisions. While models built using black-box methods such as Artificial Neural Networks or Support Vector Machines and provide better results in terms of prediction accuracy, will be welcomed in many other fields, our experience showed that physicians often hesitate to accept these results due to the lack of model understand-ability, how the involved factors are related, and how to link that to their medical experience and knowledge. Although researchers have investigated the use of many different machine learning algorithms on clinical data, and reported interesting findings [1, 2], we believe that in practice, the ability for model introspection will actually limit us to only few of the many algorithms that are used in other fields and applications, even if this comes at the expense of prediction accuracy.

Another point to consider is that in many data mining applications, it is desirable to have a prediction model with high accuracy. In clinical data mining, however, it is important to distinguish between false positive errors and false negative ones. A false negative error has bigger impact than a false positive one because it can lead an unhealthy patient to miss a proper treatment, which might be fatal. On the other hand, a false positive error can be detected and corrected at a later stage by further investigations and tests.

Clinical datasets are usually highly heterogeneous where the data are usually collected from various sources such as images, laboratory tests, patient interviews, and physicians' observations and interpretations which leads to a poor mathematical characterization. In addition, many clinical datasets are noisy, incomplete, and suffer from the problem of data imbalance, in which the data has large number of patients (cases/instances) of one class (type/category), and a small number of patients of the other class.

In this study we consider using C4.5 decision tree, widely used in clinical data mining, with different sampling methods in order to identify best solutions for tackling the imbalanced data problem commonly faced in medical data mining.

The rest of this paper is organized as follows. In the next section, we explain decision tree classification models. We then discuss data imbalance problem and provide a description of common methods to overcome it in Sect. 3. Section 4 presents the clinical datasets considered in this study. Section 5 discusses the methodology we follow to conduct our experiments. Analyzing the results and reporting our findings is in Sect. 6. Finally, Sect. 7 concludes the paper and gives directions for future works.

2 Decision Trees

A decision tree model [4] is a data structure that is capable of representing knowledge in a humanly understandable way. It consists of a set of internal nodes, each representing test conditions on the values of one data attribute. The tree emerges from one common root node and ends with many leaf nodes, where each leaf represents a final classification decision.

Being able to understand how the built model is classifying the data and to interpret that into useful domain knowledge are the main reasons why decision trees are preferable over other methods like SVM or neural networks in clinical data mining [5]. A new data instance can be classified by starting from the root of the decision tree, and moving down its branches according to its attributes test results until a leaf node is reached. The class of the leaf node represent the predicted class of the instance. Attributes selected as a node test are usually determined using some splitting criteria. However, popular splitting criteria such as information gain ratio [4, 6] and Gini measure are skew sensitive.

3 Data Imbalance

Data imbalance is a problem that is very common in clinical data mining. A data set is considered imbalanced if the number of instances of one class is considerably smaller than the number of instances in the others. In clinical data the majority class is usually the negative class and the minority class is the positive class which is the class of our main interest. Multi-class problems might also suffer from data imbalance; however, it can be easily converted into many one-versus-others problem. Many learning algorithms tend to get overwhelmed by the large number of the majority class and ignore the minority class thus provide a high total accuracy, however, it also provides a high error rate on the minority class which is usually our concern. Assuming a 90 % imbalance ratio, a classifier that classify all instance as negative will achieve a 90 % accuracy while misclassifying all positive instances of the important class. Obviously this is not the desired result and some alternation is required to overcome this problem.

Japkowicz and Stephen [7], showed that different learning algorithms have different level of sensitivity to the data imbalance problem. They showed also that decision trees is the most sensitive classifier compared to Multilayer Perceptron and Support Vector Machines. In clinical data mining, decision trees are preferable because they provide an explanation of the classification decision.

3.1 Undersampling

Undersampling achieves data balance by removing instances from the majority class. Random undersampling method is the simplest form of undersampling in which the size of the majority class is reduced by removing instances randomly as its name indicates. Random undersampling is simple and easy to implement however, a main disadvantage of data undersampling methods is that there is a possibility that we lose information contained in important majority class instances removed due to the undersampling process. A good informed-undersampling method reduces this possibility.

Informed undersampling reduces the size of the majority class in a controlled fashion in order to keep important instances from the majority class. Example of informed sampling are EasyEnsemble and BalanceCascade reported in [8]. Both methods use ensemble learning in order to explore the majority class space and select useful instances, however, ensemble learners models are usually difficult to explain and fall in the black-box learners zone.

J. Zhang and I. Mani [9] proposed four sampling methods called NearMiss-1, NearMiss-2, NearMiss-3, and Most-Distance that uses K-nearest neighbor in order to sample reduce the size of the majority class. The K-nearest neighbor of an instance is defined as the K elements whose distance between itself and the instance is the smallest. Here we provide a description of the four algorithms:

- *NearMiss-1* selects from the majority class the instances whose average distances to the three closest minority instances are the smallest. Thus the instances selected by NearMiss-1 are close to some of the minority class instances.
- *NearMiss-2* selects from the majority class the instances with the smallest average distance to the three farthest minority class. In other words, NearMiss-2 selects the majority instances close to all of the minority instances.
- *NearMiss-3* surrounds each instance form the minority class with *k* instances from the majority class. It selects a predetermined number of the closest majority instances for each minority instance.
- *Most Distance* selects the instances from the majority class that have the largest average distance to the three closest instances from the minority class.

3.2 Oversampling

As its name indicates, oversampling works by sampling more data from the minority class. Random oversampling randomly selects a set of minority class S_r, duplicates its members, and appends them to the original minority class set. This will lead to an increase in the size of the minority class by the size of S_r and a reduction in the original data imbalance distribution the process is repeated until the desired data balance reached. The problem with oversampling is that it may make the classifier susceptible to data overfitting because repeating the same instance causes the classifier to become more specific in covering these instances.

Another method of increasing the size of the minority class is synthetic sampling in which artificial data is synthesized from the original minority class. A powerful method that has shown good results in many applications is the synthetic minority oversampling technique (SMOTE) [10]. SMOTE uses feature space similarities between minority class instances in order to generate the synthesized artificial data. For each instance in the minority class in order to create a synthesized instance SMOTE randomly selects one of its K-nearest neighbor for some specified K, calculate the feature vector difference between the two instances then multiplies it by a random number in the range [0, 1] and add the resulted vector to the original minority instance to generate the new artificial instance.

3.3 Model Evaluation

It is important to validate the model performance. Usually, accuracy is the evaluation metrics used to evaluate classification models. However, accuracy assumes similar cost for false positive and false negative errors. In clinical data mining, the cost of false positive is more expensive than the cost of false negative errors, and an evaluation method that reflects this fact is required.

Evaluation metrics are usually derived from the confusion matrix shown in Table 1.

From the confusion matrix, accuracy can be calculated as the ratio of correctly classified instances: Accuracy = (TP + TN) / (P_c + N_c), and the classification error equals 1- Accuracy, i.e. Error = (FP + FN) / (P_c + N_c).

Table 1. Confusion matrix

	Positive Prediction: P_p	Negative Prediction: N_p
Positive Class: P_c	TP: True Positive	FN: False Negative
Negative Class: N_c	FP: False Positive	TN: True Negative

Sensitivity and specificity can provide better metrics in the case of imbalanced datasets. Sensitivity, defined as TP / (TP + FN) = TP / P_c, measures the proportion of positive instances that are correctly classified.

On the other hand, specificity, defined as TN / (TN + FP) = TN / N_c, measures the proportion of negative instances that are correctly classified.

A good classifier should have high values for both sensitivity and specificity. In the case of imbalanced data, a classifier that classifies all instances as negative will have high accuracy, and high specificity, but zero sensitivity.

The Area Under Curve (AUC) [11] is widely used for measuring the performance in case of imbalanced data. AUC returns the area under Receiver Operating Characteristics Curve (ROC) that provides a visual representation of the performance in regards to the true positive rate (i.e. sensitivity) and false positive rate (i.e. 1-specificity). The visual presentation is useful for showing the tradeoffs between true positive and false positive error rates, however, it is difficult to use for calculation. The AUC provides a quantitative metric for ROC.

4 Experimental Datasets

Earlier experimental studies on learning from imbalanced data have been conducted. Reference [12] discussed the use of several sampling techniques versus different machine learners and performance metrics, and reported partial results of applying combinations of these choices on 35 datasets coming from a variety of application domains. In another study [13], the researchers investigated the class-imbalance problem in medical datasets by considering different under-sampling and over-sampling techniques applied on one cardiovascular dataset. In this paper, we will investigate the effect of a group of undersampling and oversampling techniques applied on multiple clinical datasets, and under constraints suitable for data mining in the medical domain, where white-box learners and suitable metrics are of concern.

In our empirical study, we have considered 7 nonproprietary clinical datasets publically available in the following sources:

- UCI: the data repository of the Center for Machine Learning and Intelligent Systems in the University of California, Irvine, famously known as UCI Machine Learning Repository. (archive.ics.uci.edu/ml.)
- OML: an open collaborative machine learning platform (www.openml.org).

Table 2. Description of used clinical datasets

Data set ID	Description	Source URL[a]
BRC	Breast Cancer dataset, from Institute of Oncology University Medical Center Ljubljana, Yugoslavia [14]. It was donated in 1988, and it is used to predict recurrent cancer events of patients.	Breast + Cancer
BCW	Breast Cancer Wisconsin dataset, donated from University of Wisconsin Hospitals in 1992 [15]. It is used to diagnose benign and malignant breast cancers.	Breast + Cancer + Wisconsin + %28Original%29
DIB	Pima Indians Diabetes Database, donated in 1990, for predicting whether patients show signs of diabetes according to World Health Organization criteria.	Pima + Indians + Diabetes
SAH	South Africa Heart Disease dataset, taken from a larger dataset described by Rousseauw et al. in 1983.	www.openml.org/d/1498
SPF	Heart dataset of cardiac Single Proton Emission Computed Tomography (SPECT) images, donated in 2001, where features are extracted from the images and used to predict cardilogists' diagnoses of normal and abnormal patients.	SPECTF + Heart
SPT	Same classification task as SPF, with binary extracted features to form the dataset.	SPECT + Heart
TYR	Thyroid Disease dataset, donated by the Garavan Institute in1987, to diagnose patients with thyroid disease.	Thyroid + Disease

[a]Use archive.ics.uci.edu/ml/datasets/ before the value for UCI based datasets.

We have considered only datasets with binary classification problem. Table 2 lists these datasets with a brief description of each one, and an identifier to refer to later in our analysis.

The imbalance ratio, defined here as the percentage of minority class instances to majority class instances, varies from 9 % (highly imbalanced) to almost 54 % (only slightly imbalanced). The datasets have also diversity in the number of attributes, their types (continuous and categorical), and the number of instances.

Few datasets contain missing values in one or more of their attributes. In our study, we did not apply any method to fill in these values, and decided to work on complete data by removing the instances with missing values since they were only few. The TYR dataset was the only one having some attributes completely empty or redundant (these attributes were removed), and had rather big number of instances with missing values in some other attributes. The instances in the latter case have been removed, which we consider relatively acceptable given the total number of instances in this dataset. Table 3 summarizes these details.

5   Experiment Design

We have used RapidMiner (6.5) [16] to conduct our experiments. We have also used SMOTE implementation in Weka (3.16.13) software [17] to perform oversampling.

We have systematically applied each of the sampling methods, including "No Sampling", on each of the seven datasets. After performing data pre-processing, 10-folds cross-validation (stratified) was used in order to evaluate each method. In each fold, data balancing methods were applied to the training subset, while the test subset was left imbalanced. Figure 1 shows a snapshot of the cross-validation design in RapidMiner which limits the sampling application to the training set. AUC, sensitivity, and specificity were recorded for each sampling method.

Table 3. Datasets pre-processing and summaries

Data set	# of [a] instances	# of [b] attributes	Attribute types	Instances removed	# of positive	# of negative	Imbalance ratio
BRC	277	10	all nominal	9	81	196	0.41
BCW	683	10	all numeric	16	239	444	0.54
DIB	768	9	all numeric	0	268	500	0.54
SAH	462	10	8 numeric, 1 nominal	0	160	302	0.53
SPF	267	45	all numeric	0	55	212	0.26
SPT	267	23	all nominal	0	55	212	0.26
TYR	2,643	23	6 numeric, 16 nominal	1,129	212	2,431	0.09

[a]Final number of instances after removing instances with missing values.
[b]Number of attributes, including the class attribute, after pre-processing.

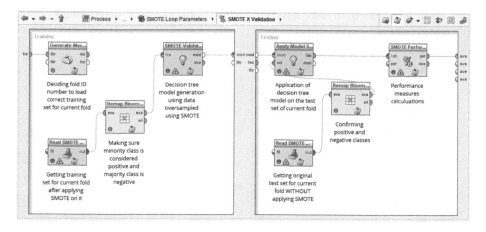

Fig. 1. Cross-validation process to evaluate SMOTE oversampling using RapidMiner.

The four sampling methods, NearMiss1, NearMiss2, NearMiss3, and Most Distance depend on calculating the distance between instances. In case of numeric attributes, Euclidean distance is used. However, when we have mixed types of attributes (numerical and categorical), Mixed-Euclidean distance is used, where for nominal attributes a distance of one is counted if corresponding values are not the same. Those algorithms are not part of RapidMiner components, and have been implemented by the authors.

For all sampling methods, we chose the parameters that rebalance the datasets to an almost equal ratio for both classes. As for the k parameter (number of nearest neighbors) for SMOTE and NearMiss3 algorithms, a fixed value of 5 has been chosen.

6 Results and Analysis

The results of our experiments are shown in Tables 4, 5, 6 and 7. For each dataset, the area under curve AUC, sensitivity, and specificity of each of the methods used rounded to two decimal places are reported.

For the Breast Cancer (BRC) dataset, Table 4. (left) shows that Most Distance method scored the highest AUC, with corresponding 0.54 sensitivity and 0.81 specificity. It shows a good improvement in sensitivity over the results obtained on the original data (indicated by **No Sampling** method) with a relatively low reduction in specificity. Table 4. (right) shows the results for the BCW dataset, and the results for the remaining datasets are summarized in Tables 5, 6 and 7.

In Table 8, we have summerized the ranking counts for each method. For example, Random Undersampling method was ranked first in only one dataset, and similarly for second, third, and fourth ranks. It also ranked fifth in three datasets.

Table 4. Performance ranks and results on BRC (left) and BCW (right) datasets.

#	Method	AUC	Sens.	Spec.
1	Most Distance	0.69±0.11	0.54±0.16	0.81±0.06
2	NearMiss1	0.68±0.09	0.53±0.15	0.74±0.10
3	**No Sampling**	0.65±0.11	0.32±0.19	0.92±0.05
4	Rand. Under.	0.65±0.09	0.40±0.20	0.74±0.18
5	NearMiss3	0.64±0.11	0.30±0.15	0.91±0.06
6	Rand. Over.	0.63±0.08	0.42±0.11	0.83±0.16
7	SMOTE	0.63±0.10	0.40±0.18	0.84±0.07
8	NearMiss2	0.62±0.09	0.69±0.11	0.46±0.11

#	Method	AUC	Sens.	Spec.
1	Rand. Over.	0.96±0.02	0.97±0.04	0.96±0.03
2	SMOTE	0.96±0.02	0.96±0.03	0.95±0.03
3	**No Sampling**	0.96±0.03	0.95±0.06	0.97±0.02
4	NearMiss2	0.96±0.03	0.95±0.06	0.96±0.02
5	Rand. Under.	0.96±0.02	0.97±0.04	0.94±0.02
6	NearMiss1	0.95±0.03	0.96±0.06	0.94±0.02
7	Most Distance	0.94±0.03	0.97±0.03	0.91±0.04
8	NearMiss3	0.91±0.05	0.96±0.04	0.86±0.08

Table 5. Performance ranks and results on DIB (left) and SAH (right) datasets.

#	Method	AUC	Sens.	Spec.
1	Rand. Over.	0.71±0.07	0.49±0.28	0.80±0.21
2	SMOTE	0.71±0.07	0.71±0.25	0.61±0.23
3	Most Distance	0.66±0.05	0.90±0.06	0.43±0.06
4	**No Sampling**	0.66±0.07	0.28±0.08	0.95±0.05
5	Rand. Under.	0.66±0.08	0.42±0.26	0.83±0.19
6	NearMiss1	0.62±0.05	0.38±0.07	0.90±0.04
7	NearMiss3	0.60±0.05	0.24±0.08	0.96±0.02
8	NearMiss2	0.50±0.00	0.53±0.07	0.50±0.09

#	Method	AUC	Sens.	Spec.
1	Most Distance	0.60±0.08	0.76±0.13	0.42±0.11
2	Rand. Under.	0.59±0.08	0.44±0.38	0.68±0.33
3	SMOTE	0.59±0.05	0.96±0.06	0.22±0.07
4	**No Sampling**	0.58±0.07	0.06±0.07	0.96±0.06
5	Rand. Over.	0.58±0.05	0.75±0.37	0.40±0.28
6	NearMiss3	0.51±0.03	0.06±0.09	0.95±0.05
7	NearMiss1	0.51±0.02	0.26±0.13	0.80±0.10
8	NearMiss2	0.51±0.02	0.35±0.11	0.59±0.12

We can see from the table that oversampling methods have ranked first and second more often than the undersampling ones, with random oversampling ranked first more than SMOTE method. Among the undersampling methods NearMiss1 and NearMiss2 methods have often scored low ranks compared to other methods.

Table 6. Performance ranks and results on SPF (left) and SPT (right) datasets.

#	Method	AUC	Sens.	Spec.
1	Rand. Over.	0.75±0.13	0.78±0.22	0.71±0.14
2	SMOTE	0.73±0.14	0.74±0.21	0.72±0.11
3	Rand. Under.	0.71±0.12	0.79±0.23	0.64±0.15
4	Most Distance	0.69±0.09	0.98±0.05	0.44±0.14
5	**No Sampling**	0.66±0.13	0.19±0.17	0.91±0.07
6	NearMiss2	0.65±0.09	0.73±0.20	0.54±0.17
7	NearMiss1	0.56±0.11	0.49±0.26	0.54±0.10
8	NearMiss3	0.50±0.00	0.22±0.17	0.76±0.11

#	Method	AUC	Sens.	Spec.
1	Rand. Under.	0.77±0.11	0.77±0.16	0.76±0.07
2	Rand. Over.	0.77±0.10	0.78±0.15	0.68±0.10
3	**No Sampling**	0.76±0.11	0.50±0.32	0.85±0.10
4	NearMiss3	0.75±0.08	0.59±0.14	0.82±0.10
5	SMOTE	0.71±0.11	0.68±0.24	0.69±0.12
6	Most Distance	0.70±0.09	0.85±0.14	0.51±0.08
7	NearMiss1	0.54±0.04	0.83±0.33	0.17±0.21
8	NearMiss2	0.52±0.03	0.65±0.43	0.32±0.34

Table 7. Performance ranks and results on TYR dataset.

#	Method	AUC	Sens.	Spec.
1	SMOTE	0.97 ± 0.01	0.97 ± 0.03	0.94 ± 0.02
2	Rand. over.	0.96 ± 0.03	0.93 ± 0.05	0.98 ± 0.01
3	NearMiss3	0.95 ± 0.02	0.91 ± 0.06	0.98 ± 0.01
4	**No Sampling**	0.95 ± 0.04	0.93 ± 0.06	0.98 ± 0.01
5	Rand. Under.	0.94 ± 0.03	0.93 ± 0.05	0.95 ± 0.02
6	NearMiss2	0.93 ± 0.03	0.90 ± 0.07	0.96 ± 0.01
7	NearMiss1	0.93 ± 0.03	0.90 ± 0.07	0.96 ± 0.01
8	Most Distance	0.60 ± 0.02	0.97 ± 0.04	0.22 ± 0.03

Table 8. Methods Rankings

Sampling method	Rank count							
	1	2	3	4	5	6	7	8
Random Undersampling	1	1	1	1	3	0	0	0
NearMiss1	0	1	0	0	0	2	4	0
NearMiss2	0	0	0	1	0	2	0	4
NearMiss3	0	0	1	1	1	1	1	2
Most Distance	2	0	1	1	0	1	1	1
Random Oversampling	3	2	0	0	1	1	0	0
SMOTE	1	3	1	0	1	0	1	0
No Sampling	0	0	3	3	1	0	0	0

7 Conclusion

In this work, we have evaluated the performance of different sampling methods on clinical data classification problem using the C4.5 decision tree due to its wide usage in clinical data mining. The methods of random oversampling and undersampling, SMOTE oversampling, NearMiss1, NearMiss2, NearMiss3, and Most Distance undersampling methods were investigated. The results showed that from the AUC point of view, random oversampling and SMOTE methods were superior to the undersampling methods.

References

1. Bellazzi, R., Zupan, B.: Predictive data mining in medicine: current issues and guidelines. Int. J. Med. Inf. **77**(2), 81–97 (2008)
2. Bellazzi, R., Ferrazzi, F., Sacchi, L.: Predictive data mining in clinical medicine: a focus on selected methods and applications. WIREs Data Mining Knowl. Discov. **1**, 416–430 (2011)
3. Nakamura, Y., Ghaibeh, A.A., Setoguchi, Y., Mitani, K., Abe, Y., Hashimoto, I., Moriguchi, H.: On-admission pressure ulcer prediction using the nursing needs score. JMIR Med. Inf. **3**(1), e8 (2015)
4. Ouinlan, J.R.: Induction of decision trees. J. Mach. Learn. **1**(1), 81–106 (1986)
5. Setoguchi, Y., Ghaibeh, A.A., Mitani, K., Abi, Y., Hasimoto, I., Moriguchi, H.: Predictability of pressure ulcers based on operation duration, transfer activity, and body mass index through the use of an alternating decision tree. J. Med. Investig. **63** (2016)
6. Quinlan, J.R.: C4.5: Programs for Machine Learning. Morgan Kaufmann, San Francisco (1993)
7. Japkowicz, N., Stephen, S.: The class imbalance problem: a systematic study. Intell. Data Anal. J. **6**, 429–449 (2002)
8. Liu, X.Y., Wu, J., Zhou, Z.H.: Exploratory undersampling for class-imbalance learning. IEEE Trans. Syst. Man Cybern. Part B Cybern. **39**(2), 539–550 (2009)
9. Zhang, Z., Mani, I.: KNN approach to unbalanced data distributions: a case study involving information extraction. In: Proceedings of International Conference on Machine Learning (ICML 2003), Learning from Imbalanced Data Sets Workshop (2003)
10. Chawla, N.V., Bowyer, K.W., Hall, L.O., Kegelmeyer, W.P.: SMOTE: synthetic minority over-sampling technique. J. Artif. Intell. Res. **16**(1), 321–357 (2002)
11. Bradley, A.P.: The use of area under the ROC curve in the evaluation of machine learning algorithms. Pattern Recogn. **30**(7), 1145–1159 (1997)
12. Hulse, J.V., Khoshgoftaar, T.M., Napolitano, A.: Experimental perspectives on learning from imbalanced data. In: Proceedings of the 24th International Conference on Machine Learning (ICML 2007), Corvalis, USA, pp. 935–942 (2007)
13. Rahman, M.M., Davis, D.N.: Addressing the class imbalance problem in medical datasets. Int. J. Mach. Learn. Comput. **3**(2), 224–228 (2013)
14. Zwitter, M., Soklic, M.: Breast cancer dataset. Data obtained from the University Medical Centre, Institute of Oncology, Ljubljana, Yugoslavia (1988)
15. Mangasarian, O.L., Wolberg, W.H.: Cancer diagnosis via linear programming. SIAM News **23**(5), 1–18 (1990)
16. RapidMiner Studio. http://www.rapidminer.com
17. Witten, I.H., Frank, E.: Data Mining: Practical Machine Learning Tools and Techniques, 3rd edn. Morgan Kaufmann (2011)

Internetworking, Security
and Internet of Things

Internet of Things (IoT) with CoAP and HTTP Protocol: A Study on Which Protocol Suits IoT in Terms of Performance

Mohammad Aizuddin Daud$^{(\boxtimes)}$ and Wida Susanty Haji Suhaili

School of Computing and Informatics, Universiti Teknnologi Brunei,
Bandar Seri Begawan, Brunei Darussalam
phpfreakz.rodriguez@gmail.com,
wida.suhaili@utb.edu.bn

Abstract. Behind Internet of Things (IoT) system, there are constrained devices and protocols that handle all the communication in the system. Constrained devices are equipped with sensor and communication capabilities to allow them to send data over the network. There are limitations on these constrained devices as they have limited resources such as processing power, memory and power consumption. In order to fit the needs for IoT systems, different types of protocols have been developed. These protocols lie in a communication protocol stacks that are from the Application layer, Transport layer, Network Layer and Network access layer. These protocols are designed to cater for the need of the systems to run smoothly with the limited resources available for the constrained devices. For this paper, the focus lies on two (IoT) protocols on the application layer: Hypertext Transfer Protocol (HTTP) and Constrained Application Protocol (CoAP). It extends how the protocol structures the message format, communication establishment and how request is handled from the client. The study is designed on different test beds based on performance factor to meet the requirement of the device's resources. The results and analysis of this study contributes to the findings of the performance where CoAP is faster than HTTP with smaller data. The study strengthens the use of CoAP for constrained devices in relation to the limited resources mentioned before thus contributing to how data are managed in any IoT environment.

Keywords: Constrained devices · CoAP · IoT · HTTP · Protocols

1 Introduction

Home Security System, Home Automation system, Disaster Management are some of the systems developed using the idea of IoT. For this project, the main focus is on how communication of these systems was influenced by the protocol that handles the IoT system. Over the internet, communication between devices is handled by protocols where in each layer a solid communication protocol mechanism for all the IoT building blocks. This is required to create efficient and reliable communication over the IoT systems [1]. Furthermore devices used for the IoT system have constrained resources which are usually associated with limited processing power, limited storage and often

© Springer International Publishing AG 2017
S. Phon-Amnuaisuk et al. (eds.), *Computational Intelligence in Information Systems*,
Advances in Intelligent Systems and Computing 532, DOI 10.1007/978-3-319-48517-1_15

runs on battery [2]. These devices are used as nodes for data collected from the sensors. Hence a RESTful approach is used being in favour of low power embedded networks [3]. Researchers in [4] compare CoAP to HTTP in terms of performance and CoAP suitability in constrained scenarios. This paper aims to extend the work to answer the questions: (1) how specific area of communication is managed and (2) how to ensure the connectivity of each of the device are efficient to meet the need of the constrained devices. Therefore the focus of this project is on which communication protocol out of these two is most favourable to IoT systems.

2 HTTP and CoAP

HTTP is an application-level protocol for distributed, collaborative, hypermedia information systems [5]. HTTP is a TCP/IP based protocol that is used to deliver data on the World Wide Web with port 80 as the default port. The 3 basic features of HTTP of connectionless, media independent and stateless make it simple but powerful protocol [6].

While CoAP is one of the latest application layer protocols developed by IETF that facilitates the integration of the embedded network with Web technologies [3]. This protocol is developed specifically for the IoT system which was developed based on the idea of HTTP protocols. CoAP runs over UDP to keep the overall implementation lightweight. It is the aim of CoAP development to keep the overall implementation lightweight since dealing with devices of limited resources.

It uses Restful Architecture which is similar to HTTP. In Restful Architecture, the commands GET, POST, PUT, and DELETE are used to provide resource-oriented interactions in client-server architecture [3]. The CoAP is developed not only for the communications and transferring data, but is also developed along with DTLS. DTLS is a chatty protocol that requires numerous message exchanges to establish a secure session [7]. CoAP uses DTLS for security transaction in the transport layer.

2.1 Constrained Devices

Constrained devices are devices that have limited processing and storage capabilities usually powered by batteries [2]. Constrained devices often used as nodes in the IoT system. Nodes are point of data collection where sensor will collect the data and pass it to the nodes and from the nodes; data will be stored and sent over the internet to the server [8]. Constrained devices also deal with the request and response message from and to the server. Efficiency of the device in processing the messages depends on the size of the message sent.

Here, CoAP have been developed to fit with the constrained devices. With the property of CoAP, IoT system are able to work efficiently with constrained devices since CoAP are designed to be lightweight which put less stress on the constrained devices processing capabilities.

2.2 Process of Communication Made Between HTTP and CoAP

The characteristics of both HTTP and CoAP protocol are based on how they establish server client communication and their message format when exchanging messages between server and client. In this Section, the focus is to compare and contrast on how HTTP and CoAP established their communication between client and server and the packet format for request and response message.

Communication Establishment. The communication for HTTP started with a sequence of handshake protocol [9]. These handshake protocols are made between client and server to establish a connection between both. A synchronization packet is sent to initiate the connection to the server [10]. This packet is sent to setup a reliable session between the client and the server. Once the SYN packet is received by the server, the server will respond with SYN-ACK packet to the client. The SYN-ACK packet is a Synchronization Acknowledgement packet. This packet is sent to acknowledge the client that sessions are allowed to be initiated [10]. Then the client will sent ACK packet, which is an Acknowledgement packet in response for the SYN-ACK packet from the server. This packet will be followed by the establishment of reliable session between the client and the server [10].

The communications in CoAP are considered to be more direct since no handshake occurs between client and server. The differences between HTTP and CoAP in transport layer are CoAP which make use of UDP and HTTP uses TCP. In order to satisfy IoT requirement, devices in IoT can only have small or limited resources. This is how CoAP contributes to the definition of lightweight communication as CoAP uses UDP, UDP's properties are low overhead [11]. This contributes to the requirement of device resources, which makes CoAP suitable for IoT. CoAP communication started with CON. CON is a confirmable message that are sent from the client. CON message carry a request message to the server. Once CON is received by the server, it will elicit the ACK message that is the acknowledgement message. The ACK message carries the response message for the request that is made by the client.

2.3 HTTP and CoAP Message Format

HTTP Message Format. In HTTP, there are two different message formats for request and response: HTTP request and HTTP response message. Once the reliable session is initiated, the client will sent HTTP request. The HTTP message size ranges from 2 kb to 8 kb which is equivalent to 2000 bytes and 8000 bytes [12].

The HTTP request message format contains Method, URL, Version, Header Lines fields and an Entity body. Method that is specified in the message format refers to the request from the client [12]. The request methods used in this project are:

(1) GET - Retrieve information from the server on specified URL
(2) POST - Submit data to be processed on specified URL
(3) PUT - Edit or add existing information in the server on specified URL
(4) DELETE - Delete existing information in the server on specified URL

In the Header Lines Section, it contains Header field name and value. This is where additional information of the request and detail of the client are stored [10] with information such as Host, Connection, User-agent and Language.

Entity Body is part of the HTTP request message format that is only used with post method and response message [10]. Header field is similar to the Header Lines Section but with additional fields such as Date, Content Length and Content Type.

CoAP Message Format. Unlike HTTP, CoAP only have one message format which is used by both request and response message. Both the request and response message are using the same format. The minimum message size for CoAP is 4 bytes and the maximum is 1024 bytes [11]. The message format for CoAP consists of Version, Type, Token Length, Code, Message ID, Options and Payload.

Version in the CoAP message format refers to the version of CoAP protocol used in [11]. Types are message types that are represented by numbers. Message types are as follow along with its number representation [11].

(1) CON (Confirmable) – 0
(2) NON (Non-Confirmable) – 1
(3) ACK (Acknowledgement) – 2
(4) RST (Reset-Message) – 3

The Code in the CoAP message format refers to the Request Method. Each method is represented using numbers. The request methods are listed below [13] which are similar to those of HTTP:

(1) GET = 1
(2) POST = 2
(3) PUT = 3
(4) DELETE = 4

The Message ID in the CoAP message format is the number to identify the message sent [11]. This ID prevents message duplication. The minimum message size for CoAP is 4 bytes and the maximum is 1024 bytes contributes to the lightweight size of CoAP compared to HTTP message format [11].

3   Implementation

The implementation of the system was based on the Client Server architecture. In Client Server architecture, the client provides the user with user interface that allows user interaction and the server deals with database and processing the request from the client [5]. In order to show the differences in terms of performance, and to make the result more reliable, settings were kept constant such as running both clients and servers in the same platform.

3.1   HTTP and CoAP Implementation

Both HTTP and CoAP server used the CoAPthon and Django REST Framework. The server was hosted by Ubuntu 14.04 using VMware player. Whilst for CoAP client, the Copper Extension of Mozilla Firefox browser and for HTTP client, Mozilla Firefox browser was used.

The client server architecture was set under an environment which connects to a network allowing them to communicate and send information over the network. The client will be the Mozilla Firefox browser; server will be the VMware which runs Ubuntu and host the HTTP and CoAP server. The protocol represents the link that governs the communication between both client and server over the network.

3.2   Network Performance Measures

Different types of network performance measures can be used. Each measure gave different types of result depending on what kind of performance measured. Such as:

1. Throughput - Measure of how much actual data can be sent per unit of time across a network.
2. Latency - Amount of time taken for data to travel from one location to another across the network.
3. Jitter - Variation in the delay of received packets.
4. Error Rate - Measure of the amount of error encountered during data transmission over a network. The higher the error, the less reliable the network connection is.

For this study, time was used as the measure, which is similar to latency. Time is referred to the time taken for the request and response process to complete and measured in seconds.

3.3   Testing

This implementation was tested on different scenarios to see how different network structure will affect the performance of the server. The scenarios were based on 5 different network structures: Wireless Home Network, Wired Home Network, UTB Network, Portable Wi-Fi Network and Wireless Hotspot Network. The testing process was done by sending Request from the client using four different methods that are GET, POST, PUT and DELETE. These methods triggered the server to send Response according to what have been requested by the client. Wireshark was used to record and analyse the packets sent between the client and the server.

For HTTP, time was measured from the first SYN message to the last ACK message. For CoAP, time was measured from the first CON message to the last ACK message. Data sizes were set for each method: GET 14 bytes, POST 9 bytes, PUT 9 bytes and DELETE 8 bytes.

4   Findings

For both protocols, the time taken was based on how communication was established. In each scenario, results were based on the test using the four methods of GET, POST, PUT and DELETE which were captured using Wireshark and the time taken was analysed from the moment communication was initiated until completed. The results obtained were placed according to the four methods for all the five scenarios. HTTP took longer time to complete request for each method than CoAP. The HTTP difference in terms of time taken can be clearly seen from the HTTP-REQUEST to HTTP-RESPONSE. The time taken patterns for both CoAP and HTTP in all 5 scenarios were similar for all the four methods. The HTTP requests for all methods were relatively higher for GET and POST method. As for CoAP, the time taken was similar with all the methods except for PUT method where it was slightly higher than the other but the differences can barely be seen.

4.1   Large Data

With small data the differences was barely seen for CoAP and similar for HTTP in all the methods, next was to test on a set large data of 1990 bytes and applied to the four methods, for both CoAP and HTTP to complete request with big data.

GET Large Data on Home Wireless Network. The Table 1 shows the time taken CoAP to complete the GET methods on large data for both request. The time taken for CoAP to complete the request is 24.273145 s.

Table 1. CoAP GET large data time taken

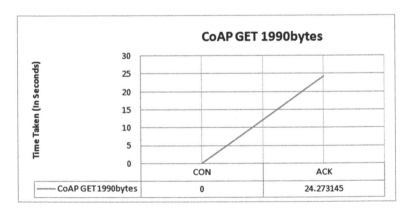

Table 2. HTTP GET large data time taken

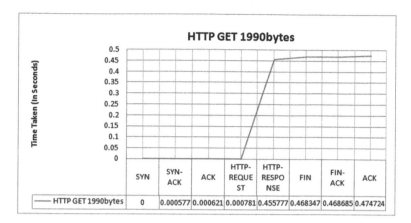

	SYN	SYN-ACK	ACK	HTTP-REQUEST	HTTP-RESPONSE	FIN	FIN-ACK	ACK
HTTP GET 1990bytes	0	0.000577	0.000621	0.000781	0.455777	0.468347	0.468685	0.474724

In Table 2, it shows the time taken HTTP to complete the GET request. The time taken for HTTP to complete the request is 0.47472 s. The pattern was similar to GET method on smaller data from the different scenarios. The one that contributed to the time taken was HTTP-REQUEST to HTTP-RESPONSE.

From previous analysis, the time taken for CoAP was relatively lower but when dealing with large data; a different set of findings was found. The time taken for CoAP is higher even compared to HTTP previous analysis. This means that large data affect the overall performance in terms of time taken for CoAP and only slight effect on HTTP.

Table 3. CoAP and HTTP overall GET time taken

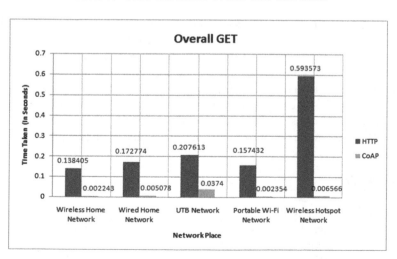

Table 4. CoAP and HTTP overall time taken large data

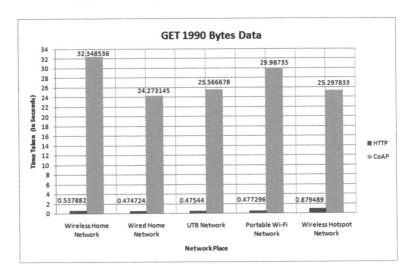

The Table 3 summarizes the time taken for GET method result for all of the scenarios.

The Table 4 shows the summary for CoAP and HTTP request on GET method for large data 1990 bytes.

The tables above show the huge gap of time taken between CoAP and HTTP for the five different network structures. The time taken to process large data by CoAP creates a large gap in terms of time compared to HTTP. Therefore, time performance for CoAP is affected by large data; this strengthens why CoAP is not suitable to process large data.

POST Large Data on Home Wireless Network. An exception on the test made was discovered when the CoAP request was done for POST and PUT method on large data of 1990 bytes. CoAP was unable to complete both requests. The error encountered was the '4.08 Request Entity Incomplete', indicating that the client was unable to send the request to the server. Data limit for CoAP was 1024 bytes where the large data are 1990 bytes. This shows CoAP client was unable to send data larger than 1024 bytes. As for HTTP, the time taken was slightly higher than the previous analysis for the rest of the methods. This indicates that large data does affect the time taken performance of HTTP.

5   Conclusion

The properties of CoAP as the chosen protocol to use when dealing with IoT have been outlined in this paper. IoT system consists of a group of constrained devices which means that every device is limited in terms of resources such as RAM, memory and processing power. With the properties of CoAP and the idea of IoT system, CoAP is efficient in delivering smaller data size as the findings suggest. This is further strengthened by comparing CoAP with the known HTTP protocol. The study shows how the light-weight feature of CoAP contribute to the short transmission time but attempt needs to be considered as this compromise the security aspect of it. With HTTP, this is taken care by the handshake before and after as well as the message format. Therefore for future enhancement for CoAP, a study on the implementation of DTLS as part of the security aspect is suggested. One known fact of HTTP is having its own ways of securing the communication. This is done by the use of HTTPS. HTTPS was introduced which is a combination of both HTTP and TLS. This combination provides an encrypted communication which is a part of security measure that has been taken by HTTP. CoAP with DTLS could also provide security to the communication. Similar test could be done to determine whether security features affect the protocol in terms of performance.

References

1. Mario, B., Candid, W.: Insecurity in the Internet of Things. In: Symantec Security Response (2015)
2. Mattern, F., Floerkemeier, C.: From the internet of computers to the internet of things. Informatik-spektrum **33**(2), 107–121 (2010)
3. Villaverde, B., Pesch, D., Alberola, R.D.P., Fedor, S., Boubekeur, M.: Constrained application protocol for low power embedded networks: a survey. In: 2012 Sixth International Conference Innovative Mobile and Internet Services in Ubiquitous Computing (IMIS), pp. 702–707 (2011)
4. Colitti, W., Steenhaut, K., De Car, N., Buta, B., Dobrota, V.: Evaluation of constrained application protocol for wireless sensor network. In: 2011 18th IEEE Workshop Local Metropolitan Area Networks (LANMAN), pp. 1– 6 (2011)
5. Fielding, R., et al.: Hypertext transfer protocol – HTTP/1.1. The Internet Society (1999). https://www.ietf.org/rfc/rfc2616.txt. Accessed 1 Sep 2016
6. Yannakopoulos, J.: Hypertext Transfer Protocol: A short course. Department of Computer Science, University of Crete, Greece (2003). http://condor.depaul.edu/dmumaugh/readings/handouts/SE435/HTTP/http.pdf. Accessed 1 Sep 2016
7. Chen, X.: Constrained application protocol for internet of things. Washington University, St. Louis (2014). http://www.cse.wustl.edu/~jain/cse574-14/ftp/coap.pdf. Accessed 1 Sep 2016
8. Ajit A., Mininath K.: Secure CoAP using enhanced DTLS for internet of things. Int. J. Innov. Res. Comput. Commun. Eng. (2014)
9. Bardford, P., Crovella, M.: A performance evaluation of hyper text transfer protocols. In: Proceedings of the 1999 ACM SIGMETRICS International Conference on Measurement and Modeling of Computer Systems, pp. 188–197 (1999)
10. Shelby, Z.: RFC 7252 - the constrained application protocol (CoAP). Internet Engineering Task Force (IETF) (2014). https://tools.ietf.org/html/rfc7252. Accessed 1 Sep 2016
11. Tod, B., Jin, Q.: The TCP handshake: practical effects on modern network equipment. Net. Protoc. Algorithm **2**(1), 197–217 (2010)

12. Fielding, R., Reschke, J.: RFC 7230 - Hypertext Transfer Orotocol (HTTP/1.1): Message Syntax and Routing. Internet Engineering Task Force (IETF) (2014). https://tools.ietf.org/html/rfc7230#section-3.2.4. Accessed 1 Sep 2016
13. Siegel, M., Sciore, E., Madnick, S.: Context interchange in a client-server architecture. J. Softw. Syst. 27(3), 223–232 (1993). http://web.mit.edu/smadnick/www/wp2/1993-07.pdf. Accessed 1 Sep 2016

NTRU Binary Polynomials Parameters Selection for Reduction of Decryption Failure

Juliet N. Gaithuru$^{(\boxtimes)}$, Mazleena Salleh, Ismail Mohamad, and Ikuesan R. Adeyemi

Universiti Teknologi Malaysia, 81310 Skudai, Johor, Malaysia
julietgaithuru@yahoo.com
http://www.utm.my

Abstract. This paper studies the NTRU public key cryptosystem to identify the most influential parameters for decryption failure confirming that decryption failure is key-dependent. The study uses binary polynomials and analyzes the correlation between the parameter sets recommended in the EESS 1v2 (2003) and Jeffrey Hoffstein et al. (2003). The observed relationships are then used to recommend an extended parameter selection criteria which ensures invertibility and reduced probability of decryption failure. We then recommend a condition for selecting an appropriately large size of q which is the least size required for ensuring successful message decryption. The study focuses on binary polynomials as it allows for a smaller public key size and for the purpose of providing better insights leading to further study into other variants of NTRU.

Keywords: Cryptography · NTRU · Asymmetric key algorithms · Decryption failure · Parameter selection · Machine learning · Classifier

1 Introduction

Securing information has become a necessity in the information age. As the volume of data transmitted online grows, so is the need to safeguard this information whether it is in the form of text, video, voice or images. This security has been provided through the use of various encryption algorithms. The Nth degree truncated polynomial ring (abbreviated as NTRU) [1] is one of the asymmetric key algorithms used, which is deemed to be future-proof as it is secure against quantum algorithm attacks due to its lattice-based structure. NTRU collectively refers to NTRUEncrypt (encryption algorithm) and NTRUSign (signature scheme). The security of NTRU is based on the approximate Closest Vector Problem (appr-CVP) in convolution modular lattices [2]. NTRU has undergone several modifications since it was invented in 1996 and patented in 1998 [3]. It has several variants, with the NTRU parameters taking either a binary, ternary or product-form.

In this paper, the study is focused on binary polynomials because binary polynomials result in a smaller public key size since they are believed to allow

© Springer International Publishing AG 2017
S. Phon-Amnuaisuk et al. (eds.), *Computational Intelligence in Information Systems*,
Advances in Intelligent Systems and Computing 532, DOI 10.1007/978-3-319-48517-1_16

for a small q parameter [4]. In addition, the study of NTRU which uses binary polynomials will provide a comprehensive understanding of the NTRU operation, the parameters and the relationships between them and how to select parameters in a way that ensures a low probability of decryption failure. This provides a good basis for educational purposes and the conclusions drawn can be applied on other variants of NTRU.

2   NTRU Parameters

NTRU operates in the polynomial convolution ring $R = \frac{Z[X]}{(X^N - 1)}$. The parameters used in the NTRU public key cryptosystem are made up of the integer parameters $N, p, q, d_f, d_g, d_r, d_m$ and the polynomials f, g, r, m, F_p, F_q. The parameters are selected such that the parameter size N is a prime integer, p is a small modulus (which could be 2, 3, or $2 + x$) and the parameter q is a large modulus which corresponds to the value of p selected; q is prime (for $p = 2$) and an integral power of 2 (for $p = 3$ and $p = 2 + x$). x stands for the indeterminate used in polynomials [5]. The private key polynomial f is selected such that it has d_f number of 1 coefficients and that it is invertible $mod\,p$ and $mod\,q$. The polynomial g has d_g number of 1 coefficients. The random polynomial r which is a one-time value used to obscure the message has d_r number of 1 coefficients. The message m has d_m number of 1 coefficients.

A list of recommended parameter sets for low, medium and high security levels for binary polynomials were proposed in [6], which were then replaced with the parameters recommended in [7,8] as depicted in the Table 1.

Table 1. Previously recommended parameter sets for $p = 2$

Parameter set	Security level (k)	N	p	q	d_f	d_g	d_r	Reference
ees251ep4	80	251	2	239	72	72	72	[7,8]
ees251ep5	80	251	2	239	72	72	72	
NTRU167.2	Low	167	2	127	45	35	18	[6]
NTRU263.2	Moderate	263	2	127	35	35	22	
NTRU503.3	High	503	2	253	100	100	65	

Binary polynomials were believed to allow for a small q parameter. The need to increase resistance against hybrid combinatorial attack created the need for the use of larger sample spaces [4]. Binary polynomials were replaced with ternary and product-form polynomials in order to improve the combinatorial search space thus improving security and efficiency [9]. List of recommended parameters for ternary and product form polynomials of NTRU have been provided in [3–5,9–11].

3   NTRU Operation

NTRU operation begins with the establishment of the integers N, p, q and the polynomials f, g, r and m. The private key is then generated by first obtaining the multiplicative inverse of $f \, mod \, p$ and $f mod \, q$ such that: $f * F_p \, mod \, p = 1$ and $f * F_q \, mod \, q = 1$. The private key is then set as the polynomial pair (f, F_p). The public key h is obtained by computing $h = p * F_q * g \, mod \, q$. The message m is encrypted using the public key h by computing $e = r * h + m \, mod \, q$. The ciphertext is decrypted by the recipient using the private key by computing $a = f * e \, mod \, q$. After which adjustment is done to a by ensuring that the coefficients of a are in the range of $\frac{q}{2}$ and $\frac{-q}{2}$. This is then followed by retrieving the decrypted message by computing $C = F_p * a \, mod \, p$.

Decryption is made possible because the polynomials p, r, g and m are chosen to have small values in the polynomial convolution ring R thus ensuring the polynomial $prg + fm$ has a high probability of having width b (which represents $prg + fm$) less than q. The coefficients of these terms are selected in a manner that ensures that their absolute value does not exceed $\frac{q}{2}$ and $\frac{-q}{2}$ [9,12].

4   Decryption Failure Approximation

A range of parameter sets have been recommended in previous work in order to ensure a low probability of decryption failure. The choices of parameter sets which will result in an overwhelming probability of decryption failure can be determined experimentally [13]. In this study we work towards this objective of experimentally identifying these parameters. Before embarking on experimentation, we evaluate the computational prediction of the probability of decryption failure as described in the subsequent section.

4.1   Recap of Previous Decryption Failure Approximation

Decryption failure refers to the inability to decrypt validly generated ciphertexts [12]. Using the recommended parameters in EESS 1v2 [8], the probability of decryption failure is once in every 2^{-12} for $N = 139$ and 2^{-25} for $N = 251$. In [12], the measure of the probability of decryption failure is based on ensuring that the polynomial product $(p * r * g + f * m)$ does not exceed q. In [14], decryption failure was considered to be message and key dependent which led to the development of a padding scheme referred to as SVES-3. Decryption failure is a threat to the security of the transmitted information. This is because the occurrence of decryption failure leaks information about the private key of the recipient [15].

In [9], the measure of the probability of decryption failure was obtained by checking if a coefficient of $(r * g + f * m)$ is greater than c where $c = \frac{(q-2)}{2p}$. Therefore the probability of decryption failure P_{dec} is given by:

$$P_{dec}(c) = Prob\left(\mid r * g + f * m \mid \geq c\right) \tag{1}$$

For ternary polynomials, an approximation of the probability of decryption failure was estimated more accurately by considering the assumptions that N is large and the coefficients of r are independent random variables taking the value 1 with probability $\frac{d_r}{N}$, -1 with probability $\frac{d_r}{N}$ and 0 with probability $\frac{(N-2d_r)}{N}$ and the same assumption holds for g, F and m.

Then, if $y_i = r_k g_l$ and $z_i = m_s f_t$ and X_j denotes a coefficient of $(r*g+f*m)$ then X_j is a sum of N terms, for some k, l, s, t given by

$$X_j = \sum_{i=1}^{N}(y_i + z_i) \tag{2}$$

Assuming N is large, the central limit theorem is applied on X_j, which is normalized to have variance as 1 ($\sigma^2 = 1$). This results in the standard normal probability density function which is given by $f(x) = \frac{1}{\sqrt{2\pi}}e^{\frac{-x^2}{2}}$. After applying the central limit theorem on X_j, translating it to complementary error function (erfc) notation and repeating the experiment of selecting a coefficient N times, the resulting probability of decryption failure is given by:

$$P_{dec} = N * erfc\frac{c}{\sigma\sqrt{2N}} \tag{3}$$

where $c = \frac{(q-2)}{2p}$ and σ^2 is the variance.

4.2   Computational Approximation of Decryption Failure for NTRU Binary Polynomials

We apply the same principle of approximation in [9] on binary polynomials in order to estimate the probability of decryption failure by assuming that:

– N is large
– The coefficients of r are independent random variables taking the value 1 with probability $\frac{d_r}{N}$ and 0 with probability $\frac{N-d_r}{N}$ and that the same assumption holds for g, f and m.

A computation of the mean of y_i and z_i (that is $E(y_i)$ and $E(z_i)$) results in:

$$E(y_i) = \frac{d_r d_g}{N^2} \tag{4}$$

$$E(z_i) = \frac{d_f d_m}{N^2} \tag{5}$$

Going on to compute the value of variance for binary polynomials results in

$$\sigma^2 = \left(\frac{N^2 d_r d_g - (d_r d_g)^2}{N^4}\right) + \left(\frac{N^2 d_f d_m - (d_f d_m)^2}{N^4}\right) \tag{6}$$

The central limit theorem is applied in arriving at the measure of the mean and standard deviation. The central limit theorem states that the average of a large number of independent identifiable random variables is approximately normally distributed [16]. Once a normal distribution is standardized, then the resulting standardized normal distribution has a mean of zero and a standard deviation of 1 unit (that is $\mu = 0$ and $\sigma = 1$).

In this case of the binary polynomials, the computed value of the mean is not equal to zero as shown by Eqs. 4 and 5 as is the characteristic of a standard normal distribution. Therefore leading to the conclusion that this method of approximation of the probability of decryption failure is not considered to be suitable for binary polynomials for the purposes of this study.

5 Studying the Relationship Between the Parameters and Their Effect on the NTRU Key Generation, Encryption and Decryption

A study of the relationship between the parameters was conducted in order to gain insight into a method of selecting parameters that will ensure a low probability of decryption failure.

5.1 Testing Parameters and Environment

Testing was done so as to identify which parameters have the greatest influence on whether decryption in NTRU is successful. An experimental analysis was done whereby all the NTRU parameters were varied one at a time while keeping the other parameters constant. The evaluation began with small parameters (low security levels), followed by medium and high security levels. The testing began by varying each of the parameters f, g and m one at a time for all possible combinations of polynomials for $N = 11$ (low security). The proportions obtained of successful key generation, encryption and decryption from the analysis done at the low security level were then used as input for computation of appropriate sample sizes for moderate and high security levels ($N = 11$ and $N = 53$ respectively). The parameter sets used for low security level and moderate security levels are based on examples published on the Security Innovation website and previous published works, while the parameter set for $N = 251$ (128-bit security) is based on parameter set in EESS1v2. The test parameters used for this study are illustrated in the Table 2.

Table 2. Test parameters for binary NTRU polynomials

Parameter set	k	N	p	q	d_f	d_g	d_r	d_m
eesTest1	3	11	2	37	4	5	4	6
eesTest2	16	53	2	67	7	27	40	35
ees251ep4 [8]	80	251	2	239	72	72	72	35

The results of the testing were used to identify the influential parameters which were then varied for $N = 53$ and the recommended parameter sets in the EESS1v2 [8] for $N = 251$ for binary polynomials as well as product-form polynomials.

The data for experimentation was selected using the following methodology:

1. The size of the population was determined. This is based on the number of all possible polynomial combinations, given by 2^N.
2. The sample size was generated by uniform random sampling without replacement. Uniform sampling method is used because it is the documented sampling method used to select secret polynomials f and g [4,9,15]. First, an initial sample size n_0 was obtained by computing $n_0 = \frac{z^2 \times p(1-p)}{e^2}$ where z is the critical value for the confidence level c, p is the proportion or distribution and e is the sampling error. The proportion, p used is based on the results obtained from the test results for low security levels, as will be discussed in Sect. 5.2.
3. Uniform random sampling was used, with a 99 % confidence interval and 5 % margin of error.
4. Given that n_0 was found to be at least 5 % of the population N and sampling is without replacement, the sample size was more accurately estimated by reducing the error in the previous computation of n_0 by applying the Finite Population Correction Factor (FPC) [17,18] by computing $n = \frac{n_0 \cdot N}{n_0 + (N-1)}$.
5. In order to ensure equal distribution of values of $f(1)$ in the population the population was split into categories where the *No. of categories*$= N_2 - N_1$. Then the number of sample values generated in each category was computed as $\frac{n}{N_2 - N_1}$. This was based on the principle that any integer $(2^t - 1)$ once its converted into a binary polynomial equation of f results in a value of $f(1) = t$ once the sum of coefficients is computed.
6. The random numbers generated were then converted to their binary polynomial equivalent and used as test data.

The tests were carried out using the Magma Computational Algebra System [19] as was used in previous studies by Hermans et al. [20] on speed records for NTRU run on a GPU and also the cryptanalysis study of countermeasures imposed on NTRU to enable it withstand multiple transmission attacks [21] and a study on algebraic attack on NTRU [22]. The experiments were run on two computer systems; one with a Windows 8.1 64-bit operating system with Intel core i5 processor with 4 GB RAM while the second computer system had a Windows 8.1 64-bit OS Intel core i7 processor with 8 GB RAM.

5.2 Testing for Identification of Influential Parameters: Key Determinants of Decryption Failure

Testing was done in order to identify influential parameters for the successful decryption of messages in the NTRU algorithm by varying the polynomial parameters f, g, r, m and the integer q. The results obtained from these tests are as shown in Table 3.

Table 3. Initial test results for identification of influential NTRU parameters

Parameter	Percentage with successful key generation		
	$N = 11$	$N = 53$	$N = 251$
f	51.15	53.74	51.09
f (product form)	100	100	100
g	100	100	100
r	100	100	100
m	100	100	100
q	large q	large q	large q

The results in Table 3 revealed that parameters g, r and m had no effect on the occurrence of decryption failure. Parameter f had the most influence on decryption failure, coupled with the size of the parameter q. In the case where the parameter f is of the product form, the polynomial f given by $f = 1 + pF$ was always invertible thus resulting in successful key generation thus had no effect on the likelihood of the occurrence of decryption failure. This is attributed to the fact that the selected polynomial should be invertible $mod\,p$ and $mod\,q$ so as to be an acceptable parameter f. The multiplicative inverses can be obtained using the Euclidean algorithm. Once invertibility is confirmed, then the polynomial inverses F_p and F_q can be generated and subsequently the private key parameters (f, F_p) obtained. The experiments showed that once an invertible polynomial of f was obtained but which resulted in decryption failure, variation of the size of the large prime q upwards resulted in successful decryption.

This led to the conclusion that the parameters f and q are the most influential parameters for successful decryption in NTRU. This confirmed the assertion by Howgrave-Graham et al. in [14] that decryption failure is largely key dependent. Upon identification of the parameters f and q as the influential parameters for successful decryption in NTRU, further evaluation was carried out into these parameters as described in the subsequent section.

The tests revealed that out of all the polynomial combinations of f tested, 3×10^3 polynomials, only 51.15 % of the polynomial combinations were invertible for low security level at $N = 11$. Therefore 48.85 % of the polynomials were not invertible therefore resulting in unsuccessful key generation. For moderate security level at $N = 53$, only 53.74 % of the polynomials were invertible while for high security level only 51.09 % of the polynomials were invertible. These measures indicate the probability that a randomly chosen polynomial in the ring R will be invertible.

In [13], the probability that a randomly selected element in the ring of convolution polynomials is invertible is given by:

$$\left(1 - \frac{1}{p}\right)\left(1 - \frac{1}{p^n}\right)^{\frac{(N-1)}{n}}. \qquad (7)$$

where n is the smallest integer such that $p^n = 1\,(mod\,N)$. Using the Eq. 7 to compute the probability of a randomly selected polynomial being invertible and comparing this measure to the experimental observation results in a disparity ranging from 0.01 to 0.04 as shown in the Table 4.

Table 4. Probability that a randomly chosen polynomial is invertible

	$N = 11$	$N = 53$	$N = 251$
Computed probability	0.4995	0.4999	0.5000
Experimental observation	0.5115	0.5374	0.5109
Difference	0.0120	0.0375	0.0109

The slight disparity may be attributed to the use of sampling. These measures confirm that the sampling procedure used in this study results in an appropriately large data size from which reasonable conclusions can be drawn. The observations can therefore be considered to be representative of the entire population, with an acceptable degree of accuracy.

5.3   Studying the Private Key Polynomial f

Following the identification of the private key polynomial f as one of the influential parameters, an in-depth study of the parameter was done. An evaluation of the polynomials which resulted in successful key generation but had decryption failure showed that varying the size of the modulus q to a larger prime number resulted in successful message decryption. Furthermore, the polynomials of f which were invertible thus resulting in successful message decryption had the following properties:

1. Odd values of $f(1)$ have successful key generation.
2. $f(1) \neq q$.
3. $f(1) \neq N$.

5.4   Using Machine Learning to Analyze the Relationship Between the Polynomial f and Large Modulus q

Following the observation from this study that varying the size of the modulus q to a larger prime number resulted in successful message decryption, it begs the question what is the minimum size of q required to ensure a successful message decryption if all the other NTRU parameters are selected accordingly. Initial observations showed that the modulus q was larger than the integer N.

Given that the private key polynomial f had the highest influence on the occurrence of decryption failure along with the large modulus q, further analysis was conducted to determine the correlation between these parameters so as to identify the underlying conditions which if varied could favourably affect the

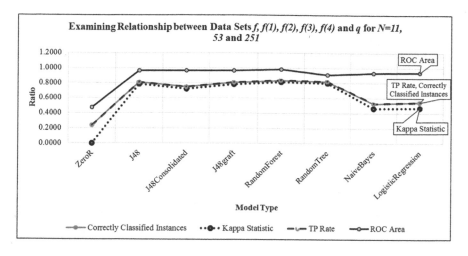

Fig. 1. Results of examining the relationship between data attributes of f and q.

chances of a successful decryption. The evaluation of this relationship was done using machine learning techniques by using classifiers provided in WEKA [23]. The evaluation was three fold:

1. First, an evaluation was done to establish whether it is possible to classify the attributes based on the classes of q, in other words given f is it possible to predict the corresponding value of q.
2. Secondly, of these attributes used for the evaluation in step 1, which ones have the most predictive capability.
3. Thirdly, if indeed it is possible to classify the value of f based on its corresponding minimum value of q required for successful message decryption in NTRU, then what is the predictive relationship.

In the first step, in order to address the classification of attributes, additional attributes of f were introduced for the evaluation. The polynomial $f(x)$ was evaluated when $x = 1, 2, 3, 4$ resulting in the attribute list: $f(1)$, $f(2)$, $f(3)$, $f(4)$. The minimum size of q required for successful message decryption served as the class name. Then several machine learning models were run using 10-fold cross-validation for the attribute list $f(1)$, $f(2)$, $f(3)$, $f(4)$ and q in order to explore the possibility of classifying attributes of f based on the minimum size of q required for successful message decryption. The results of the classification are as depicted in the Fig. 1.

As illustrated in the Fig. 1, the baseline accuracy using the ZeroR showed that there were 23.73 % correctly classified instances while the random forest tree classifier showed 83.4328 % correctly classified instances. This confirmed the existence of a classification of attributes of f based on the corresponding minimum value of q required for successful message decryption. Thus pointing to the likelihood of a predictive relationship between the data sets of f and q.

The second step was then conducted in order to select the attributes, in this case NTRU parameters, which have greater predictive capability. The attributes selected as having greater predictive capability would then be used to formulate the predictive relationship in the next step. The attributes $f(1)$, $f(2)$, $f(3)$, $f(4)$ with q as the class name were evaluated using various attribute selectors. The Greedy Stepwise, Best first, SubsetSizeForwardSelection models showed that $f(1)$ and $f(2)$ had the highest predictive capability. The LinearForwardSelection model showed that $f(1)$ and $f(3)$ had the highest predictive capability while the Ranker ranked the attributes as $f(1)$, $f(2)$ followed by $f(3)$ and finally $f(4)$. Therefore, the attributes $f(1)$ and $f(2)$ were then selected as the attributes with the most predictive attributes of the value of q.

In the final step, the attributes with the greatest predictive capability, $f(1)$ and $f(2)$ were used to predict the value of q. Several models were used and the results compared in an effort to find an appropriate predictive model, whose results are shown in Table 5.

Table 5. Comparison of function classifiers for establishment of a predictive model for minimum size of q for successful NTRU decryption

Functions classifier type	Correlation coefficient	Mean absolute error	Root mean squared error	Deduced model-Min. size of q for successful dec
Linear regression	0.8313	55.7003	67.9571	$2.5754 * f(1) + 83.9038$
Pace regression	0.8308	55.9965	68.0585	$84.1958 * f(1) + 2.5618$
Simple linear regression	0.8286	56.398	68.4487	$2.53 * f(1) + 85.01$
LeastMedSq	0.0551	1165979.865	21036891.04	$0.482 * f(1) + 276.489$

The comparison of several classifiers in order to deduce an appropriate model as shown in Table 5 showed that the linear regression model resulted in the highest correlation coefficient of 0.8313 followed by the pace regression model at 0.8308, the simple linear regression model and finally the LeastMedSq. In addition, the linear regression model showed the lowest mean absolute error, root mean square error and relative absolute errors. Therefore, the resulting deduced model from the linear regression classifier was selected as the most appropriate model.

This process is considered relevant for the exploration of the probable relationship between the extracted attributes of f and q so as to reveal the underlying intrinsic relationship (if any) based on the function of f. By extracting f as an attribute based on a derivation function of f, the introduction of machine learning process could extract patterns which would be otherwise unnoticeable to the human eyes. Consequently, based on the observed relationship, a probable inference on the causation of decryption failure could be further explained. Exploratory process of various supervised machine-learning techniques was carried-out using the WEKA tool. The predictive relationship (the probability of the existence of relationship between the $f(x)$ and q) observed was measured with reference to a probabilistic baseline accuracy using ZeroR algorithm. The ZeroR algorithm is a highest-class prior probabilistic classifier. It classifies all instances in a dataset into the class of the highest prior probability. Among all explored classifiers, regression model presented higher class-prediction accuracy and capability to present a reliable predictive tendency. Parameters considered in the classification process include accuracy, RMSE, Kappa statistics, F-measure, and Area Under the receiver operating characteristic Curve (AUC). Further exploration was then considered using various functions of regression model, result as shown in Table 5. This resulted in the establishment of the predictive model that: For successful message decryption in NTRU, the minimum size of q should be selected as a prime number larger than $(2.5754 * f(1) + 83.9038)$.

6   Conclusion

This study provides an insight into the parameter selection criteria for NTRU using binary polynomials. The study first evaluates the most influential parameters for decryption failure, proving that decryption failure is key-dependent. We then study NTRU parameters in order to establish a relationship which provides a predictive indicator of which parameter combinations will result in successful message decryption. This study shows that in order to ensure the selection of polynomial of f that is invertible, it should have an odd number of ones, that is $f(1)$ should be an odd integer when the sum of coefficients is computed. In addition, $f(1)$ should not be equal to N and q should be greater than the parameter size N. Then we establish that the criteria for establishing the appropriately large value of q is that in order to ensure successful message decryption q should be selected such that it is the next prime number greater than $(2.5754 * f(1) + 83.9038)$ thus $q > (2.5754 * f(1) + 83.9038)$. It should be noted that binary polynomials were originally introduced when NTRU was invented in 1998 and later recommended parameters use ternary polynomials which were subsequently followed by the release of parameter sets of the product form, which eliminate the need to find the multiplicative inverse of f. In this study, we focus on the earlier version of NTRU recommended parameters which use binary polynomials in order to provide for a smaller public key size and providing an avenue for further study into other variants of NTRU.

References

1. Qingjun, C., Yuli, Z.: Subliminal channels in the ntru and the subliminal-free methods. Wuhan Univ. J. Nat. Sci. **11**(6), 1541–1544 (2006). http://dx.doi.org/10.1007/BF02831816
2. Whyte, W., Hoffstein, J.: NTRU. In: Whyte, W., Hoffstein, J. (eds.) Encyclopedia of Cryptography and Security, pp. 858–861. Springer, Boston (2011). http://dx.doi.org/10.1007/978-1-4419-5906-5_464

3. Hoffstein, J., Pipher, J., Silverman, J.H.: NTRU: a ring-based public key cryptosystem. In: Buhler, J.P. (ed.) ANTS 1998. LNCS, vol. 1423, pp. 267–288. Springer, Heidelberg (1998). http://dx.doi.org/10.1007/BFb0054868
4. Hoffstein, J., Pipher, J., Schanck, J.M., Silverman, J.H., Whyte, W., Zhang, Z.: Choosing parameters for ntruencrypt. Report, Cryptology ePrint Archive, Report 2015/708 (2015)
5. onsortium for Efficient Embedded Security: Efficient embedded security standard (EESS) EESS 1, version 3.0, 31 March 2015. https://github.com/NTRUOpenSourceProject/ntru-crypto
6. Silverman, J.H.: Wraps, gaps, and lattice constants. NTRU Report 11 (2001)
7. Hoffstein, J., Silverman, J.H., Whyte, W.: Estimated breaking times for NTRU lattices. In: version 2, NTRU Cryptosystems (2003). Citeseer (1999). http://www.ntru.com/cryptolab/tech_notes.htm#012
8. IEEE: Efficient embedded security standards (EESS), EESS 1: implementation aspects of ntruencrypt and ntrusign, version 2.0, 20 June 2003
9. Hirschhorn, P.S., Hoffstein, J., Howgrave-Graham, N., Whyte, W.: Choosing NTRUEncrypt parameters in light of combined lattice reduction and MITM approaches. In: Abdalla, M., Pointcheval, D., Fouque, P.-A., Vergnaud, D. (eds.) ACNS 2009. LNCS, vol. 5536, pp. 437–455. Springer, Heidelberg (2009). doi:10.1007/978-3-642-01957-9_27
10. Ieee draft standard specification for public- key cryptographic techniques based on hard problems over lattices. IEEE Unapproved Draft Std P1363.1/D12, p. 1, October 2008
11. Hoffstein, J., Jill Pipher, W.W.: More efficient parameters keys and encoding for hybrid resistant ntruencrypt and ntrusign. Report, NTRU Cryptosystems Inc., Security Innovation (2009)
12. Howgrave-Graham, N., Nguyen, P.Q., Pointcheval, D., Proos, J., Silverman, J.H., Singer, A., Whyte, W.: The impact of decryption failures on the security of NTRU encryption. In: Boneh, D. (ed.) CRYPTO 2003. LNCS, vol. 2729, pp. 226–246. Springer, Heidelberg (2003). http://dx.doi.org/10.1007/978-3-540-45146-4_14
13. Pipher, J.: Lectures on the ntru encryption algorithm and digital signaturescheme: Grenoble, June 2002. Report, Brown University, Providence, RI 02912, June 2002. http://www.math.brown.edu/~jpipher/grenoble.pdf
14. Howgrave-Graham, N., Silverman, J.H., Singer, A., Whyte, W., Cryptosystems, N.: Naep: Provable security in the presence of decryption failures. IACR Cryptology ePrint Archive 2003, 172 (2003)
15. Howgrave-Graham, N., Silverman, J.H., Whyte, W.: Choosing parameter sets for NTRUEncrypt with NAEP and SVES-3. In: Menezes, A. (ed.) CT-RSA 2005. LNCS, vol. 3376, pp. 118–135. Springer, Heidelberg (2005). http://dx.doi.org/10.1007/978-3-540-30574-3_10
16. Athreya, K.B., Lahiri, S.N.: Central limit theorems. In: Athreya, K.B., Lahiri, S.N. (eds.) Measure Theory and Probability Theory, pp. 343–382. Springer, New York (2006). http://dx.doi.org/10.1007/978-0-387-35434-7_12
17. Lincoln University: Sample size (2006). http://library.lincoln.ac.nz/global/library/learning/mathsandstats/qmet103/sample-size.pdf
18. Levy, P.S., Lemeshow, S.: Sampling of Populations: Methods and Applications. Wiley, Hoboken (2013)
19. Magma Group: Magma computational algebra system, version 2.21-2, Sydney (2015)

20. Hermans, J., Vercauteren, F., Preneel, B.: Speed records for NTRU. In: Pieprzyk, J. (ed.) CT-RSA 2010. LNCS, vol. 5985, pp. 73–88. Springer, Heidelberg (2010). doi:10.1007/978-3-642-11925-5_6
21. Xu, J., Hu, L., Sun, S., Xie, Y.: Cryptanalysis of countermeasures against multiple transmission attacks on NTRU. IET Commun. **8**(12), 2142–2146 (2014)
22. Bourgeois, G., Faugère, J.C.: Algebraic attack on ntru using witt vectors and Gröbner bases. J. Math. Crypt. **3**(3), 205–214 (2009)
23. Hornik, K., Buchta, C., Zeileis, A.: Open-source machine learning: R meets Weka. Comput. Stat. **24**(2), 225–232 (2009)

Energy Efficient Operational Mechanism for TDM-PON Supporting Broadband Access and Local Customer Internetworking

S.H. Shah Newaz[1], Alaelddin Fuad Yousif Mohammed[1(✉)],
Mohammad Rakib Uddin[2], Gyu Myoung Lee[3], and Jun Kyun Choi[1]

[1] Korea Advanced Institute of Science and Technology (KAIST),
Daejeon, South Korea
{newaz,alaelddin,jkchoi59}@kaist.ac.kr
[2] Universiti Teknologi Brunei, Gadong, Brunei Darussalam
rakib.uddin@utb.edu.bn
[3] Liverpool John Moores University (LJMU), Liverpool, UK
G.M.Lee@ljmu.ac.uk

Abstract. Considering ever increasing importance of traffic sharing among the users connected with different Optical Network Units (ONUs) of a TDM-Passive Optical Network (TDM-PON), to date, different TDM-PON architectures have been introduced. These architectures facilitate traffic flow among the ONUs directly along with traffic flow between an ONU and the Optical Line Terminal (OLT), which is the centralized intelligence of a TDM-PON system. The TDM-PON that facilitates direct communication among different ONUs under a single TDM-PON system is namely called TDM-PON Internetworking architecture. In this paper, we come up with a novel operational mechanism for one of the well-known TDM-PON Internetworking architectures in order to improve energy saving performance. An ONU in the TDM-PON Internetworking architecture based on which we propose a novel Internetworking operational mechanism can have two modes: LAN-PON (sharing traffic with other ONUs directly without any assist from the OLT) and broadband access (facilitates OLT and ONU communication). In order to facilitate operating under these two modes, an ONU uses two low-cost Optical Switches (OSWs). We refer to this Internetworking architecture as OSW based solution. Our novel energy efficient operational mechanism allows an ONU in OSW based solution to use Energy Saving Mode (ESM). The simulation results state that the operational mechanism introduced in this paper can contribute in increasing energy saving performance of ONUs noticeably.

Keywords: TDM-PON · Internetworking · Optical switches · Sleep mode

1 Introduction

Passive optical network (PON) technology is an access network technology which has tremendously contributed in expansion of communication network.

© Springer International Publishing AG 2017
S. Phon-Amnuaisuk et al. (eds.), *Computational Intelligence in Information Systems*,
Advances in Intelligent Systems and Computing 532, DOI 10.1007/978-3-319-48517-1_17

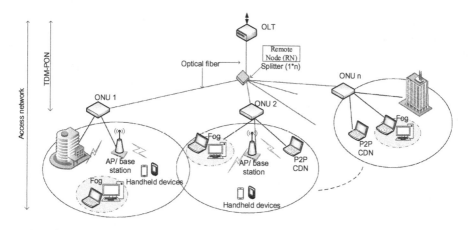

Fig. 1. TDM-PON deployment supporting different applications running at customer premises.

The reasons that motivated many network operators to adopt PON technology are: (i) higher scalability, (ii) more bandwidth capacity, (iii) more cost-effective services and (iv) lower energy consumption compared to other access network technologies (e.g. WiMAX, WiFi) [1,2]. We have witnessed tremendous growth of data rate of PON over the last several years. In 2011, Huawei tested 40 G PON prototype [3].

Rich Internet applications and high quality streaming services (e.g. 4 K/8 K video) have propelled the high bandwidth demand in the access segment of network. Additionally, with the increasing number of contents, Content Delivery Network (CDN) providers are experiencing higher Capital Expenditure (CAPEX) and Operational Expenditure (OPEX). This is because they need to continuously upgrade their content storage facilities along with expanding CDN infrastructure. One of the possible ways to improve content availability while not increasing CAPEX and OPEX is to use P2P-CDN (combine CDN and Peer-to-peer) [4,5]. In P2P-CDN, users cache the popular contents and can exchange the contents with other users (peers) [4,5]. In case of distributed computing paradigm, computational tasks are assigned to the locally available computational resources, namely Fog servers [6]. These Fog servers could be users' desktop computers or any hand-held devices. These aforementioned discussion clearly indicates that there is a growing need to have traffic forwarding in a local area domain. Hence, authors in [1] argue that it would be a very possible scenario in which an end user (an end user refers to any devices with Internet connection) connected with an Optical Network Unit (ONU) might need to communicate with another end user which is served by another ONU within the same TDM-PON system. Therefore, arguably, not only ONU and Optical Line Terminal (OLT) communication but also direct communication among ONUs has become increasingly important. Figure 1 demonstrates a TDM-PON deployment scenario along with several possible applications (e.g. P2P CDN, Fog servers).

To date, many researchers proposed their Internetworking architecture among the ONUs. Each of those has some advantages and limitations as well. Typically, TDM-PON architectures standardized by IEEE and ITU-T (e.g. IEEE 802.3ah) support inter-ONU communication under the same TDM-PON system by using a router attached to an OLT [7]. The major limitation of such Internetworking architecture comes from the fact that the bandwidth required for the OLT and ONUs communication (broadband access) could be insufficient in a TDM-PON system. This is because LAN emulation could lead to narrowing the scope of broadband access traffic forwarding. Apart from that, authors in [7] argue that the router installed in the OLT, which facilitates emulation protocol for LAN, is expensive; thereby, imposing additional CAPEX to the network operators. In the meantime, a few methods have been proposed to provide ONUs Internetworking functionality in different ways [7,8]. In [7], authors propose a physical-layer solution, which is attractive since this method is not only bandwidth efficient but also potentially cost effective. Two low-cost optical switches (OSWs) in each ONU are used to virtually divide a single TDM-PON system into two independent networks so as to facilitate direct inter-ONU communication (referred as LAN-PON traffic) and conventional communication between an ONU and the OLT (referred as Broadband Access (BA) traffic). In this paper, we refer to this architecture as OSW based solution. In OSW based solution, a pair of optical wires connect an ONU with a star coupler (SC) and the SC is connected with the OLT through a single optical fiber. The OSWs in an ONU switch mode based on requirement in order to facilitate LAN-PON traffic and BA traffic forwarding. The mode in which an ONU can forward LAN-PON traffic is named as LAN-PON mode, whereas in BA mode an ONU is capable of forwarding BA traffic. The main drawback of OSW based solution relies on the fact that the OSW in an ONU requires around 2 to 10 ms to switch from one mode to another mode, thereby making frequent mode change cumbersome. In OSW based solution, an ONU uses the same wavelength for LAN-PON traffic and BA traffic forwarding.

In [9], authors came up with a tunable transmitter (Tx) and receiver (Rx) based solution in order to forward LAN-PON traffic and BA traffic. This proposed mechanism uses a tunable Tx and Rx so as to establish inter-ONU communication through a dedicated wavelength. That is, unlike OSW based solution, an ONU in the architecture introduced in [9] uses two wavelengths to forward LAN-PON traffic and BA traffic. This architecture improves bandwidth utilization significantly both for broadband access (conventional PON communication) and ONUs Internetworking (LAN-PON communication). In addition, authors in [9] introduced a time slot based transmission scheme where each ONU's transmission opportunity is separated into two equal time slots allowing an ONU to forward LAN-PON traffic and BA traffic through two different wavelengths. It needs to highlight that the switching time of a tunable transmitter is in nanosecond range whereas an OSW's switching time is in millisecond range [7]. However, a tunable Tx and Rx are very expensive compared to a normal Tx and Rx, thereby increasing CAPEX in the tunable Tx and Rx based solution than that of the OSW based solution. Therefore, arguably, the proposed solution in [9] may not be a feasible solution to the PON operators.

In [1], based on router based Internetworking architecture, we came up with an energy efficient LAN-PON and BA traffic forwarding technique. Considering the rapid growth of traffic forwarding capacity in PON, we argued in [1] that router based Internetworking would be a promising architecture to support both LAN-PON traffic and BA traffic forwarding. We consider OSW based solution could be another promising solution as it requires low cost switches to be deployed in ONUs, thereby not increasing CAPEX significantly. Additionally, unlike the router based Internetworking architecture, in OSW based solution, downlink and uplink bandwidth do not need to be sacrificed for LAN-PON traffic forwarding. Furthermore, OSW based Internetworking architecture is flexible and secured [7].

In this paper, we propose a novel LAN-PON traffic and BA traffic forwarding mechanism in OSW based solution in order to minimize energy consumption in ONUs as much as possible without imposing noticeable traffic delay. The simulation results show the importance of our proposed idea. The rest of the paper is organized as follows. Section 2 explains the system model. In Sect. 3, the proposed operational procedures are introduced. Section 4 evaluates the performance of the proposed operational procedures and finally Sect. 5 draws conclusion and future research directions.

2 System Model

We propose a novel energy efficient operational procedures based on the OSW based solution, which is introduced in [7]. This section narrates how the LAN-PON Internetworking architecture presented in [7] works to facilitate LAN-PON traffic and BA traffic forwarding. In addition to that, in this section, we present the architectural assumptions associated with the OLT and an ONU so as to implement energy saving functionalities in this LAN-PON Internetworking architecture.

The OLT consists of two major parts: (i) digital circuitry and (ii) analog circuitry. The OLT MAC resides in the digital circuitry part and it is in charge of packet processing, forwarding (both incoming and outgoing), ONU status tracking, etc. Whereas, the analog circuitry in the OLT is in charge of receiving and transmitting traffic through optical signal. In analog circuitry part, there are two major components: a Continuous-Mode optical Transmitter (CMT) and a Burst-Mode optical Receiver (BMR) which are connected through a Coarse Wavelength Division Multiplexer (CWDM), as we can notice from Fig. 2.

The Star Coupler (SC), which is a passive device, is located at the remote node. This plays a crucially important role to guide optical signal in order to forward traffic among the ONUs and OLT. A pair of SC ports connect each of the ONUs (i.e. a pair of optical fibers are connected with each of the ONUs), as shown in Fig. 2.

The analog circuitry part of an ONU in the OSW based solution [7] consists of a CWDM, two OSWs, a Burst-Mode transmitter (BMT), and a BMR along with all the necessary components that a conventional ONU has. Additionally, we consider that the analog part of an ONU in our proposal uses a counter and a Sleep Control Logic (SCL), as considered in [2]. The role of the counter and SCL is to trigger any activities inside an ONU (e.g. changing states of an ONU). Furthermore, we assume that an ONU in our proposal can have Energy Saving Mode (ESM). Under this mode, there are three states: sleep state (both Tx and Rx are turned off), doze state (only Tx is turned off) and active state (every components are powered on in an ONU allowing the ONU to be fully functional).

As mentioned before, an ONU in the OSW based solution [7] can be operated under two modes: LAN-PON mode and BA mode. When the ONU is in BA mode, OSW 1 and OSW 2 move into cross state (cross state refers to the dotted connection) and bar state (bar state refers to the solid connection), respectively [7]. This combination of OSWs allows the ONU's BMR to receive downstream traffic from the OLT and the ONU's BMT to send upstream traffic using wavelength λ_{UL} [7]. On the other hand, when the OSW 1 and OSW 2 are set to bar state and cross state, respectively, the ONU is operated under LAN-PON mode [7]. That is, the ONU can send and receive traffic to/from other ONUs under the same TDM-PON system. To know more in detail how the state change of OSWs facilitates an ONU's mode change, please refer to the paper [7]. At this point, it needs to mention here that regardless the mode of an ONU (i.e. BA mode, LAN-PON mode), the ONU can have three states which are: sleep state, doze state and active state while using the ESM.

3   Proposed Energy Efficient Operational Mechanism

Similar to the operational procedures stated in [7], in our proposal, we consider that ONUs that have traffic to share among themselves move into LAN-PON mode. Whereas, the remaining ONUs in the same TDM-PON system are operated under BA mode. We propose that an ONU should maintain its ESM regardless its current mode at a given time. The subsequent part of this section narrates how our novel energy efficient operational mechanism works.

ONU Group Management. The OLT in a TDM-PON system maintains a Dynamic Bandwidth Allocation (DBA) cycle for both uplink and downlink transmission for the ONUs in BA mode. The length of the DBA cycle, T_{cycle}, is a function of OSW's state changing time (T_{OSW}). In this paper, we consider that $T_{cycle} = 2 \times T_{OSW}$.

When an ONU has traffic to forward to a particular ONU, it movies into LAN-PON mode after $T_{cycle}/2$ in a T_{cycle}. Before it moves into LAN-PON mode, it should exchange traffic with the OLT. To do so, at the beginning and middle of each T_{cycle}, ONUs send their uplink bandwidth requirement to the OLT along with mentioning whether they have LAN-PON traffic to forward or not. The OLT uses DBA algorithm to calculate uplink transmission slot for each of the ONUs in BA mode. Once uplink slots are measured for ONUs, the OLT notifies the ONUs (see Fig. 2). At the same time, the OLT mentions the downlink transmission slot

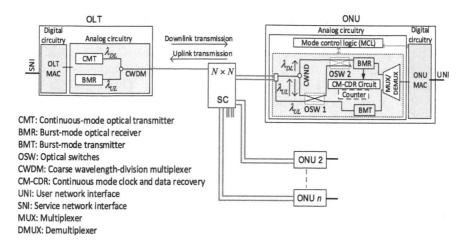

CMT: Continuous-mode optical transmitter
BMR: Burst-mode optical receiver
BMT: Burst-mode transmitter
OSW: Optical switches
CWDM: Coarse wavelength-division multiplexer
CM-CDR: Continuous mode clock and data recovery
UNI: User network interface
SNI: Service network interface
MUX: Multiplexer
DMUX: Demultiplexer

Fig. 2. OSW based Internetworking architecture along with ONU's energy saving capability.

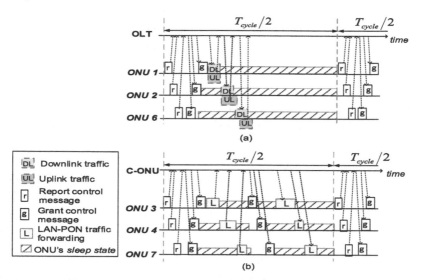

Fig. 3. Traffic forwarding under proposed operation procedures; (a) BA mode and (b) LAN-PON mode.

to each of them after taking into consideration average downlink traffic arrival rate for each of the ONUs. The OLT allocates uplink and downlink slots during the first half of a T_{cycle} to those ONUs, which have LAN-PON traffic to share. This allows the ONUs having LAN-PON traffic to forward to move into LAN-PON mode during the immediate $T_{cycle}/2$ duration.

LAN-PON Traffic Forwarding Policies. Before a group of ONUs move into LAN-PON mode, the OLT nominates one of those ONUs as a Coordinator-ONU (C-ONU). The criteria that the OLT uses for selecting a C-ONU is out of the scope in this paper. The role of the C-ONU in LAN-PON mode is almost the same as the OLT in BA mode in a TDM-PON system. That is, all the ONUs in LAN-PON mode send bandwidth request to the C-ONU using report control message. Upon receiving bandwidth requests, the C-ONU measures an uplink slot for each of the ONUs in LAN-PON mode and notifies them using grant control message (see Fig. 2). In turn, each ONU forwards uplink traffic to the C-ONU destined to any ONUs in the TDM-PON system. Note that, for an ONU in LAN-PON mode, the destination ONU (traffic recipient ONU) may not be in the same mode at a given time (i.e. the destination ONU could be in BA mode while the source ONU is in LAN-PON mode).

After receiving traffic from all the ONUs in LAN-PON mode, the C-ONU measures a downlink transmission slot for each of the ONUs and informs them. If any of the received traffic at the C-ONU is destined to an ONU which is not in LAN-PON mode during that time. Then C-ONU buffers those traffic and forwards to the OLT once it moves into BA mode. After receiving traffic from the C-ONU during the assigned reception slot, an ONU should turn off its Tx and Rx while staying in LAN-PON mode. This is possible in our solution because an ONU in this mode can receive all the traffic from different ONUs at once from the C-ONU, thereby allowing the ONUs to stay active state only during their traffic reception slots. In contract to the procedures explained above, the solution presented in [7] forces all the ONUs in LAN-PON mode to stay active always (each ONU forwards traffic to the destination ONU by itself), resulting in wasting energy significantly.

Energy Saving Functionality. In our proposal, an ONU turns of its components whenever possible in order to reduce its energy consumption. An ONU in both LAN-PON mode and BA mode, transmits and receives traffic during the assigned transmission slots. Apart from that, the ONU needs to stay in active state for a while in order to communicate with the OLT during each T_{cycle}. We consider that an ONU moves into sleep state turning off its Tx and Rx whenever it does not have any transmission and/or reception related activities. Figure 3 presents the overall procedures of our energy efficient TDM-PON supporting broadband access and local customer Internetworking.

4   Results and Discussion

To evaluate the performance of our proposed operational procedures, we rely on our outstanding TDM-PON OPNET based simulation model which has been used in our previous research efforts (e.g. [12]). In this paper, similar to [11], we assume that an ONU consumes 4.69 W when the ONU is in active state, and 0.7 W when it is in sleep state. We suppose that an ONU requires 2 ms to transit

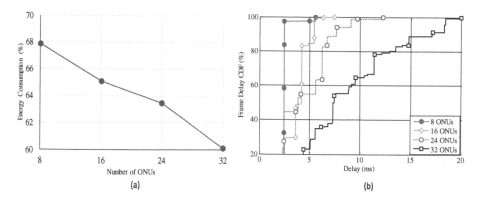

Fig. 4. Energy consumption and frame (traffic) delay performance evaluation.

from sleep to active state, similar to [10,14]. We assume that the sleep duration of an ONU is 50 ms. Additionally, we consider that $T_{OSW} = 2$ ms.

In this paper, we are interested to evaluate the delay and energy performances under 8, 16, 24, and 32 ONUs in a TDM-PON system. It has to note here that we compare the energy performance of an ONU in our proposed operational mechanism in front of the ONU introduced in OSW based solution, in which an ONU always remains active (no energy saving functionalities) regardless the presence or absence of traffic [7].

Figure 4(b) shows the delay performance of LAN-PON frames when our proposed operational mechanism is in place. In this figure, we represent the delay as Cumulative Distributed Function (CDF). Results show that the higher the number of ONUs in LAN-PON mode, the more traffic delay. Particularly, we can see from Fig. 4(b), when there are 8 ONUs in a LAN-PON, around 98 % of frames have delay around 6 ms, whereas; only around 30 % of frames have around 6 ms delay when there are 16 ONUs in the LAN-PON mode. Furthermore, the worst delay performance is when there are 32 ONUs in the LAN-PON mode. The reason behind this is that when there is a small number of ONUs (e.g. 8 ONUs), the chance for ONUs in the LAN-PON to get enough grant to send all the frames in its buffer is very high. However, an ONU in this case has around 6 ms delay due to sleep to active state transition time and switching time from BA mode to LAN-PON mode (i.e. OSW's state changing time). Moreover, when the number of ONUs increases (e.g. 16 ONUs) in LAN-PON mode, an ONU may not be able to send all its traffic within the allocated slot assigned by the C-ONU, which is in turn, results in increasing traffic delay. It needs to highlight here that long traffic delay may not be acceptable for many delay sensitive applications [13]. A number of research efforts report that traffic forwarding delay in a TDM-PON system could lead to reduce TCP traffic throughput noticeably [2,15]. Therefore, future research should put effort in finding the optimal number of ONUs that should be in LAN-PON mode at a given time, so that, traffic delay requirement is not violated.

Note that the longer the sleep interval length of an ONU, the higher the traffic delay, but lower the energy consumption of an ONU [2,13]. In this paper, we present the initial results of an ONU's energy consumption in Fig. 4(a). This figure shows that the energy consumption of an ONU decreases when number of ONUs increases. The reason behind this is that, under large number of ONUs, an ONU can have a chance to sleep for long time. However, this results in increasing traffic delay for that ONU.

5   Conclusion

In this paper, we put effort to come up with an energy efficient operational mechanism for TDM-PON system that allows both broadband access (BA traffic) and local customer Internetworking (LAN-PON traffic). Here, we have presented very initials results of our research. Among the several issues, in our future research, we are planning to: (i) investigate how to improve latency performance in both LAN-PON and BA mode when T_{OSW} is relatively long, (ii) study how we can incorporate QoS aware traffic forwarding mechanism under OSW based Internetworking architecture and (iii) devise an algorithm to select an appropriate C-ONU when more than one ONUs are in LAN-PON mode.

References

1. Newaz, S.H.S., Mohammed, A.F.Y., Lee, G.M., Choi, J.K.: Energy efficient and latency aware TDM-PON for local customer internetworking. In: IFIP/IEEE International Symposium on Integrated Network Management (IM), Ottawa, ON, pp. 1184–1189 (2015)
2. Newaz, S.H.S., Cuevas, A., Lee, G.M., Crespi, N., Choi, J.K.: Adaptive delay-aware energy efficient TDM-PON. Comput. Netw. **57**, 1577–1596 (2013)
3. Huawei Unveils World's First 40G PON Prototype. http://pr.huawei.com/en/news/hw-103292.htm
4. Zhang, G., Liu, W., Hei, X., Cheng, W., Kankan, U.X.: Understanding Hybrid CDN-P2P video-on-demand streaming. IEEE Trans. Multimedia **17**(2), 229–242 (2015)
5. Chen, L., Zhou, Y., Jing, M., Richard, T.B.: Thunder crystal: a novel crowdsourcing-based content distribution platform. In: Proceedings of the 25th ACM Workshop on Network and Operating Systems Support for Digital Audio and Video (NOSSDAV) (2015)
6. Zao, J.K., et al.: Pervasive brain monitoring and data sharing based on multitier distributed computing and linked data technology. Front. Hum. Neurosci. **8**, 1 (2014). Article 370
7. Tran, A.V., Chae, C.-J.: Bandwidth-efficient PON system for broadband access and local customer internetworking. IEEE Photonics Technol. Lett. **18**(5), 670–672 (2006)
8. Seo, Y.-J., Kim, J.G., Kang, M.: Dynamic bandwidth allocation algorithm for a PON system with local customer internetworking. In: International Conference on Optical Internet (COIN), pp. 1–2 (2008)

9. Kim, J.G., Chae, C.-J., Kang, M.-H.: Mini-slot-based transmission scheme for local customer internetworking in PONs. ETRI J. **30**(5), 282–289 (2008)
10. Kubo, R., Kani, J., Fujimoto, Y., Yoshimoto, N., Kumozaki, K.: Adaptive power saving mechanism for 10 Gigabit class PON systems. IEICE Trans. Commun. **E93–B**(2), 280–288 (2010)
11. GPON power conservation, ITU-T Recommendations – supplement 45 (G.sup 45), May 2009
12. Mohammed, A.F.Y., Newaz, S.H.S., Uddin, M.R., Lee, G.M., Choi, J.K.: Early wake-up decision algorithm for ONUs in TDM-PONs with sleep mode. IEEE/OSA J. Opt. Commun. Netw. **8**(5), 308–319 (2016)
13. Newaz, S.H.S., Jang, M.S., Mohammed, A.F.Y., Lee, G.M., Choi, J.K.: Building an energy-efficient uplink and downlink delay aware TDM-PON system. Opt. Fiber Technol. **29**, 34–52 (2016)
14. Shi, L., Mukherjee, B., Lee, S.S.: Energy-efficient PON with sleep-mode ONU: progress, challenges, and solutions. IEEE Netw. **26**(2), 36–41 (2012)
15. Alaelddin, F.Y.M., Newaz, S.H.S., Lee, J., Uddin, M.R., Lee, G.M., Choi, J.K.: Performance analysis of TCP traffic and its influence on ONUs energy saving in energy efficient TDM-PON. Opt. Fiber Technol. **26**, 190–200 (2015). Part B

Performance Analysis of MANET Under Black Hole Attack Using AODV, OLSR and TORA

Fatin Hamadah M.A. Rahman and Thien Wan Au[✉]

School of Computing and Informatics, Universiti Teknologi Brunei, Jalan
Tungku Link, Gadong, Brunei Darussalam
fatinh.rahman@gmail.com, twan.au@utb.edu.bn

Abstract. Black Hole attack is one of the many attacks that can occur on a MANET. It works on the network layer by dropping all incoming packets instead of forwarding them to the destinations. MANET conveys communication in a multi-hop manner from the source node to the destination. But without any security means, the effects of such attack on MANET can disrupt the performance and operations of the network. This paper describes the effect of the attack on MANET that is using AODV, OLSR and TORA routing protocols with the introduction of IPSec protocol under the influence of Black Hole attack. The simulation of the attack is achieved using Riverbed Modeler Academic Edition. Based on the Black Hole attack simulations, MANET suffered the lowest throughput when it is using TORA and the highest using OLSR routing protocol.

Keywords: MANET · Black Hole attack · IPSec · OLSR · TORA

1 Introduction

A network attack can be classified as insider or outsider. When an attacker, which is not part of the network, performs the attack outside of the network, the attack is considered as an outsider attack. Meanwhile, in an insider attack, the attacker happens to have penetrated inside the network by means such as impersonation. These attacks can be further classified into active or passive attack. In active attack, the attacker performs conspicuous disruption to the data in the communication by means such as data modification. Meanwhile in passive attack, the attacker does not disrupt the transferred data between nodes, but mostly eavesdrop on the network traffic to gather valuable information instead. Detecting passive attack is more complex compared to the active attack as the passive attack is usually done in discreet where damage is not transparent. These attacks prove that strong encryption is needed in order to prevent them from occurring. Examples of such common attacks on MANET are Jamming attack, Repudiation attack, Wormhole attack and Black Hole attack. The focus of this study is directed towards the understanding of Black Hole attack on MANET using various routing protocols. The upcoming Sect. 2 describes the background study on Black Hole attack and Sect. 3 elaborates on the simulations carried out to perform the attack. Section 4 presents the results obtained from the simulations. Further findings are discussed in Sect. 5 and finally, Sect. 6 concludes the whole study.

© Springer International Publishing AG 2017
S. Phon-Amnuaisuk et al. (eds.), *Computational Intelligence in Information Systems*,
Advances in Intelligent Systems and Computing 532, DOI 10.1007/978-3-319-48517-1_18

2 Background Study

The Black Hole attack is one of the many routing protocol attacks occurring on the network layer where the authentication aspect of the network is at stake. In Black Hole attack, the attacker node deceives and advertises other node that it has the shortest route to reach a destination [1]. From Fig. 1 for instance, node D wants to send a packet to node B and begin its route discovery process by sending the RREQ packet to its neighbors. Node A responds with RREP packet to node D, claiming that it has the shortest route to the destination. Therefore, node D will ignore other RREP packets and start sending packet to the malicious node A instead. This node A then drops any of the incoming packets similar to that of a black hole in the solar system. Lowering the buffer size of the malicious nodes can make them appear to have the shortest route. Since the buffer size of the malicious nodes in this attack is lowered from the default size, it can also result in packet loss.

Other studies have been conducted to see the effects of the attack on MANET. In one study, a throughput comparison was made between the AODV and OLSR routing protocols under the Black Hole attack [2]. Their study found that the OLSR still gained the highest throughput value compared to AODV, regardless with the introduction of the attack. They have also doubled the network size, from 16 to 30 nodes, and the result remains the same in this scenario as well. Singh also obtains similar findings in their study [3]. They have concluded that detection of the attack is difficult, by looking at the performance of the network. Prevention measures such as encipherment, and digital signature were also suggested to prevent such attack from occurring. Meanwhile, the study by Jasvinder and Sachdeva has simulated the Black Hole attack on MANET with an increasing number of malicious nodes [4]. All the nodes are using the AODV routing protocol. As the malicious nodes increase, the performance of the network degraded in terms of delay and throughput. All these studies did not imply a security mechanism that was not added in their studies. Since Black Hole attack occurs on the network layer, this layer should be protected. One way that can help to achieve this is

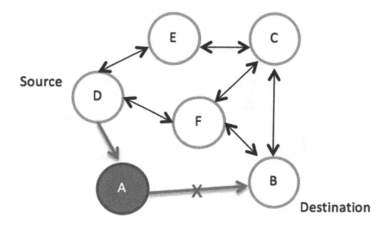

Fig. 1. Black Hole attack

by using IPSec protocol that works on the network layer and furthermore, not one of the studies conducted have applied the IPSec protocol specifically.

IPSec protocol is not a single protocol, but it is rather a protocol suite that encrypts a packet data and the header information [5]. Although the use of IPSec is optional in IPv4, the usage of IPSec protocol is made mandatory in IPv6 whereby using 128-bit address in IPv6 enables easy deployment of the IPSec. It has two modes of operation, namely the transport mode and tunnel mode. In transport mode, authentication and encryption only occurs at the payload of the IP packet. Meanwhile, the tunnel mode provides authentication and encryption of the entire IP packet. It is an open standard protocol that contains other subsequent components. Two of IPSec's main protocols are Authentication Header (AH) and Encapsulating Security Payload (ESP). The AH provides data authentication and encryption for packets between two systems, but it does not provide data confidentiality of packets. Whereas the ESP is a security protocol that provides encryption of the IP packet where it authenticates the inner IP packet and ESP header [6].

3 Simulation

3.1 Network Layout

Figure 2 below shows the design layout of the simulations. The nodes are placed randomly in a predefined area. Details of the components of the simulations are presented in Table 1.

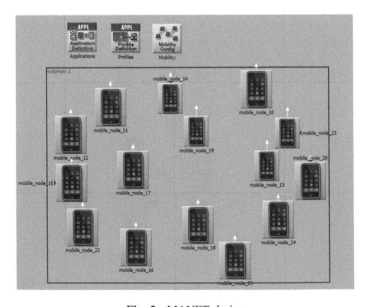

Fig. 2. MANET design

Table 1. Node attributes and values

Attribute	Value
Routing protocol used in Simulation 1	AODV
Routing protocol used in Simulation 2	OLSR
Routing protocol used in Simulation 3	TORA
No. of normal nodes	14
No. of malicious nodes	1
IP Addressing	IPv4
Model	wlan_iphone
Standard	IEEE 802.11a
Data rate	54 Mbps
Trajectory	Vector
Simulation area	500 m × 500 m

3.2 Parameter Configurations

The nodes represent the individuals that make up the network. Fifteen nodes are arranged randomly where they all share the common configurations. In the attack, a different routing protocol in three separate but identical simulations is used to see how they perform against one another.

For the IPSec configuration, the IP Security demand is used between all the nodes in a full mesh manner. The transport mode is preferred, as communication is only within peer-to-peer. The Destination and Source Port of the IP Security demand is set to 'voice' and the value '8' denotes Best Effort that is set for the Type of Service. This IPSec protocol is applied between all the nodes in a full mesh manner in order to get more accurate results. The malicious node is IPSec-enabled, thus it can send and receive the IPSec packets from the normal nodes. In the Application Definitions Attribute, the VoIP application is created where the voice is set to use IP telephony. To define the type of profile that is applied to the nodes, the Profiles Attribute is used. The 'User' profile is created and configured where the Application is also integrated into this profile. To enable mobility, the nodes' trajectory is set to vector and the Random Waypoint Model is used for the mobility model.

3.3 Black Hole Attack Configurations

In order for the malicious node to advertise itself and deceive other nodes that it has the shortest path, it needs to be able to show its availability of fresh routes. To accomplish that, the malicious node must have a low buffer size. The buffer size relates to the maximum size of the higher layer data buffer in bits. Once the buffer limit is reached, the data packets arriving from higher layers are discarded until some packets are removed from the buffer so that the buffer has some free space to store these new packets. Hence, the malicious node must have a low buffer size to show that it is always available to process other nodes' requests and later on drops the request packets it has received, to ensure the validity of the attack. The buffer size of the malicious node is

Table 2. Routing protocols with different buffer size for malicious node

Buffer size (bits)	AODV	OLSR	TORA
32000	X	X	X
16000	✓	X	✓
12000	✓	X	✓
8000	✓	X	✓
4000	✓	✓	✓
1000	✓	✓	✓
800	✓	Simulation aborted	✓
600	Simulation aborted		✓
500			Simulation aborted

Legend: X - No packet drop   ✓- Packet drop occurs

decreased gradually from the default 256000 bits to see from which value the packet drop will occur. In this study, the starting point of the packet drop varies with the three routing protocols. The packet drop in AODV and TORA starts at approximately 16000 bits and for the OLSR protocol it occurs from around 4000 bits. However, using buffer size lower than a certain value would abort the simulation and different routing protocols have different minimum buffer size that it can handle. Having a buffer size of below 500 bits could be too low for a node to function in the simulations. Theoretically, a typical packet header is around 40 bytes minimum, which is equivalent to 320 bits. A whole packet, including the header and the payload could be more than that. Therefore, to ensure consistency in all 3 routing protocols in the attack, the most reasonable buffer size range for the Black Hole attack is between 900 to 4000 bits as shown in Table 2 below. From that range, buffer size of 1000 bits is selected for the malicious node.

As a precautionary measure to prevent misconfiguration and to ensure that the attack is valid, the Discrete Event Simulation (DES) log is checked after every simulation run to ensure any warnings and errors are taken into consideration, thus improving the reliability of gathered results. Figures 3 and 4 have shown that the malicious node does indeed experience packet loss due to insufficient buffer capacity.

From the simulation, several metrics are obtained. The performance metrics are also used as a measuring tool to help with the analysis:

- *The throughput metric* represents the total number of bits forwarded from wireless LAN layers to higher layers in all WLAN nodes of the network

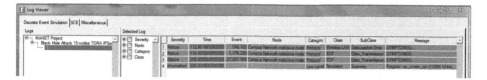

Fig. 3. DES log

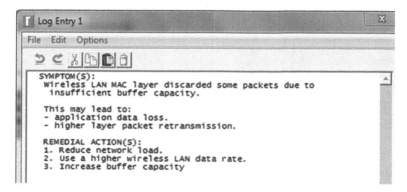

Fig. 4. Log entry for the malicious node

- *Delay* represents the end-to-end delay of all the packets received by the wireless LAN MACs of all WLAN nodes in the network and forwarded to the higher layer. This delay includes medium access delay at the source MAC, individual reception of all the fragments and transfer of the frames via AP if access point functionality is enabled.
- *Retransmission attempts* metric gives the total number of retransmission attempts by all WLAN MACs in the network until either packet is successfully transmitted or it is discarded as a result of reaching short or long retry limit.

4 Results

The scenarios ran for duration of 5 min where the average values of the performance metrics in terms of throughput, delay and retransmission attempts are examined. Figure 5 shows that the implementation of IPSec protocol has slightly improved the throughput of all 3 routing protocols, although the difference cannot be seen clearly for the OLSR with and without IPSec. As expected, the OLSR routing protocol has the highest throughput value and on the contrary, TORA has the lowest. Both AODV and OLSR have shown an increase in throughput value with the use of IPSec protocol. However, only the TORA routing protocol with IPSec has a slightly lower throughput in comparison to TORA without IPSec.

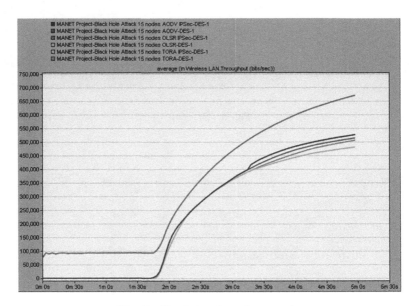

Fig. 5. Simulation throughput results

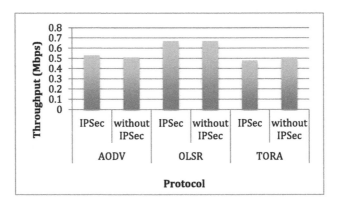

Fig. 6. Throughput results

The x-axis of graphs shown in Figs. 6, 7 and 8 represents the routing protocol being used i.e. AODV, OLSR and TORA, and whether or not the routing protocol is using IPSec for security. Meanwhile, the y-axis of the same Figs. 6, 7 and 8 represents the measured throughout the simulation in Mbps, delay in millisecond (ms), and retransmission attempt in number of packets respectively.

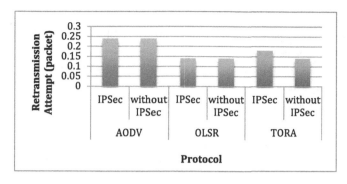

Fig. 7. Delay results

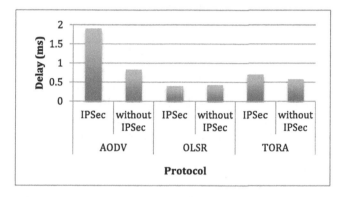

Fig. 8. Retransmission attempt results

5 Findings and Analysis

Further investigations are made on the throughput values of the three routing protocols towards varying buffer size. Recognizing the fact that the packet drops occur at 16000 bits for AODV and TORA, and 4000 bits for OLSR, as the buffer size decreases, the throughputs of the simulations with IPSec have shown slight increase or are similar to that in simulations without IPSec as shown in Table 3. While different protocols have

Table 3. Throughput results in Kbps based on varying buffer sizes

Buffer size (kb)	AODV (Kbps)		OLSR (Kbps)		TORA (Kbps)	
	No IPSec	With IPSec	No IPSec	With IPSec	No IPSec	With IPSec
16	625	525	775	690	550	525
12	625	550	775	690	530	525
8	570	580	775	680	500	500
4	514	525	775	775	504	505

Table 4. Performance comparison

Protocol	AODV	OLSR	TORA
Throughput (%)	+1.69	+0.03	−2.72
Delay (%)	+106.60	+2.23	+19.60
Retransmission attempts (%)	−0.55	+0.34	+81.82

Note: 1. A −ve sign means a decrease in value with the implementation IPSec

2. A +ve sign means an increase in value with the implementation IPSec

shown different reactions with varying buffer size values, Table 3 also shows that OLSR obtained the highest throughput value and TORA has the lowest throughput value with regard to the different buffer sizes. As the malicious node in Black Hole attack usually deals with RREQ and RREP messages with its surrounding nodes, therefore the network using OLSR protocol is least affected by the attack because OLSR focuses on using Hello messages between the nodes for route discovery. Furthermore, being a table-driven protocol, OLSR stores and updates the routing information in its routing table permanently. The networks using AODV protocol and TORA are mostly affected by the attack as they use RREQ messages for route discovery. Hence, this explains how OLSR has managed to maintain high throughput values. While AODV and TORA may use Hello messages as well, it is used for maintenance purposes.

The IPSec protocol is expected to improve the performance as the attack occurs at the network layer, where it is proven true in terms of the throughput of this attack. Although the results are not really significant, it provides enough evidence that the lower buffer size can help improve the simulations with IPSec. From Table 4, in terms of throughput, the use of IPSec has produced slight increase of 1.69 % and 0.03 % for AODV and OLSR. The OLSR routing protocol has shown better results with only 2.23 % delay increase in value with the implementation of IPSec, compared to the AODV and TORA with much higher increase of 106.6 % and 19.6 %. On the other hand, for the retransmission attempts of the TORA protocol has highly increased to 81.82 %, which could explain its lowly gained throughput.

Acknowledging that this is an insider attack, the malicious node is also considered to be a part of the network. The alterations of IPSec settings on the malicious node are not made, as the focus of the project is to see the effect of the attack. With the use of IPSec especially with the component of Internet Key Exchange that is responsible in ensuring end-to-end authentication, the malicious node is has to obey the protocols policy. Instead of dropping the packets, the malicious node forwards the packets to the destination despite having a small buffer size. Hence, the throughput is improved. The reduced buffer size explains the high delay, as the malicious node needs time to process the packet before it can work on a new packet to their respective destination.

Although IPSec works on the network layer, it does little and has minute control over the routing of the network. The actual routing protocols, in this case, AODV, OLSR and TORA are the ones that determine the routes towards the destination. After route towards the destination is established, only then the IPSec works to ensure data authentication, integrity and confidentiality of the data, depending on the AH and ESP protocols used in IPSec. Using IPSec to protect the network layer is not entirely effective to act as a defense as efficient routing protocols also play an important part to defer this type of attack.

6 Conclusions and Future Work

IPSec protocol can secure a network but the performance will most likely be compromised either in terms of throughput, delay, retransmission attempts or other performance metrics. The study has demonstrated that in terms of throughput, OLSR performs better under Black Hole attack compared to AODV and TORA routing protocols. Nonetheless, AODV performed worst in terms of delay and TORA performed the worst in terms of retransmission attempt. Different routing protocols have different starting points of packet drop with varying buffer sizes. The packet drop starts to occur at 16000 bits for AODV and TORA, and 4000 bits for OLSR. As the buffer size decreases, the throughputs of the simulations with IPSec have shown slight increase or are similar to that in simulations without IPSec. However, the usage of IPSec protocols this study are limited, thus we cannot clarify whether security goals of authentication, integrity and confidentiality are achieved. Therefore, in the future, further studies on IPSec in a network can be explored specifically to highlight whether the security goals can be achieved by using different combinations of the AH and ESP protocols.

References

1. Godwin, P.J., Srivinasan, R.: A survey on MANET security challenges, attacks, and its countermeasures. Int. J. Emerg. Trends Technol. Comput. Sci. 3(1), 274–279 (2014)
2. Ullah, I., Anwar, S.: Effects of black hole attack on MANET using reactive and proactive protocols. Int. J. Comput. Sci. Issues 10(3), 152–159 (2013)
3. Singh, H., Singh, G., Singh, M.: Performance evaluation of mobile ad hoc network routing protocols under black hole attack. Int. J. Comput. Appl. 42(18), 1–6 (2012)
4. Jasvinder, Sachdeva, M.: Effects of black hole attack on an AODV routing protocol through the using of Opnet simulator. Int. J. Adv. Res. Comput. Sci. Softw. Eng. 3(8), 657–664 (2013)
5. Eastom, W.: Computer Security Fundamentals. Pearson, New York (2012)
6. Forouzan, B.A.: Data Communications and Networking. McGraw-Hill, New York (2007)

Management Information Systems and Education Technology

Enhancement of Learning Management System by Integrating Learning Styles and Adaptive Courses

Ean Heng Lim[1($\boxtimes$)], Wan Fatimah Wan Ahmad[2],
and Ahmad Sobri Hashim[2]

[1] Department of Information Systems,
Universiti Tunku Abdul Rahman, Kampar, Malaysia
ehlim@utar.edu.my
[2] Department of Computer Information Sciences,
Universiti Teknologi Petronas, Teronoh, Malaysia
{fatimhd,sobri.hashim}@petronas.com.my

Abstract. Learning management systems (LMS) such as LattitudeLearning, BIStrainer, Blackboard, Google Classroom and Moodle are commonly adapted by many education institutions. Most of the existing LMS are focusing on assisting teaching and learning related matters and does not take consideration in regards to the differences that existed between each individual learners. The main problem is that learners have different motivation, cognitive traits, and learning styles. This paper is examining the effect of learning styles by integrating adaptive courses into the LMS to suit according to the learner's learning styles. The results revealed that learners exhibited different preferences in LMS environment based on different learning style. This paper focuses on taking account of the learner learning styles by incorporating adaptivity into the learning model. Based on this approach, the current Moodle based LMS has been implemented. By extending LMS with adaptivity, it will enable a support to teachers and learners. The research results are important to ensure that the courses include the features which fit to different learning styles. This will further identify the needs and characteristics of learners by responding to the learners and present them with the enhanced adaptive LMS based on the learners' needs.

Keywords: Learning Management System (LMS) · Personalization · Learning style detection · Adaptivity

1 Introduction

The way or method a learners used in learning has been a concerned for researchers for many years [2]. Many researches has been carried out about learning methods and its influence towards learners in the classroom and prior knowledge is one of the main and reliable individual difference forecasters of accomplishment [3]. The issue with the current available LMS is that most of it offers learner the similar learning materials. Personalized LMS attempt to help learners to adapt to the learning content to suit the needs of different learners learning styles. Currently, majority of the LMS focusing too

© Springer International Publishing AG 2017
S. Phon-Amnuaisuk et al. (eds.), *Computational Intelligence in Information Systems*,
Advances in Intelligent Systems and Computing 532, DOI 10.1007/978-3-319-48517-1_19

much on the content development and ignore the learners learning styles. Adaptivity should be introduced to increase the effectiveness of LMS. This paper discusses the adaptive LMS based on the learner's preferences identified through a set of questionnaire. The reason LMS is gaining popularity in educational institutions or in corporate world is because it allows the display of information in any available multimedia medium (graphic, animation, sound, text or video), on any subject, anytime and anywhere [1]. Due to the increasing range of learners, technology advancement and rapid changes in learning tasks has led to significant challenges. All of these have led to the complexity of defining the context of use LMS more than before [15].

1.1   Learning Management System in General

LMS has been evolving rapidly in terms of educational contents, technological resources and interaction methods which is integrated into a LMS application. A lot of LMS have been developed and used in education institution which includes [16, 17] Blackboard and WebCT. There are also education institution which adopt open source LMS [18] such as Moodle, dotLRN and Sakai. The latter is frequently adopted by education institutions due to cost factor. The advantage gained in this cost factor is this type of technology which can assist in learning and teaching can be acquire without paying any license fees [19]. The availability of a variety of LMS in the market makes education institution difficult to select them. All of it is down to making a choice that will satisfy partly or even all of its requirements [20]. However, it would be a difficult task in choosing any of these LMS to fit well in a structured e-learning implementation plan and strategies of an education institution [21, 22].

From the statistics above, it clearly indicate that the usage of LMS are gaining more and more popularity. Since the LMS will be able to adapt to a learner, it may also be called as personalization. Personalized LMS require a preemptive learning strategy in order to control the learning content, speed and scope. Studies on learning styles are driven by theories which claim that different learners have different preferences to adapt when learning. So, assumption can be made that by integrating learning styles into the learning environment, it will make the learning process much easier for learners and at the same time enhances their learning efficiency.

In contrast, if learning styles are not in the favor of the learner by the learning environment, then the learners may find it difficult throughout learning process. Hence, personalized LMS address this particular issue. Personalized LMD purpose is to provide learners with courses which will fit based on the learners learning styles. There are also some drawbacks though supporting personalization will bring huge advantage to LMS. For example, personalized LMS lack of integration and supports only a few functions of web-enhanced education [27]. Developing a personalized LMS and incorporate learning styles into the learning processes is a difficult and costly process which involves many participants in the process.

2 Problem Statement

First and foremost, the researchers are looking for a method to identify the different learning styles of a learner in order for the system to assign the course material accordingly. Therefore, a learner learning styles need to be identified first in order for a LMS to detect the learning style itself. In order to identify learning styles of a learner, an automated approach based on the survey device from Felder-Soloman's Index of Learning Styles is proposed. The various learning styles of learners is compared and assessed based on a literature review approach. This is to categorize the different learning preferences within the learning style dimensions. Based on the findings, an automated process has been integrated into the LMS which enable the identification of a learner learning styles as the current existing LMS does not take into consideration of learner's characteristics but treat all learners in an equal manner.

When the learning styles of a learner are identified, the LMSs can be enhanced to help them produce and deliver adaptive courses. In this research, an exploratory study is conducted to see the effect of integration of learning styles into a LMS. It is to serve as a commonly accepted framework or template for LMS design and usage in teaching and learning.

3 Literature Review

Learning strategies are normally used to support provisional success where learners apply in a certain condition. However, learning strategies may vary from time to time and condition. If learners are frequently utilizing these learning strategies then learning styles can be detected from these strategies [5]. Based on [6], a learning strategy is define as "the way a learners chooses to tackle a specific learning task in the light of its perceived demands" and learning style "as a broader characterization of a learner's preferred way of tackling learning tasks generally". Moreover, [6] argued that different learning styles are the fundamental of learning strategies.

3.1 Learning Styles

Learning styles are referring to the approaches and methods which help to ease learning process. Generally, learning styles theories are established based on the hypothesis that different learner has different learning style. There are many researches carried out that are looking to identify and categorize different learning styles. [4] has group five families of learning style models. First family includes the visual, auditory, kinesthetic, and tactile where it affects the learning styles and preferences of learners. The second family imitates features of the cognitive structure which may include the patterns of abilities. The third family indicates that to a learner learning styles are closely related to a learner personality. The last family stated that learning styles is an enhancement from learning preferences. Learning styles theories propose that all individuals can be classified according to a learning style.

The issue here is that there is lack of researches which is able to provide appropriate combinations of features and functions in a LMS and learning styles. The features in a LMS can be utilized and manipulated depending on the difference of teaching strategies in order to match with the difference of learning styles. For instance, an online assignment submission can be used in multiple ways. It may be used to submit a written assignment or learner may use it to upload the finish assignment as a reference for other learner. This type of method suit well for a learner with active learning style. Other features such as online chat may also be utilized to give a platform to learners to interact with one another. This type of communication is closely resemble to the structure of communication with the corresponding chat is a suitable method for learner who has a tendency on sequential style.

Some review on the existing learning styles are conducted and presented regarding the present adaptive LMS with learning styles.

3.2   Learning Styles Model by Kolb

The Kolb Learning Style Inventory is based on a four stage experiential learning theory. The four-stage cycle includes:

   i. create a solid information which serves as the fundamental
  ii. observation and reflection
 iii. a "theory" for actions which can be determined
 iv. the theory serves as a guide to create new experiences [28].

In stage one, in order to create solid information, the learner are required to involve in a new experience. In stage two which is the observation and reflection, the learner observe the others learners activity or improves the observations on the new experience. In stage three, abstract conceptualization, the learner creates theories to describe the observations. Lastly in stage four, it requires the learner to use the concepts in order to answer the problems or develop assessments.

To conclude, the Kolb model emphasize on the way a leaner identifies and process the information [29]. It evaluates the learner's abilities together with two ranges which are the creation of solid information to observation and reflection, and abstract conceptualization to reflective observation.

3.3   The Felder-Silverman Learning Style Model (FSLSM)

Felder-Silverman Learning Style Model (FSLSM) was developed by Richard Felder and Linda Silverman in 1988. This model concentrates on the learning styles aspects of the learners. It divides learners in four main categories. The Sensory/Intuitive and Visual/Verbal are referring to how learners observe or see information. The Active/Reflective and Sequential/Global are referring to the learners understanding [7].

The Felder-Soloman's Index of Learning Styles (ILS) instrument contains forty four questions. Each category contains eleven questions and it can be easily done on the web [8].

[9] presented the FSLSM's differences to other learning style models including describing learning style in more detail, representing also balanced preferences, and describing tendencies. [10–12] said that their reliability and validity data verified an argument that the ILS is a proper method to assess learning styles but both studies are suggested that the continuing research on this is needed. [13] concluded that the ILS is the best method to be utilized in order to allow learners strengths to be evaluated in relation to their learning preferences rather than comparing with other individuals.

4 Research Methodology

The adaptive LMS will be designed based on three main steps. The first step is to create and define the main phases in developing the adaptive LMS. The second step will be categorizing the learners based on their learning styles. Lastly, adaptivity will be applied into the course contents. A case study is conducted to provide an assessment on learning styles and integration results into LMS.

A questionnaire will be established to have a better understanding of the learners' habits of learning. Through this questionnaire, the researchers recognize the learning styles. Thereafter, the learners will be provided different learning materials accordingly. By analyzing the questionnaire, the researchers may find several patterns where different learners may have different preferences when using LMS. The results is vital to ensure that the LMS is able to produce an adaptive courses that which is suitable to accommodate different learning styles which can be divided into extravert, thinker, judger and intuitive.

In an effort to find out an efficient, automatic method for identifying learners' learning styles, the researchers utilize the literature-based approach that study and analyze the learners' behaviors in order to generate the teaching strategy. FSLSM model is often used as it has been proved to be effective in most adaptive LMS [14–19]. The researchers will construct its own LMS to serve the study and to be used conveniently in the future.

Thirty learners are involved in the lab test to test the adaptive LMS conducted. The initial login requires the learners to answer the survey based on the Felder-Soloman's Index of Learning Styles (ILS) instrument. The system will then determine the learners belong to which group of learning style before presenting them the appropriate learning materials. By identifying the learning styles of learners, the researchers will be able to choose the relevant type of teaching strategies and the most suitable learning materials based on the following teaching strategy.

Interviews are performed after the lab test process. A survey tool is developed before using in the main phase of this research. Demographic factors will also be considered. All these proposed ethnography methods purpose is to verify the current levels LMS and identify the problems faced by users when using the LMS system.

In the adaptive LMS course developing phase, the researchers will be testing the effectiveness and features before making an assessment to measure the efficiency of the course before adaptation. Overall this test will be divided into three category which is the pre-test, post-test, and survey to observe the learner's attitudes towards the adaptivity of courses. Learners will be divided into experimental and control group where both groups have to take the pre-test. Lastly, the post-test will take place to see whether there are substantial variances existed by the experimental and control group.

An independent t-test and paired sample t-test will be conducted to verify the difference between experimental and control groups in the pre-test and post-test.

5   Discussion

An exploratory study is conducted to provide an assessment on learning styles. It is also to seek additional signs of relationships between the four dimensions of FSLSM and adaptive LMS. A good LMS system with adaptive features can be an added advantage to assist the learners in learning process as it may guide them through an adaptive course by presenting material suitable for learners [30].

Based on the results collected, it shows that both active and reflective appeared to have no substantial differences in the experiment conducted. It was confirmed by using the multi-variate one way between the subject analysis of variance. The results disclose that there is no substantial personality effect was found. The researchers can conclude based on the experimental results where it indicates that the personality alone will have no substantial effects on learning performance when the learners were using a traditional LMS.

On the other hand, the reflective type of learner seems to be using their time equally compared to the active type of learner in the learning course. In other words, the former tend to be progressing in a more continuous manner. The reflective learner also shows that they become faster when utilizing the adaptive LMS whereas the active learner becomes slower which may demonstrate a less efficient process.

For the other two dimensions, it shows no sign of substantial correlations.

This suggests that the personality effect could be more significant in adaptive e-learning systems than the traditional e-learning system. The results of the exploratory study again confirm the existence of relationships between learning styles and working memory capacity, even when using a small sample size. Therefore, results endorse the conduction of a study with a larger sample size. A larger sample size yields to more reliable results by using more representative data. Furthermore, it makes more detailed analyses possible.

References

1. Jun, W., Gruenwald, L., Park, J., Hong, S.: A web-based motivation-supporting model for effective teaching-learning. In: Fong, J., Cheung, C.T., Leong, H.V., Li, Q. (eds.) ICWL 2002. LNCS, vol. 2436, pp. 44–55. Springer, Heidelberg (2002)
2. Radwan, N.: An adaptive learning management system based on learner's learning style. Int. Arab J. e-Technol. 3(4), 228–234 (2014)
3. Pinto, J., Ng, P., Williams, S.: The effect of learning styles on course performance: a quantile regression analysis. Northern Arizona University the W. A. Franke College of Business, Working Paper Series—08(2), pp. 1-14 (2008)

4. Jonassen, D.H., Grabowski, B.L.: Handbook of Individual Differences, Learning, and Instruction. Lawrence Erlbaum Associates, Hillsdale, New Jersey (1993)
5. Coffield, F., Moseley, D., Hall, E., Ecclestone, K.: Should we be using learning styles? what research has to say to practice. Learning and Skills Research Centre, University of Newcastle upon Tyne, London (2004)
6. Pask, G.: Styles and strategies of learning. Br. J. Educ. Psychol. **46**, 128–148 (1976)
7. Entwistle, N.J., Hanley, M., Hounsell, D.: Identifying distinctive approaches to studying. High. Educ. **8**(4), 365–380 (1979)
8. Felder, R.M.: Learning and teaching styles in engineering education. J. Eng. Educ. **78**(7), 674–681 (1988)
9. Soloman, B.A., Felder, R.M.: Index of Learning Styles (1997). http://www.engr.ncsu.edu/learningstyles/ilsweb.html
10. Brown, B.L.: Learning Styles and Vocational Education Practice. ERIC Clearinghouse on Adult, Career, and Vocational Education, Columbus (1998). http://www.calpro-online.org/eric/pab.asp
11. Zywno, M.S.: A contribution to validation of score meaning for felder-soloman's index of learning styles. In: Proceedings of the 2003 American Society for Engineering Education Annual Conference and Exposition, pp. 23–51 (2003)
12. Livesay, G.A., Dee K.C., Nauman E.A., Hites, L.S.: Engineering learners learning styles: A statistical analysis using felder's index of learning styles. In: Proceedings of the American Society for Engineering Education, Montreal, Quebec, pp. 2144–2151 (2002)
13. Felder, R.M., Spurlin, J.: Applications, reliability and validity of the Index of Learning Styles. Int. J. Eng. Ed. **21**(1), 103–112 (2005)
14. Van Zwanenberg, N., Wilkinson, L.J., Anderson, A.: Felder and silverman's index of learning styles and honey and mumford's learning style questionnaire: how do they compare and how do they predict? Educ. Psych. **20**(3), 365–381 (2000)
15. Zaharias, P.: Usability and e-learning: The road towards integration. ACM eLearn Mag. **2004**(6), 4 (2004)
16. Leeming, D., Pitia, P., Haccehy, G.: Introduction to Learning Management System. Workshop Report, Distance Learning Centers Project, Pfnet Internet Centre, Honiara (2005)
17. Paulsen, M.F.: Experience with learning management system in 113 european institutions. Educ. Technol. Soc. **6**(4), 134–148 (2003)
18. Aberdour, M.: Open Source Learning Management Systems. An Epic White Paper (2007)
19. Cavus, N., Uzunboylu, H., Ibrahim, D.: Assessing the success of learners using a learning management system and together with a collaborative tool in web based teaching of programing languages. J. Educ. Comput. Res. **36**(3), 301–321 (2007)
20. Nielsen, J.: Finding usability problems through heuristic evaluation. In: Proceedings of the ACM CHI 1992 Conference (Monterey, California), pp. 373–380 (1992)
21. Lin, H.-F.: Measuring online learning systems success: applying the updated delone and mclean model. CyberPsychol. Behav. **10**(6), 817–820 (2007)
22. Rubin, J.: Handbook of usability testing: how to plan, design, and conduct effective tests. Wiley, New York (1994)
23. Hollingsed, T., Novick, D.G.: Usability Inspection methods after 15 years of research and practice. In: 2002 SIGDOC International Conference on Web-Based Learning, Texas, USA, pp. 44–55. Springer, Berlin (2007)
24. Zhang, D., Zhou, L., Briggs, R.O.: Instructional video in e-learning: assessing the impact of interactive video on learning effectiveness. Inform. Manage. **43**(1), 15–27 (2006)
25. Hiltz, S.R., Turoff, M.: What makes learning networks effective? Commun. ACM **45**(4), 56–59 (2002)

26. Paramythis, A., Loidl-Reisinger, S.: Adaptive learning environments and e-learning standards. Electron. J. e-Learn. **2**(1), 181–194 (2004)
27. Brusilovsky, P.: KnowledgeTree: A distributed architecture for adaptive E-Learning. In: Proceedings of 13th International World Wide Web Conference, WWW 2004, May 17–22, pp. 104–113. ACM Press, New York (2004)
28. Zanich, M.L.: Learning styles/teaching styles. Unpublished manuscript, Indiana University of Pennsylvania, Teaching Excellence Center, Indiana (1991)
29. Kelly, C.: The theory of experimental learning and ESL. Internet TESL J. **3**(9), 1–5 (1997)
30. Hauger, D., Köck, M.: State of the art of adaptivity in e-learning platforms. In: 15th Workshop on Adaptivity and User Modeling in Interactive Systems, held in the context of Lernen-Wissensentdeckung- Adaptivität 2007 (LWA 2007). University of Hildesheim, Hildesheim, Germany, pp. 355–362 (2007)

InterviewME: A Comparative Pilot Study on M-Learning and MAR-Learning Prototypes in Malaysian English Language Teaching

Lim Kok Cheng[1(⊠)], Ali Selamat[2], Rose Alinda Alias[2],
Fatimah Puteh[2], and Farhan bin Mohamed[2]

[1] Department of Software Engineering, Universiti Tenaga Nasional, Jalan
IKRAM-UNITEN, 43300 Kajang, Selangor, Malaysia
`kokcheng@uniten.edu.my`
[2] Faculty of Computing, Universiti Teknologi Malaysia, 81310 Johor Bahru,
Johor, Malaysia
`{aselamat,alinda,m-fatima,farhan}@utm.my`

Abstract. The Malaysian English Language Teaching (ELT) of English as a Second Language (ESL) has been a long debated topic from implementation of methodologies, strategies and models to the tools used to deliver English education to Malaysian learners. This paper highlights the two emerging tools namely Mobile Learning (M-learning) and Mobile Augmented Reality Learning (MAR-learning) in ELT. The aim of this study is to conduct a comparative study verifying if MAR-learning is significantly more motivating and satisfying than M-learning in ELT. This paper will first present the current trends in both technologies, followed by the development of the first MAR-learning prototype before being evaluated comparatively with the current existing M-learning application. The last section of this paper will highlight the results of the study in compliance with the generated hypotheses.

Keywords: English language teaching · English as second language · Mobile learning · Mobile augmented reality learning · Motivation satisfaction

1 English as a Second Language in Malaysia

Malaysia is a country of diverse race, culture, practices, languages and accents where as according to the Government of Malaysia's Official Portal in [1], Malaysian English (ME), also known as Malaysian Standard English, is a form of English derived from British English. Malaysia was coined earlier as a richly multilingual country that can be categorized generally as diglossic or polygossic [2]. Despite having multiple ethnic languages, ME is widely used in business, along with Manglish, which is a colloquial form of English with heavy Malay, Chinese, and Tamil influences [1].

In a study conducted by [3], Malaysian pre-university students have very high motivation and positive attitudes towards learning English and that they are more instrumentally motivated. However, despite the positive motivation, studies done by [4–7] on the other hand has contradicting results on English proficiency among Malaysians especially in communication, speaking and pronunciation. Although being

© Springer International Publishing AG 2017
S. Phon-Amnuaisuk et al. (eds.), *Computational Intelligence in Information Systems*,
Advances in Intelligent Systems and Computing 532, DOI 10.1007/978-3-319-48517-1_20

one of the highest ranked countries in English proficiency, results has also shown that Malaysian graduates do not have English abilities up to the current industrial expectations. English is recognized as an important second language in view of its status as the lingua franca of the world, essential for economic advancement and international communication [8]. The discussion of declining English standards revealed varying viewpoints such as difficulties and reasons that students faced in learning the four language skills i.e. speaking, listening, reading and writing; and the lack of confidence, which hampered their language improvement [9]. Musa, Koo and Azmn in [10] added on that the overall picture is discouraging and is indicative of the need to change the ways in which English language literacy is taught to Malaysian learners. In learning a second language or a foreign language, research has established that it is utmost important that learners receive maximum support in terms of supportive and conducive learning environment as well as adequate, meaningful language experience [10]. Syllabuses are delivered to students in a variety of methodologies and techniques in conventional classroom condition and also through the benefits of Information and Communication Technology. In recent years, ICT has contemporarily progress from supplementary to major academic role, especially in blended learning, where many researches have shown proofs of effectiveness in using the tools of ICT [8]. However, the exploration and utilization of ICT in Malaysian ELT is still at the very beginning and is foreseen to prosper within many years to come.

1.1 Listening and Speaking

Speaking and Listening is regarded as the two parameters with huge proficiency differences. In a study by [6], the author shows that fresh graduates scored lowest in speaking compared to other components of English skills. The result is perhaps co-related with the current policy of ELT in Malaysia where the 'literacy' aspect seems to be focusing on the reading and writing skills. This is because emphasis of these skills is measured in the national examinations [7]. It is of no surprise that speaking and listening have been a challenge in Malaysia. From the findings by Wahi in [11] on selected multi-racial students in Malaysia, considering their socio-economic background and rural settings where English was used minimally compared with their respective mother tongue dialects, the students declared that conversing in English was regarded as "odd" and "abnormal" in their domestic contexts [11]. In [11], the author mentions that in a non-native English-speaking environment, it is difficult for students who are non-native speakers of English to speak it accurately and fluently. That difficulty is compounded by lack of exposure to good models of English and opportunities to use English [11, 12].

Another study presented by [10], which highlighted one more reason for speaking and listening to be more of a challenge for Malaysian students is due to classroom practices that are mainly characterized by answering reading comprehension questions and essay writings with limited listening and speaking exercises. Realizing the current needs in ELT, Juhary in [13] mentioned that Malaysia has come to a point where second and third languages become part of the requirements to be employed especially in the multinational and international companies. Therefore it is within desperate conditions for competent graduates to acquire above average English communication skills, aligned with the study of Ngah et al. in [14] that highlights the importance of communicative competence among graduates to secure employment upon graduation.

Listening and speaking are two English skills directly related to the Willingness to Communicate (WTC) in English. As mentioned earlier, works by [11] shows that Malaysian student generally lacked in WTC due to cultural background. Adding to the findings of [11], [10] also highlights the unwillingness to communicate in English as one of the major causes of limited English proficiency among Malaysian learners. Not only lacking in WTC using English, [10] also discusses the lack of motivation and high anxiety to learn English. Relating to findings of [10], [15] presented results showing that language learning communication strategies directly affect motivation, self-perceived communication competence, and WTC in English. Therefore, it can be symmetrically assumed that self-perceived communication competence here may co-relate with motivation level and anxiety of English learners.

Therefore, technologies such as M-learning and MAR-learning may intervene the learning strategies of listening, speaking and WTC. Considering the benefits of these technologies mentioned above, it would be a broad research area motivated to improve English listening and speaking performances by blending technologies, pedagogical revision and ELT strategies.

1.2   Mobile Learning and ELT in Malaysia

With the fastest growing web community to be mobile visitors, the development of educational technologies recently has tended to be mobilized, portable, and personalized [16]. Mobile learning (M-learning) is also an appeared education model, which can be very beneficial for students with providing the opportunity of education independent of time and environment [17]. Although e-learning has much more advantages than traditional education methods, some deficiencies of its own have lead science world to new pursuits such as the development of mobile technologies and the need for movement of the technology in education to new dimensions have revealed the new notion M-learning [17]. Coining M-learning as here and now learning, [18] introduce a framework explaining M-learning through three principles: engaging, authentic and informal. More and more scholars believe that M-learning will be beneficial through learning anywhere and anytime in future education [19–23]. Mobility and spontaneity that mobile devices offer are observed in the present study. Because of the two features, mobile participants could engage themselves in reading online material ubiquitously, either on or off campus [24].

Considering immense positive remarks and proofs on the effectiveness of M-learning, it is not surprising that ELT in Malaysia progress in a parallel fashion with M-learning. Soleimani, Ismail and Mustaffa in [25] for instance show evidence of positive perception of Mobile Assisted Language Learning (MALL) among a local university's post-graduates in ESL. From a study done by [26], who reviewed the works done by 10 different researches on Malaysian ELT or Mobile Assisted Language Learning (MALL), it is concluded that research area in Mobile related ELT is fast moving, but is still in embryonic stage due to ample rooms for diversity in the research trends.

Despite the positive measures of using mobile technology in ELT, there are several factors that need to be analyzed further considering the blending the technology with English language pedagogy. Some drawbacks highlighted in [27], where the distractions associated with mobile phone usage in an educational setting are also of concern. For example, the ringing of mobile phones in public and at inappropriate times can be a nuisance and an annoying distraction to others [27]. Other challenges highlighted would be related to the ergonomics and affordances of mobile devices including greater difficulty in reading text, presenting graphical information and increased challenges in interactivity with the mobile interfaces [27–29].

1.3   Mobile Augmented Reality in ELT

Mobile Augmented Reality Learning (MAR-learning) is the latest technology combining the mobile criteria of M-learning and visualization effectiveness of (AR-learning). According to [30], wireless mobile devices, such as smart phones, tablet PCs, and other electronic innovations, are increasingly ushering AR into the mobile space where applications offer a great deal of promise, especially in education and training. However, even from an international context, little research has been conducted using MAR-learning in ELT. Liu in [31] for example has explored the combination of AR M-learning and ubiquitous (U-learning) in an English learning environment named HELLO (Handheld English Language Learning Organization). In an experiment done on 3 teachers and 64 seventh grade students comparing HELLO and traditional learning methods, [31] found that students HELLO performed better in listening and speaking. Liu and Tsai in [32] applied global positioning and AR Techniques in mobile assisted English Learning in their study. In an experiment involving 5 undergraduate participants, Liu and Tsai in [32] discovered that mobile AR assisted the participants with English vocabulary and expressions needed for descriptive writing. He et al. in [33] has attempted to use AR technology to design and develop mobile-based English ELT software. In an experiment involving 40 pre-school children, He et al. concluded that mobile-based AR learning software is helpful to students who are non-native speakers for learning vocabularies [33].

Other studies related to mobile AR includes scholars' works in general education. Fitzgerald et al. in [34] suggested a taxonomy of classifying AR in mobile environment which can be used to categorized different research aspects in mobile learning. However, despite many positive feedbacks from mobile AR studies, there are still many challenges and research gaps to be explored. There are two major areas, highlighted by [34] in mobile AR, which consists of technical challenges and pedagogical challenges. Pedagogical challenges in details highlight interesting facts where study has shown conflicts between technology experience and learning experience, which renders learning objectives are altered to fit around an AR device limitations [34]. Furthermore many results from the above mentioned studies did not highlight clearly or at least normalize biases on a technological "wow" factor which makes learning experience engaging and interesting, but not the content of the knowledge itself. The works of [35, 36] have highlighted the limitation of "wow factor". This is definitely a challenge where the interest of a user can be only temporary and might defeat the purpose of the learning experience, which is to achieve understanding of knowledge rather than the excitement of the technology interfaces. It is therefore still a big question if MAR-learning will be worth the technological investment despite having more properties as compared to M-learning.

2 InterviewME MAR-Learning Application Prototype

Deriving from the literature from the previous section, M-learning and MAR-learning has been two of the emerging technologies in ELT. This research therefore would like to measure suitable technology by conducting a pilot experiment comparing M-learning and MAR-learning.

In order to narrow down the context of the content for this pilot test, a major chapter in the university's English Communication course called "interviews" is selected as a test bed content for this test. This chapter is selected due to the component of the content itself that is designed to improve ESL in speaking, listening and WTC. While M-learning for this chapter is already available via the current MylinE server [37] (Fig. 1), MAR-learning has never been existed in Malaysian ELT within the extent of current literatures.

Therefore an MAR-learning prototype named "InterviewME" is developed for the purpose of this comparative study. In order to develop the MAR-learning prototype, elements of English learning strategies from the Oxford's Strategy Inventory for Language Learning (SILL) has been adapted [38]. Shah et al. [38] has conducted a study by examining the Language Learning Strategies (LLS) of English for specific purpose students at a Malaysian public university. The study interestingly yields results of different Oxford SILL used by sampled Malaysian students in English learning. Being one of the pioneer research tackling pedagogy issues, the knowledge from the work of [38] can be fundamental in blending English LLS with current content delivery tools such as ICT. Each element in the Oxford's SILL learning strategy is mapped to MAR-learning feasibilities inclusive of media and mixed reality utilities shown in

Fig. 1. MylinE English learning application in M-learning [37]

Table 1. InterviewME design mapping with Oxford's SILL [38] and Specth's matrix [40] for prototype development

Oxford strategies [38]	Definition [38]	Oxford SILL mapping	Specht matrix mapping [40]
Memory Strategies	Creating Mental Linkages Applying Images and Sounds Reviewing Well Employing Action	InterviewME provides standard Malaysian Communication English Syllabus, which includes interconnecting modules, which incorporates media utilization of not only images and sounds, but also animated and interactive character plus menus. The arranged learning instruction interconnect ESL learners from module to module. Furthermore, the existence of hints within the interview games aimed to create mental linkages between the learner and items or terms from prior experience interconnecting the learning process	Performance, Understanding, Reflection
Cognitive	Practicing Receiving and Sending Messages Analyzing and Reasoning Creating Structure for Input and Output	Module repetition with multiple different scenarios is aimed to trigger cognitive responses in practicing	Independent Context, Identity
Compensation	Guessing Intelligently Overcoming limitations in speaking and writing	With given hints and choices of answers during the interview game	Illustration, Exploration
Metacognitive	Centring your learning Arranging and planning your learning Evaluating your learning	[40] showed evidence where MAR-learning is related to metacognitive learning process. InterviewME is developed with procedures of inter-linked learning arrangements with simple game-based evaluation at the end of each module	Understanding, Reflection, Relation

(*continued*)

Table 1. (*continued*)

Oxford strategies [38]	Definition [38]	Oxford SILL mapping	Specht matrix mapping [40]
Affective	Lowering your Anxiety Encouraging yourself Taking your emotional temperatures	Given the advantage of AR proved to improve phobia and anxiety by several researchers. Besides AR has also been proven to increase learning motivation and learning engagement [42,43]	Location, Environment
Social	Asking questions Cooperating with others Empathising with others	Since this strategy is much related to social engagement and the comparative study focuses more on motivation, satisfaction and technology preference. This strategy will be mapped in the near future in the next prototype development phase where communication will be made available via InterviewME	Collaboration, Reflection (once prototype improvement is implemented)

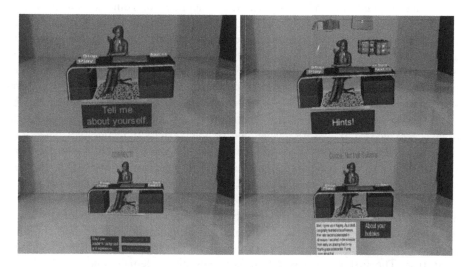

Fig. 2. InterviewME MAR-learning prototype

Table 1. The prototype also incorporates the criteria of a high-fidelity prototype where de Sa and Churchill in [39] presented evidence that it benefits probing, concept validation, feature validation, usability testing and user experience evaluation. Besides, development of InterviewME prototype adapts the matrix guidelines of Specht in [40], where 3 matrix categories namely Dynamic 3D objects, Augmented Books and Instructional AR Manipulation are chosen relative to the content deliverance of

InterviewME (Table 1). As far as the content is concern, InterviewME has 100 % similarity academic content with one of MylinE's module called "interviews", but of course presented with all MAR properties suggested by [40, 41].

Device used to represent InterviewME is an Apple Iphone 6 where whole screen is utilized as spatial cues for displays of 3-Dimensional interviewer character, interactive buttons, quizzes, sounds and hints augmenting the real environment (Fig. 2).

3 Methodology

The comparative studies of M-learning (Mobile MylinE) versus MAR-learning (InterviewME) uses the Real World Evaluation Strategy by [44] where control experiment are carried out in common students' area in a university. This is to ensure that the participants can carried out the test at a location which they are familiar and comfortable with (Fig. 6). The process starts with screening of 23 students who volunteered. 10 undergraduates are selected based on few criteria. First they have to be enrolled in the compulsory English communication subject where interview skills are part of their syllabus. Both male and female genders are equally represented. All 10 graduates are chosen with several criteria including having personal smartphones and have experienced learning application at least once through the mobile environment. Many researches argue on sample sizes where 30 participants are considered the general rule of thumb depending on varieties of application. Tullis and Albert in [45] however find this theory a myth and believe smaller sample sizes of 8 or 10 can still be meaningful. Adding to that, from the experiences of [45], 5 participants per significantly different class of user is usually enough to uncover the most important usability issues. This is supported by Frick's review where based on the cumulative binomial probability formula, led to the statement that testing only 4 or 5 users will uncover 80 % of the usability problems [46]. Students were chosen based on gender as well since [47, 48] show evidence where gender played certain roles in determining individual acceptance on mobile AR, visual and spatial capacity in technology. Therefore, 5 male and 5 female students were profiled to limit the gender bias in this experiment. Since all students will be giving opinions on both M-learning and MAR–learning experiences, students will then be separated into 2 different groups. Both groups will be given similar device (an Apple Iphone 6) to experience M-learning and MAR-learning. Even though Furio in [49] shows no significant differences in participants' performance while using devices with different screen size and weight, this study aims to eliminate any possible device handling biases in order to focus only on users' motivation and satisfaction. The first group has 2 male students and 3 female students, while the second group has 3 male students and 2 female students. The first group will experience the M-learning first, then the MAR-learning, while contrary, the other group will experience MAR-learning first before M-learning. After each session with one of the technology, these students will be given a set of modified Intrinsic Motivation Inventory (IMI) questionnaires to measure students' motivation and satisfaction in 4 areas namely Enjoyment and Interest (E&I), Perceived Competence (PC), Perceived Pressure and Tension (PP&T) and Perceived Choice (PCh), which according to [50] are very much related to motivation and satisfaction metrics. Since Interview training

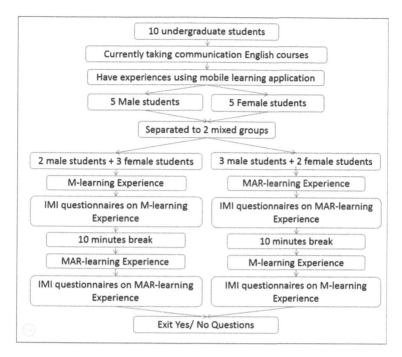

Fig. 3. Experiment flow in M-learning and MAR-learning comparative study

requires verbal competency, which is also a part of the contributing factor to WTC [10, 15], motivation is seen to be an element co-related to WTC as well. Therefore IMI is selected as the measuring tool based on its objective to gauge users' self-reported motivation in multiple categories. The 4 categories discussed above is selected based on reference works done in [42, 43, 50–52], where they are directly related to users' satisfaction, which is one of the 3 pillars in ISO 9241-11 usability model [53]. In order to ensure users' consistency in answering the IMI questionnaires, repetitive questions is embedded in two of the categories to eliminate if any, answers that does not match each other in the given Likert Score. As to limit the effects of central tendency bias discussed in [54], an ipsative 4 (even numbered) Likert scale is used where 1 represents "Strongly Disagree", 2 represents "Disagree", 3 represents "Agree" and 4 represents "Strongly Agree" [55–57]. In each technology experience, several discrete steps are standarized to practice similar procedures on every participant. The steps are presented in Fig. 3.

In every M-learning experience or MAR-learning experience, a sub experiment flow is conducted shown in Fig. 4.

3.1 Hypotheses

In order to compare M-learning (Mobile MylinE) and MAR-learning (InterviewME), the study aims to verify 4 hypotheses based on the literature study on suitability of M-learning and MAR-learning in ELT. From the literature and methodologies

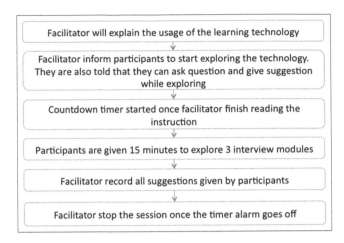

Fig. 4. Experiment flow in M-learning and MAR-learning comparative study

discussed above, it is believed that both M-learning and MAR-learning have impact on motivation and satisfaction. However, the impact is seen to be more significant in MAR-learning due to the blending of virtuality and reality, which is an extension of better visualization as compared to M-learning. Therefore, the 4 hypotheses are discussed below in Table 2.

Table 2. Research hypotheses

Hypotheses	Definition
Hypothesis 1 (H1)	Due to blended media and reality in MAR-learning, this study believes that MAR-learning will be significantly better than M-learning in enjoyment and interest
Hypothesis 2 (H2)	Due to the nature of interaction with reality in MAR-learning according to matrix by [40], this study believes that MAR-learning will be significantly better than M-learning in perceived competence
Hypothesis 3 (H3)	Due to the fact that selected participants are frequent mobile device users and have experienced mobile learning at least once, this study believes that there will be no significant difference comparing the perceived pressure and tension of MAR-learning and M-learning
Hypothesis 4 (H4)	Due to the lack of MAR-learning option in current language learning application and due to the interaction styles of MAR-learning, this study believes that MAR-learning will be significantly better than M-learning in perceived choice

4   Results and Discussion

Before analyzing the IMI data collected from the participants, to ensure there was no gender bias discussed in [47, 48], an unpaired t-test is performed to differentiate the motivation and satisfaction of male and female participants. From the results collected, there is no significant difference in all IMI categories comparing the 2 genders. The results therefore confirmed the non-existence of gender bias in this comparative study. The t-values are presented in Table 3, where degrees-of-freedom is 9 with 95 % of confidence level (df = 9, p < 0.05).

Table 3. Unpaired t-value for comparing male and female participants in IMI scores

IMI categories t value	M-learning	MAR-learning
Enjoyment and interest	1.09	0.00
Perceived competence	0.224	0.00
Perceived pressure and tension	0.784	0.632
Perceived choice	0.894	0.00

The results for all 4 IMI categories will be discussed next where the IMI total scores for each category is presented in Table 4. There are a total of 10 questions where paired t-test is used for all categories considering evaluation of M-learning and MAR-learning per participant.

Table 4. Number of questions and Likert points in IMI scores

IMI categories	Number of questions	Total Likert points
Enjoyment and interest	4	16
Perceived competence	3	12
Perceived pressure and tension	2	8
Perceived choice	1	4

In a paired t-test verifying H1, the results show that there were significant differences in enjoyment and interest comparing the Likert scores between M-learning and MAR-learning. (t = 4.13, df = 9, p < 0.05). Therefore, it verifies hypothesis H1 where MAR-learning creates more enjoyment and interest than M-learning (Table 5, Figs. 5 and 6).

Table 5. Paired IMI scores for each participant in enjoyment and interest (E&I)

E&I	P1	P2	P3	P4	P5	P6	P7	P8	P9	P10
M-learning	6	11	12	11	15	16	12	12	12	12
MAR-learning	16	15	16	16	15	16	16	14	16	16

In another paired t-test verifying H2, the results show that there were significant differences in perceived competence comparing the Likert scores between M-learning and MAR-learning. (t = 4.29, df = 9, p < 0.05). Therefore, it verifies hypothesis H2 where participants perceive themselves to be more competent in MAR-learning than M-learning (Table 6), (Figs. 5 and 6).

However, in another paired t-test verifying H3, the results show that there were no significant differences in perceived pressure and tension comparing the Likert scores between M-learning and MAR-learning. (t = 2.09, df = 9, p < 0.05). Hypothesis H3 is therefore rejected even though participants seem to have less pressure and tension experiencing MAR-learning than M-learning (Table 7), (Figs. 5 and 6).

Table 6. Paired IMI scores for each participant in perceived competence (PC)

PC	P1	P2	P3	P4	P5	P6	P7	P8	P9	P10
M-learning	7	9	9	7	9	11	9	8	7	7
MAR-learning	11	11	12	11	9	10	12	10	12	10

Table 7. Paired IMI scores for each participant in perceived pressure and interest (E&I)

PP&T	P1	P2	P3	P4	P5	P6	P7	P8	P9	P10
M-learning	6	7	6	6	8	7	7	7	6	8
MAR-learning	7	8	8	8	8	7	8	7	8	8

Table 8. Paired IMI scores for each participant in perceived choice (PCh)

PCh	P1	P2	P3	P4	P5	P6	P7	P8	P9	P10
M-learning	2	3	3	4	4	4	4	3	3	4
MAR-learning	4	4	4	4	4	4	4	4	4	4

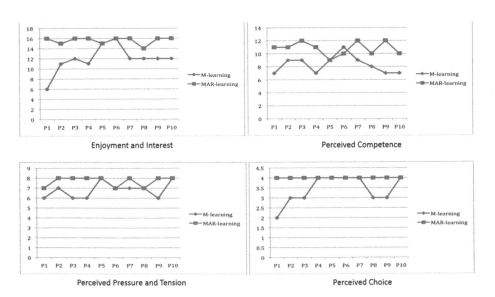

Fig. 5. Chart representation of IMI Results in E&I, PC, PP&T and PCh

Finally in the last paired t-test verifying H4, the results show that there were significant differences in perceived choice comparing the Likert scores between M-learning and MAR-learning. ($t = 2.71$, $df = 9$, $p < 0.05$). Therefore, it verifies hypothesis H4 where participants perceive themselves to be more willing to explore MAR-learning than M-learning (Table 8), (Figs. 5 and 6).

Fig. 6. Experimental setup, location and evaluation process

5 Conclusion

The results in the comparative study of M-learning and MAR-learning have given new perspective in MAR-learning. Evidence from this pilot test shows that MAR-learning significantly outperform M-learning due to AR components which help enhancing motivation and satisfaction of participants using it. Even though the above experiment is objectified to measure only motivation and satisfaction, the analyzed result aspire to create new leads to research in MAR-learning in ELT, especially in a context of multi-racial country like Malaysia.

References

1. The Government of Malaysia's Official Portal, Language (2015). https://www.malaysia.gov.my/en/aboutmalaysia?subCatId=3208956&type=2&categoryId=3208945
2. Mahir, A., Normazla, Jarjis, S., Kibtiyah, M.: The use of Malay Malaysian English in Malaysian English: key considerations. In: The Second Biennial International Conference on Teaching and Learning of English in Asia: Exploring New Frontiers. (TELiA2 2007), Langkawi, Malaysia, pp. 1–9 (2007)
3. Muftah, M., Rafik-Galea, S.: Language learning motivation among Malaysian pre-university students. Engl. Lang. Teach. **6**(3), 92–103 (2013)

4. Pillai, S.: Speaking English the Malaysian way - correct or not? Engl. Today **24**(4), 42–45 (2008)

5. Thirusanku, J., Yunus, M.M.: Status of English in Malaysia. Asian. Soc. Sci. **10**(14), 254–260 (2014)

6. Nair, G.K.S., Rahim, R.A., Setia, R., Husin, N., Sabapathy, E., Jalil, N.A.A., Seman, N.A.: Malaysian graduates English adequacy in the job sector. Asian Soc. Sci. **8**(4), 143–147 (2012)

7. Yamat, H., Fisher, R., Rich, S.: Revisiting English language learning among Malaysian children. Asian Soc. Sci. **10**(3), 174–180 (2014)

8. Thang, S.M., Mustaffa, R., Wong, F.F., Noor, N.M., Mahmud, N., Latif, H., Aziz, M.S.A.: A quantitative inquiry into the effects of blended learning on english language learning: the case of Malaysian undergraduates. Int. Educ. Stud. **6**(6), 1–7 (2013)

9. Hiew, W.: English language teaching and learning issues in Malaysia: learners' perceptions via facebook dialogue journal. Res. World **3**(1), 11–19 (2012)

10. Musa, N.C., Koo, Lie, Y., Azman, H.: Exploring English language learning and teaching in Malaysia. GEMA Online J. Lang. Stud. **12**(1), 35–51 (2012)

11. Wahi, W.: English language literacies of undergraduate students in Malaysia's culturally and linguistically diverse environment. Casualties of national language policies and globalisation? Educ. Res. Perspect. (online) **42**, 329–362 (2015)

12. Davies, A.: Native Speaker. Blackwell Publishing Ltd., Oxford (2003)

13. Juhary, J.: English language teaching: the reflective practices of an oral communication class. Engl. Lang. Teach. **7**(4), 136–146 (2014)

14. Ngah, E., Radzuan, N.R.M., Fauzi, W.J., Abidin, N.A.Z.: The need for competent work ready english language learners. Procedia Soc. Behav. Sci. **29**, 1490–1499 (2011)

15. Yousef, R., Jamil, H., Razak, N.: Willingness to communicate in English: a study of Malaysian pre-service English teachers. Engl. Lang. Teach. **6**(9), 205–216 (2013)

16. Chen, C., Chung, C.: Personalized mobile English vocabulary learning system based on item response theory and learning memory cycle. Comput. Educ. **51**(2), 624–645 (2008)

17. Korucu, A.T., Alkan, A.: Differences between M-learning (mobile learning) and e-learning, basic terminology and usage of M-learning in education. Procedia Soc. Behav. Sci. **15**, 1925–1930 (2011)

18. Martin, F., Ertzberger, J.: Here and now mobile learning: an experimental study on the use of mobile technology. Comput. Educ. **68**, 76–85 (2013)

19. Ally, M., Prieto-Blázquez, J.: What is the future of mobile learning in education? RUSC Univ. Knowl. Soc. J. **11**(1), 142–151 (2014)

20. Hockly, N.: Mobile learning. ELT J. **67**(1), 80–84 (2013)

21. Stephens, M.: Learning everywhere. Libr. J. **137**(7), 48 (2012)

22. Cavus, N., Al-Momani, M.M.: Mobile system for flexible education. Procedia Comput. Sci. **3**, 1475–1479 (2011)

23. Keengwe, J., Bhargava, M.: Mobile learning and integration of mobile technologies in education. Educ. Inf. Technol. **19**(4), 737–746 (2014)

24. Lin, C.: Learning English reading in a mobile-assisted extensive reading program. Comput. Educ. **78**, 48–59 (2014)

25. Soleimani, E., Ismail, K., Mustaffa, R.: The acceptance of mobile assisted language learning (MALL) among post graduate ESL students in UKM. Procedia Soc. Behav. Sci. **118**, 457–462 (2014)

26. Mohamad, M., Muniandy, B.: Mobile-assisted language learning in Malaysia: where are we now? Practice **47**, 55–62 (2014)

27. Velikovsky, A., Shammass, S.: Mobile education. In: Mobile Technology Consumption: Opportunities and Challenges, pp. 16–29 (2011)

28. Ting, Y.L.: The pitfalls of mobile devices in learning: a different view and implications for pedagogical design. J. Educ. Comput. Res. **46**(2), 119–134 (2012)
29. Blakemore, D., Svacha, D.: Mobile handheld devices in education. In: Mobile Technology Consumption: Opportunities and Challenges, pp. 32–45 (2011)
30. Lee, K.: Augmented reality in education and training. TechTrends, **56**(2), 13–21. Springer, US (2012)
31. Liu, T.: A context-aware ubiquitous learning environment for language listening and speaking. J. Comput. Assist. Learn. **25**(6), 515–527 (2009)
32. Liu, P., Tsai, M.: Using augmented-reality-based mobile learning material in EFL English composition: an exploratory case study. Br. J. Educ. Technol. **44**(1), E1–E4 (2013)
33. He, J., Ren, J., Zhu, G., Cai, S., Chen, G.: Mobile-based AR application helps to promote EFL children's vocabulary study. In: Proceedings of IEEE 14th International Conference on Advanced Learning Technologies (ICALT 2014), Athens, Greece, pp. 431–433 (2014)
34. Fitzgerald, E., Adams, A., Ferguson, R., Gaved, M., Mor, Y., Thomas, R.: Augmented reality and mobile learning: the state of the art. Int. J. Mob. Blended Learn. **5**(4), 43–58 (2013)
35. Beauchamp, G., Parkinson, J.: Beyond the 'wow' factor: developing interactivity with the interactive whiteboard. Sch. Sci. Rev. **86**(316), 97–103 (2005)
36. Murray, L., Barnes, A.: Beyond the "wow" factor evaluating multimedia language learning software from a pedagogical viewpoint. System **26**(2), 249–259 (1998)
37. MyLinE: online resources for learning in English (2015). http://myline.utm.my/moodle282
38. Shah, M.I.A., Ismail, Y., Esa, Z., Muhamad, A.J.: Language learning strategies of English for specific purposes students at a Public University in Malaysia. English. Lang. Teach. **6**(1), 153–161 (2013)
39. De Sá, M., Churchill, E.: Mobile augmented reality: exploring design and prototyping techniques. In: Proceedings of the 14th international Conference on Human-computer Interaction with Mobile Devices and Services (MobileHCI 2012), San Francisco, CA, USA, pp. 221–230 (2012)
40. Specht, M., Ternier, S., Greller, W.: Mobile augmented reality for learning: a case study. J. Res. Cent. Educ. Technol. **7**(1), 117–127 (2011)
41. Lotherington, H., Jenson, J.: Teaching multimodal and digital literacy in L2 settings: new literacies, new basics, new pedagogies. Annu. Rev. Appl. Linguist. **31**, 226–246 (2011)
42. Yamabe, T., Nakajima, T.: Playful training with augmented reality games: case studies towards reality-oriented system design. Multimedia Tools Appl. **62**(1), 259–286 (2013)
43. Hartl, A., Grubert, J., Reinbacher, C., Arth, C., Schmalstiegp, D.: Mobile user interfaces for efficient verification of holograms. virtual Reality (VR) IEEE, pp. 119–126 (2015)
44. Fitzpatrick, R.: Strategies for evaluation of software usability, articles, department of mathematics, statistics and computer science. Dublin Institute of Technology, Ireland (1999). http://www.comp.dit.ie/rfitzpatrick/papers/chi99%20strategies.pdf
45. Tullis, T., Albert, B.: Measuring the User Experience, Interactive Technologies Series. Morgan Kaufmann, San Francisco (2008)
46. Frick, T.W., Dodge, T., Liu, X., Su, B.: May. How many subjects are needed in a usability test to determine effectiveness of a web site. In: Meeting of the Association for Educational Communication and Technology, Anaheim, CA (2004). http://education.indiana.edu/~frick/aect2003/frick_dodge_liu_su.Pdf
47. Olsson, T., Kärkkäinen, T., Lagerstam, E., Ventä-Olkkonen, L.: User evaluation of mobile augmented reality scenarios. J. Ambient Intell. Smart Environ. **4**(1), 29–47 (2012)
48. Lin, P.C., Chen, S.I.: The effects of gender differences on the usability of automotive on-board navigation systems–a comparison of 2D and 3D display. Transp. Res. Part F Traffic Psychol. Behav. **19**, 40–51 (2013)

49. Furió, D., González-Gancedo, S., Juan, M., Seguí, I., Costa, M.: The effects of the size and weight of a mobile device on an educational game. Comput. Educ. **64**, 24–41 (2013)

50. Mekler, E.D., Brühlmann, F., Tuch, A.N., Opwis, K.: Towards understanding the effects of individual gamification elements on intrinsic motivation and performance. Comput. Hum. Behav. 1–10 (2015)

51. Eishita, F.Z., Stanley, K.G.: The impact of sensor noise on player experience in magic window augmented reality aiming games. In: Chorianopoulos, K., Divitini, M., Hauge, J.B., Jaccheri, L., Malaka, R. (eds.) ICEC 2015. LNCS, vol. 9353, pp. 502–507. Springer, Heidelberg (2015). doi:10.1007/978-3-319-24589-8_47

52. Grubert, J., Pahud, M., Grasset, R., Schmalstieg, D., Seichter, H.: The utility of magic lens interfaces on handheld devices for touristic map navigation. Pervasive Mob. Comput. **18**, 88–103 (2015)

53. International Standards Organization (1998) ISO 9241–11: Ergonomic requirements for office work with visual display terminals (VDTs); Part 11-Guidance on Usability (2015)

54. Peer, E., Vosgerau, J., Acquisti, A.: Reputation as a sufficient condition for data quality on amazon mechanical turk. Behav. Res. Methods **46**(4), 1023–1031 (2014)

55. Salgado, J.F., Anderson, N., Tauriz, G.: The validity of ipsative and quasi-ipsative forced-choice personality inventories for different occupational groups: a comprehensive meta-analysis. J. Occup. Organ. Psychol. **88**(4), 797–834 (2015)

56. van Eijnatten, F.M., van der Ark, L.A., Holloway, S.S.: Ipsative measurement and the analysis of organizational values: an alternative approach for data analysis. Qual. Quant. **49**(2), 559–579 (2015)

57. Joubert, T., Inceoglu, I., Bartram, D., Dowdeswell, K., Lin, Y.: A comparison of the psychometric properties of the forced choice and likert scale versions of a personality instrument. Int. J. Sel. Assess. **23**(1), 92–97 (2015)

A Preliminary Evaluation of ICT Centers Performance Using COBIT Framework: Evidence from Institutions of Higher Learning in Brunei Darussalam

Afzaal H. Seyal$^{(\boxtimes)}$, Sheung Hung Poon, and Sharul Tajuddin

School of Computing and Informatics, Universiti Teknologi Brunei,
Bandar Seri Begawan, Brunei Darussalam
{afzaal.seyal, sheunghung.poon,
sharul.tajuddin}@utb.edu.bn

Abstract. The present study is undertaken among Information and Communication Technology Centers (ICTCs) among four higher education institutions (HEIs) in Brunei Darussalam to evaluate and highlight their performance in achieving IT Governance using a performance measuring COBIT framework. The COBIT also assesses the maturity level indicators from the range of 0 to 5. The results of this preliminary study show that not all the items measuring five domains of COBIT is applicable to these ICTCs and weighted average maturity level of all these ICTs range from 1.40 to 1.72. This further highlight that the five domains of COBIT measuring IT Governance are at initial stage and the top management of these HEIs should adopt a framework and provide additional resources to their ICTs in bringing IT Good Governance. Based on the study results some recommendations are made to enhance the performance.

Keywords: IT governance · Performance measures · COBIT · Higher learning institutions · ICT center · Brunei darussalam

1 Introduction

In the second decade of 21st century, the business use of Information and Communication Technologies (ICTs) has become matured by entering into next stage of assessment where organizational performance and investment in ICT was measured in terms of increasing organizational efficiency and bringing good governance. With the extensive use of ICT for Business Process Reengineering (BPR) and simplification of the tasks the business organizations are striving their best to assess and evaluate their ICT investment and optimize ICT to support their organizational' s strategic focuses. Business organizations are of much concern in determining as how IT is used to bring good governance at the workplace. IT Governance (ITG) is a term that indicates specifying decision rights as well as the provision of an accountability framework that encourage desirable behavior in the use of IT [1]. Haes and Gremergen [2] described that ITG is implementable via a framework of three main key elements; Structures, Processes and Relational Mechanism. Although, most of the applications of ITG

© Springer International Publishing AG 2017
S. Phon-Amnuaisuk et al. (eds.), *Computational Intelligence in Information Systems*,
Advances in Intelligent Systems and Computing 532, DOI 10.1007/978-3-319-48517-1_21

focuses in business organizations, many universities have also shown great interest in the implementation of these procedures to manage their IT [3]. The last decade saw an increasing demand as how IT is governed in higher educational institutes especially with the availability of the corporate management tools.

We understand that assessing IT management practices among educational institutions is still at grass root level and need to be addressed holistically. However, this lack of applied procedures is due to initial gap into planning mechanism. The academic institutions because of the different strategic focuses lacks in necessary tools and theoretical frameworks to measure the performance of their ICTs resources. In other words, IT performances measures need to be aligned with main strategic focuses and business goals set for evaluation and measurement by using an appropriate framework [4]. Similarly, top management and leadership of these educational institutions are now more optimistic in determining as how important for these institutions to understand as why new investments have been identified and linked within the strategy in particular and strategic plan in general. In pursuit of enhancing productivity the academic institutions are formulizing the institution-wide strategy and using a proactive approach of not only formulizing but also creating separates IT and IS strategies for competitive advantages. With this approach, the educational institutions are using ICT strategies to reinforce alignment to the overall institutional strategy and to device a mechanism to review and update the strategic process regularly.

Therefore there is dire need that the performance of ICT in general and ICT centers in particular be measured through periodic audits as to find out whether good governance has being achieved. Equally, it is important to measure whether ICT centers are performing well above the threshold level to meet the institutional goals in delivering the services and providing competitive advantage. In addition, there has been less effort in term of research that focus on the issues of implementing ITG in higher institutions [5]. This is despite the realization of the importance of IT in the educational industry. Whereas, it is believed that the effectiveness of an ITG structure becomes the sole important predictor of benefits from the use of IT [6, 7]. We at this stage must ask one question on how we can improve ITG adoption and implementation in HEI. There are several frameworks available in the management science such as ITIL (IT Infrastructure Library), COSO (Committee of Sponsoring Organizations of the Treadway Commission, 1992), and a comprehensive framework COBIT (Control Objectives for IT related technology) that was developed by Information systems audit and control association (ISACA- www.isaca.org) and IT Governance Institute (ITGI) in 1992 however; the first version was only released in 1996 [8]. A COBIT steering committee branch under the IT Governance institute continuously researches and helps evolve the COBIT body of knowledge with the evolvements of businesses that now has more focus on IT aspects. The ITIL or COBIT governance framework list down an outstanding set of best practices for various IT-related processes and particularly in supporting IT governance. In general COBIT provides a framework to ensure that as how IT is aligned with the business, how IT resources should be used sensibly and responsibly as well as how IT risks should be addressed and managed appropriately. COBIT therefore provide a comprehensive and easy to understand framework covering IT organization, IT users, IT professionals, IT governance, IT risk and IT processes.

1.1 COBIT Structure

COBIT area of focus in IT Governance is supported by 34 total IT processes divided among four main domains such as; Plan and Organize (PO), Acquire and Implement (AI), Deliver and Support (DS) and Monitor and Evaluate (ME).

1.2 Why COBIT?

The basic reason for selecting COBIT is because of its appropriateness' to the education context [9, 10] and also because of it measuring IT Resources Decisions and to identify IT resources (systems, people, data, and technology) and to align with planning step in educational framework. A survey carried out by Price Waterhouse and Coopers in 2005 for ITGI revealed that 75 % of the organizations currently using COBIT and found that COBIT was very useful or somewhat useful. 15 % were unclear and 10 % had negative response. With this evidence this pioneering study that was conducted in April-May 2016 using COBIT framework among institutions of higher learning in Brunei Darussalam. The study has the following objectives:

- To find out how the formal IT governance is adopted and implemented in the ICTCs among higher education institutions in Brunei.
- To evaluate the IT governance and to measure IT performance in their respective ICT centers in fulfilling overall strategic alignment.
- To further determine the organizations' current maturity level and to compare with the maturity level of the best practice in the industry.

2 Review of Literature

There are extensive studies highlighting the performance measurements at the corporate level especially in the context of developed world. However, researchers in Asia-Pacific and especially among ASEAN group have conducted studies from performance management and IT good governance. Based on COBIT, an academic architecture Information system model in higher education was created by Mutyarini and Sembiring [11]. As they believed COBIT as a framework not only deliver but measures effectively the performance of the Information System architecture.

Besides these several other studies focused on the suitability of COBIT in measuring IT performances [12–14]. Gheorghe [4] proposed an audit methodology of the IT governance at the organizational level. The evaluation of the risk assessment with IT governance remained as key process in planning the audit mission. The identification of the risk was achieved in seven main domains such as: IT strategic planning process, IT organizational management, IT delivery process, IT investment management, IT project management, IT risk management and IT performance process. Lapao [15] studied the challenges and barriers of introducing IT Governance into one of the hospital in Portugal. The study assessed the relationship between corporate governance and the role of IT governance in managing new service deployment by using both COBIT and ITIL assessment to further identify IT governance weaknesses. The study concluded that poor IT Governance remained an important barrier to hospital management especially in the IT service management. In another study, it was highlighted that although the management of IT in higher learning institutions has been practicing some techniques and methodologies but unfortunately these are not structured [16] Maria, Fibriani and Sinatra [17] conducted another study in one university in Indonesia to measure the IT performance in achieving business goals. They assessed the maturity level of IT processes using COBIT. Results showed appropriate management of IT was achieved and as well as various IT processes are concerned they supported the business goals in a standardized, well communicated and documented approach. However, the service aspect of the user was poorly managed and need a lot of improvement to achieve business goals. Ajami and Al-Qirim [5] developed an ITG framework to support HEI in Abu Dhabi to govern their IT project. The framework focused on evaluating decisions concerning the alignment and the compatibility of IT with overall strategies and goals of HEIs. They used a combination of COBIT and six sigma framework. Sadikam, Hardi and Wachyu [18] studied the IT alignment processes in one of the Indonesian Universities with the focus of the strategic alignment with organizational strategy focusing on IT Governance, using COBIT framework. Result shows that implementation of IT Governance in the university is still at the early stage of its development. They proposed some improvement to uplift the process of IT Governance within the university.

2.1   Maturity Level Model

One of the trending process assessment in various technological and organizational areas is the development of capability maturity models. These models provide individuals and organizations the opportunity to self-assess the maturity of various aspects of the processes they employ against benchmarks. One of the well-known models belongs to the Capability Maturity Model Integration (CMMI) family. This model was developed by Humphrey at Carnegie Mellon University. Making use of Maturity Models provide insight to an organizations' development processes. There is a lack of maturity model in the discipline of Enterprise Architect. However, the one model that is widely used is Capability Maturity Model (CMM) which was originally defined by the Software Engineering Institute (SEI) [19]. In this model, the reliability of the controlled activities on IT systems are further explained into six levels (0–5). The 0 is for non-existence of management processes to 5 with best practices is followed. These levels was classified by ISACA from the consensus of opinions of many experts and best practices in generic information technology and it also has been used as an international standard.

According to the study by IT Governance Institute (ITGI 2009) around 50 % of the organizations have already implemented the IT Governance system, 18 % are in the process of implementation and remaining 32 % are in wait-and-see or at planning stage. An average global value of IT Governance maturity level of 2.67 on a scale that range from 0 to 5 of those organizations that have an IT Governance framework in place was calculated by the same study. However, the universities have not yet reached their level of maturity. Yanosky and Borrenson, [20] found the average maturity level of universities on a global level of 2.30. In addition, Llorens and Fernandez, [21] reviewed that an average maturity level for Spanish Universities is 1.44. Similarly, Harmse [22] in his study of aligning business with IT within in University of Pretoria to study the IT Governance and measured the processes. The total mean of 1.37 weighted average shows that maturity of IT processes are at initial stage and informal Enterprise Architect (EA) processes are underway. It is therefore evident that use of a particular framework and maturity level are related in shaping up the good governance.

3   Research Methodology

This is a qualitative in nature and follows the case study approach. Benbasat *et al.* [23] mentioned the case research method is useful for addressing the "how" question, i.e. in the exploratory stage of knowledge building. This is relevant within the context of study on IT Governance among institutions of higher learning. The case study therefore provides rich insight of the exploratory nature of the research.

This study measures the maturity level of various IT process by using COBIT framework which are being practiced in this study of four institutions of higher learning in Brunei Darussalam. Data used in this study consists of primary data that was obtained from interviews with the Directors and Chief Information Officers of these four ICT centers. This research adopts Yin's [24] multiple case (comparative) design in studying three single units of analysis (holistic). In keeping with participants' requests for anonymity, the universities will be referred to as Universities A, B, C, and D in this research. The data was gathered through open and semi-structured interviews with the CIO's in all four universities. Relevant documents obtained from both interviewees and the universities' websites was also referred to. The primary data that was obtained through interview and examining the organizational chart, various operational plan, and personal statistics. The prior appointment was fixed with the Directors/CIO of the data centers of these universities and responses were recorded and later transcribed and analyzed. Recording was done with their permission as per techniques described Yin [24]. One exceptional Director/CIO did not allow recording the interview.

The two out of the four institutions labelled as "A" and "B" was established in mid 1980s whereas, "C" and "D" are comparatively new. All these higher learning institutions are public sector organizations that were established to cater for advance teaching, learning and research in science and humanities education, technical education, religious education and teaching education. The maturity level was calculated by asking question as "how" in their opinion and up to what extent they are applying the particular process from 0-nonexistence to 5-optimized.

4   Results and Discussion

The measurement of IT performance in this research includes the use of various information systems developed along with the use of Internet resources to support operational activities. Various IT resources such as information infrastructure, application, and people should be appropriately managed and controlled in order to attain the desired IT goals. In certain universities, when IT will be implemented and developed, the IT goals are set by the university. The IT goals are presented in Table 1. In order to achieve the IT goals, the control of the IT processes in ICTCs is measured by COBIT framework. The IT processes consists of four domains, namely planning and organization (PO), Acquisition and implementation (AI), Delivery and Support (DS) and Monitoring (ME) for the total of 34 sub domains. Table 1 also presented the results of relationship between the IT goals and the COBIT IT process.

The result further highlights that all four ICTCs are using some of the processes of COBIT to the extent showing the suitability of COBIT framework to measure performance among HEIs. We then analyze the performance measures of each institution of higher learning ICT centers, it is cleared from the study results that all four universities are following some steps of developing the IT/IS strategy and is doing some measures to stream line the processes and to develop and apply IT Governance. Although these ICTCs for these four HEIs are not fully using all the thirty-four processes of the COBIT framework because of the various processes such as budgetary and financial aspects including cost effectiveness are beyond their direct control. Similarly the HR requirement is not in the direct control of their directors. The items were reworded from the original ones to make more clarity and simple way rather than asking in a more complex way. Result further reveals some interesting findings as none of these ICTCs have IT steering committee, only half of them have developed their ICTC Strategic plan and half of them an effective way to identify and prioritize the ICT projects. The most interesting findings emerged with the significant and sensitive area of ICT Security Awareness and backup plan. Under the present business environment where the data centers are vulnerable to various threats ignoring such a sensitive issue may result with chaos, and malfunctioning of the data centers resulting organizational efficiency of these HEIs would be jeopardized.

IT Governance covers five principal domains that include: IT Strategic Alignment, service delivery, risk management, resource management and performance management (http://www.itgi.org) that was provided in Table 2. The summary data in Table 2 outlines all the processes under five principal domains with majority of the maturity levels pointing to 1 and 2 showing adhoc to regular pattern assessment of these processes. These maturity indicators do not show an ideal picture as the global average for the universities is 2.30 and for Spanish universities is 1.37 [22]. Our results therefore are better than the Spanish universities performance level. This might be because Spanish universities using IT Governance framework shows the performance of the whole university and not of one particular faculty or center.

There is strong need to focus on addressing ethical and legal issues and compliance on health and safety policies in addition to developing the security awareness program including developing and implementing business impact analysis for business continuity plan including disaster recovery plan. These plans need to be periodically updated as organizations such as institutions of higher learning are in constant state of change in general and upgrades in software, hardware and new approach of various Web-based systems in particular require updating the plan on regular basis.

Table 1. Evaluation of Final Checklist of Questionnaire model for IT Governance among respondents

No	Question	A	B	C	D
1	Does the IT strategy plan align with organizational (business) strategic plan?	✓	✓	✓	✓
2	Does there any ICT users' policy document and copy is provided to all the users?	✓	✓	✓	✓
3	Does the ICT user policy as a medium of espousing value of information?	✓	✓	✓	✓
4	Does the ICTC follow University's objectives goals, strategies and mission (OGSM)?	✓	✓	✓	✓
5	Is IT staffed adequately, with right skills and competencies?	✓	✓	✓	✓
6	Does the ICTC have effective internal processes to choose alternative strategies that have long-term impact?	X	✓	X	✓
7	Does the ICTC establish measures to track the progress of executing chosen strategies?	✓	X	✓	X
8	Does the ICTC and organization have an IT steering committee?	X	X	X	X
9	Does the ICTC have its own strategic plan?	✓	X	✓	X
10	Does ICT delivered services on time and offers the quality expected (through help desk)?	✓	✓	✓	✓
11	Does ICTC have effective way to identify and prioritize potential projects?	X	✓	✓	X
12	Does the ICTC measure and evaluate the result of the project?	X	✓	✓	✓
13	Does ICTC have authority to cancel the project that fails to meet expectations?	✓	✓	✓	✓
14	Does ICTC prepare Service Level Agreements (SLAs) for all the outsourced projects?	✓	✓	✓	✓
15	Do IT projects have a clear budget and timeline?	X	X	X	X
16	Does the board obtain regular progress reports on major projects?	X	✓	X	X
17	Does the ICTC identify the Security Awareness Requirement?	X	✓	X	X
18	Does the ICTC prepare a guidelines and scope of Security Awareness Program?	X	✓	X	X
19	Does the ICTC deployed staff to assist in planning for Security Awareness Program?	X	✓	X	X
20	Does the ICTC have regular backup of all updates files, sensitive data and software?	X	✓	X	X
21	Does ICTC have high level management commitment to support Security Program?	X	✓	X	X
22	Does the organization take a regular inventory of its IT resources?	✓	✓	✓	✓
23	Is a risk management policy, assessment and mitigation practice followed for IT?	X	X	X	X

(Continued)

Table 1. (*Continued*)

No	Question	A	B	C	D
24	Are the security and business continuity processes regularly tested?	X	X	X	X
25	Are sufficient IT resources and infrastructure available to meet required enterprise's architecture?	✓	✓	✓	✓
26	Does ICTC develop, prepare and maintain Ethics, Legal and Professional issues and guideline?	X	✓	✓	X
27	Does ICTC have health, and safety policy and has notified the users?	X	X	✓	X

(Key: ✓ = Yes and X = No)

Table 2. Summary result of IT Audit activities & COBIT measures in four ICTCs

	Perspective Business Goals	Case			
	IT Strategic Planning Process	A	B	C	D
PO1	Obtain reliable and useful information for strategic decision making including developing Strategic Plan for ICTC	1	2	3	1
	IT Risk Management				
PO9	Manage IT-related business risk	2	2	2	2
DS5	Manage IT Security awareness	0	2	0	0
AI3	Manage IT backup and continuity plan	1	2	0	0
	Performance Management				
PO10	Manage project, product and/or innovation	2	2	2	2
ME4	Improve corporate governance and transparency.	1	1	1	1
	Service Management				
DS8	Manage service help desk and incidents	3	3	3	3
DS1	Define and manage service-level agreement	2	2	2	2
DS3	Establish service continuity and availability	2	2	2	2
DS6	Achieve cost optimization of service delivery	1	1	1	1
DS13	Improve and maintain business process functionality	1	1	1	1
ME3	Provide compliance with external laws, regulations and contracts	2	2	2.5	2
ME2	Provide compliance with internal policies and transparency	2	2	2	2
	Resource Management (IT staff management & Organization and Structure				
PO7	Improve and maintain operational and staff productivity	2	2	2	2
AI5	Acquire and maintain skilled and motivated people	1	1	1	1
	Total Maturity level points	**1.53**	**1.80**	**1.63**	**1.46**

5 Conclusion

The current study that is conducted in Brunei Darussalam is an attempt to empirically investigate the impact of IT Governance framework COBIT to evaluate and enhance the implementation of Information Technology Governance (ITG) in Bruneian Higher Education Institutions (HEIs). The study investigates the 24 processes of audit, responsibility and accountability out of 34 COBIT processes. A qualitative interview approach was carried out to identify the implementation of IT Governance framework. The study fulfilled its objectives and answered the corresponding questions. The study further calculates the maturity levels and found below the international standard followed by global best practices organizations. The evaluation is carried out by calculating the maturity level of IT process control in the context of achieving business goals. Our results show that the IT in all four ICTCs' has been managing from adhoc to regular pattern as per maturity level indicators. IT *processes*, however, support the business goals in a standardized approach. Unfortunately, in assessing and measuring the *service* aspect for the users, the ICTs should apply necessary adjustments and improvements on the priority basis. The relevant authorities especially the top management of these Higher Learning Institutions should address the security issues and awareness program need accordingly by allocating resources.

References

1. Weill, P., Ross, J.W.: IT Governance: How Top Performers Manages IT Decision Right for Superior Results. Harvard Business School Press, Boston (2004)
2. Haes, S.D., Grembergen, W.V.: Analyzing the relationship between IT governance & business IT alignment maturity. In: Proceedings of the 41st Hawaii International Conference on System Science (HICSS) (2005)
3. Dewey, B.I., DeBlon, P.B., Committee, E.C.I.: Top-10-IT issues. EDUCAUSE Rev. **41**, 58–79 (2006)
4. Gheorghe, M.: Audit methodology for IT governance. Informatica Economica **4**(1), 32–42 (2010)
5. Ajami, R., Al-Qirim, N.: Governing IT in higher education institutions. In: Advanced Science and Technology Letters, vol. 36 (2013)
6. ITGI. IT Governance Institute: An Executive View of IT Governance. Ills. Rolling Meadow (2009)
7. Weill, P.: Don't just lead govern: how top performing firms govern IT. MIS Q. Executive **3** (1), 1–17 (2004)
8. Pande, P., Holpp, I.: What is Six Sigma?. Berkshire, McGraw-Hill, Maidenhead (2001)
9. Al-Atiqi, I.M., Deshpande, P.B.: Transforming higher education with six sigma. In: INQAAHE International Network of Quality Assessment Agencies in Higher Education Biannual Conference in Abu Dhabi (2009)
10. Gomes, R.J.: IT governance using COBIT implemented in a higher public educational institutions: a case study. In: Proceedings of the 3rd International Conference on European Computing Conference, pp. 41–52 (2009)
11. Mutyarini, K., Sembiring, J.: Sistem informasi untuk institusi perguruan tinggi di Indonesia. In: Conference Proceeding of ICT for Indonesia, Bandung (2006)

12. O'Donnell, E.: Discussion of director responsibility for IT governance: a perspective on strategy. Int. J. Account. Inf. Syst. **5**, 101–104 (2004)
13. Anugrah, B.S.: Pengukuran Tingkat Kematangan Tata Kelola Teknologi Informasi Menggunakan Model Kematangan COBIT di PT Bank Mandiri, Tbk, unpublished thesis. ITS, Surabaya (2008)
14. Setiawan, A.: Evaluasi Penerapan teknologi informasi di perguruan tinggi swasta Yogyakarta dengan menggunakan COBIT Framework. In: National Seminar of Information Technology Application, Yogyakarta (2008)
15. Lapao, L.V.: Organizational challenges & barriers to implement IT governance in a hospital. Electron. J. Inf. Syst. Eval. **4**(1), 37–45 (2011)
16. Maria, E., Haryani, E.: Audit model development of academic information system: Case study on academic information system of Satya Wacana. J. Art Sci. Commer. Res. World **2**(2), 1–13 (2011)
17. Maria, E., Fibriani, C., Sinatra, L.: The measurement of information technology performance in Indonesian higher educational institution in the context of achieving institution business goals using COBIT framework version 4.1 (case study: Satya Wacana Christian University Salatiga). Int. Refereed Res. Arts Sci. Commer. **3**(3), 9–19 (2012)
18. Sadikam, M., Hardi, H., Wachyu, H.H.: IT governance self-assessment in higher education based on COBIT case study: Universiti of Mercu Buana. J. Adv. Manage. Sci. **2**(2), 83–87 (2014)
19. Schekkerman, J.: Extended Architecture Maturity Model Support Guide v 2.0. Support Guide. IFEAD Institute for Enterprise Architecture Developments, The Netherlands (2006)
20. Yanosky, R., Caruso, J. B.: Key findings process and politics: IT governance in higher education. Educause Center for Applied Research (ECAR) (2008)
21. Llorens, F., Fernandez, A.: Conclusiones del Taller de Gobierno de las T1 en Las universides. Seminario, Gobierno, de. Las Tie n las universidades Espanolas-sectional TIC de la CRUE, Universidad Politecnica de Madrid (2008) (www.upm.es/eventos/gobrernoTI-SUE)
22. Harmse, L.: The Alignment of Business with IT within Tertiary Education Institution through IT Governance: A Study Conducted at The University of Pretoria. Unpublished Dissertation for Bachelors of Industrial Engineering (2010)
23. Benbasat, I., Goldstein, D., Mead, M.: The case research strategy in studies of information systems. MIS Q. **11**(3), 368–388 (1987)
24. Yin, R.K.: Case Study Research: Design and Methods, 3rd edn. Sage Publishing, Thousand Oaks (2003)

A Cognitive Knowledge-based Framework for Adaptive Feedback

Andrew Thomas Bimba[1](✉), Norisma Idris[1], Rohana Binti Mahmud[1], and Ahmed Al-Hunaiyyan[2]

[1] Faculty of Computer Science and Information Technology,
University of Malaya, 50603 Kuala Lumpur, Malaysia
bimba@siswa.um.edu.my,
{norisma,rohanamahmud}@um.edu.my
[2] Computer and Information Systems Department,
Public Authority for Applied Education and Training, Kuwait City, Kuwait
hunaiyyan@hotmail.com

Abstract. Adaptive learning environments provide personalization of the instructional process based on different parameters such as: sequence and difficulty of task, type and time of feedback, learning pace and others. One of the key feature in learning support is the personalization of feedback. Adaptive feedback support within a learning environment is useful because most learners have different personal characteristics such as prior knowledge, learning progress and learning preferences. In a computer-based learning environment, feedback is considered as one of the most effective factors which influence learning. Although, there are various tools that provide adaptive feedback in learning environments, some problems still exist. One of the problems we are looking into is *How to design effective tutoring feedback strategies?* We propose a cognitive knowledge based framework for adaptive feedback, which combines the three facets of knowledge (pedagogical, domain and learner model) in a learning environment, using concept algebra.

1 Introduction

Adaptive learning environments provide personalization of the instructional process based on different parameters such as: sequence and difficulty of task, type and time of feedback, learning pace and others [1, 2]. It is a more personalized, data-driven and technology-enabled approach to learning, which has a potential to improve student retention, measure student learning and improve pedagogy [3]. An adaptive learning system alters its behavior based on how a learner interacts with it. These alterations are decided based on the learner's characteristics which are represented in the learner model [4]. It involves the accurate tracking of learner's activity, monitoring their individual characteristics and providing timely adaptive feedback according to effective pedagogical principles [5]. In a computer-based learning environment, feedback is considered as one of the most effective factors which influence learning [6–8]. Tailoring feedback according to learner's characteristics and other external parameters is a promising way to implement adaptation in computer-based learning environment

© Springer International Publishing AG 2017
S. Phon-Amnuaisuk et al. (eds.), *Computational Intelligence in Information Systems*,
Advances in Intelligent Systems and Computing 532, DOI 10.1007/978-3-319-48517-1_22

[5] Adaptive feedback unlike generic feedback is dynamic, as learners work through instructions different learners will receive different information [8]. A computer-based learning environment represents knowledge in form of models. The three key models in a computer-based learning environment are the pedagogical model, domain model and learner model. For a computer-based learning system to provide appropriate feedback to a student, it has to have a knowledge of effective approaches to teaching (pedagogical model), the subject been learned (domain model) and the learner (learner model) [8]. Knowledge base modeling and representation techniques such as expert knowledge base and ontologies have been used in representing pedagogical, domain and learner knowledge [9, 10]. However, ontologies are most frequently used in modeling pedagogical, domain and learner knowledge because, it could afford a consistent structure through a means of describing and processing concepts, terms and the relationship between them [11].

Feedback is frequently provided in a typical classroom setting, however most of the information is poorly received, because feedback is presented to groups and so often students do not believe such feedback is relevant to them [7]. Currently, the gap between students who excel the most and those who excel less is a challenge that teachers, school administrators, and government officials face frequently [12]. Although, there is an increase in the development of adaptive feedback-based learning environments [6, 13, 14], some problems still remain unresolved. One of the questions faced is *How to design effective tutoring feedback strategies?* [5]. Several researchers have recommended frameworks with methods for implementing adaptive feedback [15, 16], however, these methods are not fully grounded in thorough educational theories and their empirical evaluation with students has been lacking. The design, implementation and evaluation of adaptive feedback strategies is quite challenging, due to various variables such as learning strategies and style, prior knowledge, motivational state and cognitive skills, which can hinder or facilitate the effectiveness of feedback in learning. A background on adaptive feedback in learning environments and knowledge modeling will be discussed in Sect. 2. In order to provide effective feedback, we propose a cognitive knowledge-based approach, which will be discussed in Sect. 2.2. A mechanism to evaluate our proposed technique will be briefly introduced in Sect. 3 and we will conclude in Sect. 4.

2 Background

The two main approaches to teaching and learning are the deductive and inductive techniques [17]. Traditionally, using the deductive approach, Science and Engineering are taught by introducing a topic on general principles; then using the principles, models or equations are derived; afterwards, an illustration of the application of the models are provided; assignments with similar derivations are given for practice and finally student's abilities are tested in exams [17]. There is little attention on what real-world phenomena these models explain. No explanation on the practical problems they can solve and why the students should care about it. A preferable substitute is the inductive approach. Instead of starting with the general principles, the lesson commences with real-world occurrences such as a case study to analyze, experimental data to interpret, or a complex real-world problem to solve [17]. As the students try to solve the problem, they create a need for rules, procedures, facts and guiding principles, only at this point are they provided with the needed information or guided to discover it. However, in practice, teaching is not purely inductive or deductive. Learning involves moving in both directions, where student infer rule and principles from new observations and test the principles through deducing consequences and applications that are verifiable experimentally. The inductive approach is learner-centered, enabling students to take responsibility of their own learning. Instructional methods which adopt the inductive approach include problem-based learning, inquiry learning, project-based learning, case-based teaching, discovery learning and just-in-time teaching [17]. These methods are in-line with the *constructivist* idea that knowledge is constructed and re-constructed in an individual's brain, rather than simply absorbing what is given by the teacher [17–20].

In a computer-based learning environment the pedagogical model represents the knowledge and expertise of teaching. Specific knowledge represented in the pedagogical model includes, effective teaching techniques (deductive and inductive); the various instructional methods (lectures, problem-based learning, inquiry learning etc.); Instructional plans that define phases, roles and sequence of activities [21]; feedback types, depending on a learner's action; assessment to inform and measure learning [12]. The domain model represents knowledge of the subject been learned. It mainly consists of concepts such as how to add, subtract, multiply numbers; newton's law of motion; how to structure an argument; different approaches to reading etc. [12]. The size of the domain knowledge which represents concepts can differ between computer-based systems, according to the domain size, application area and the choice of the designer [22]. In complex settings, concepts can have relationships between each other, resulting in a conceptual network representing the domain's knowledge. Common relationships used by most systems include *prerequisite* (where the learner has to know the first concept before studying the next related concept), *is-a* (where a concept is an instance of another concept) and *part-of* (where a concept is part of another concept) [22, 23]. The learner model represents the student's knowledge of the domain, individual characteristics and personal interactions with the computer [12, 22].

The learner model contains information about the user's current knowledge of the domain; user characteristics such as preferences, learning style, cognitive style etc. and; interactions with the system such as student's current activity, previous achievements and difficulties, concepts mastered, emotional state, time spent on task, kinds of errors made, misconceptions, response to feedback etc. [12, 22, 24]. This information can then be used by the domain and pedagogical models to assess the learner and decide the next most appropriate interaction [12]. Various knowledge modelling techniques have been used to model domain knowledge. Cognitive Tutors [10] used production-rules, while Ontologies have been used by Munoz-Merino (2011) [25]. A semantic network is used by Web PVT to represent relationships among concepts of the domain knowledge [23]. Most computer-based learning environments do not clearly state the knowledge base modeling technique used in the pedagogical, domain and learner models. In the next sections we will discuss the characteristics of adaptation and feedback, the types of adaptive feedbacks, previous works on adaptive feedback based on adaptive characteristics, knowledge modeling in a cognitive knowledge base and how it can represent pedagogy, domain and learner models to support adaptive feedback.

2.1   Approaches to Adaptive Feedback in Learning Environments

Brusilovsky (1996) defined systems that model student's learning style, prior knowledge, goals and preference as adaptive [26]. According to Chieu (2005), there are five main adaptation techniques which are related to the key components of constructive learning environment [27]. He highlighted them as follows:

- Adaptive presentation of learning contents. The course designer should define which learning contents are appropriate to a specific learner at any given time, for example simpler situations and examples for a *novice* learner than for an *expert* one.
- Adaptive use of pedagogical devices. The course designer should define which learning activities are appropriate to a specific learner, for instance simpler tasks to a *novice* learner than to an *expert*.
- Adaptive communication support. The course designer should identify which peers are appropriate to help a specific learner, for example learners with more-advanced mental models help learners with less-advanced ones.
- Adaptive assessment. The course designer should identify which assessment problems and methods are appropriate to determine the actual performance of a specific learner, for instance simpler tests for a *novice* learner than for an *expert*.
- Adaptive problem-solving support. The tutor should give appropriate feedback during the problem-solving process of a specific learner, for example to show the learner his or her own difficulties and provide him or her with the way to overcome those difficulties.

These adaptation techniques rely on a *learner model*; an essential component which, among other student relevant data, keeps data about the student's knowledge of the subject domain under study.

Adaptive learning involves multiple disciplines such as Educational Psychology, Cognitive Science and Artificial Intelligence. This complexity prompted the structuring of research on adaptation along the methodological questions distinguishing means, target, goal and strategy [28].

- Adaptation Means: what information about the learner such as knowledge level, cognitive style, learning style, gender, student's current activity, previous achievements and difficulties, misconception is known and used for adaptation?
- Adaptation Target: what aspect of the instructional system (pedagogy and domain model) is adapted based on the learner model?
- Adaptation Goal: what are the pedagogical reasons for the system to adapt to the learner model? Is the system aiding inductive or deductive learning; is the system adapting to a specific instructional method based on the learner model?
- Adaptation Strategy: what are the steps and techniques used to adapt the system to the learner model, and how active or reactive are the learners and system to the adaptation process?

In a learning environment, feedback is seen as the teacher's (artificial or real) response to the student's action. There are four main characteristics of feedback: function, timing, schedule and type [29]. Although, other researchers [15] suggested other characteristics of feedback, we adhere to the characterization by Carter (1984)

Table 1. Characteristics of Feedback.

Characteristics of feedback	Explanation
Function	Feedback can be provided in relation to the instructional goals and objectives. For example, feedback is provided based on, cognitive functions such as promoting information processing, motivational functions such as developing and sustaining persistence or provide correct response
Timing	Feedback can be given with respect to timing. It could be in advance, appearing before an action; it could be immediate, appearing immediately after an action or delayed, appearing at a longer time after the action has been made. The feedback is intended to advise, notify, recommend, alert, inform or motivate the learner about some concerns
Scheduling	Feedback can also be made available at scheduled instances. For example, when the learner exceeds a certain time threshold, expertise level, after solving certain questions or after every subtopic
Type	There are various feedback types resulting from function, timing and scheduling. For example, verification feedback, avoidance feedback, correction feedback, informative feedback, cognitive feedback, emotional feedback, scheduled feedback, dynamic feedback, immediate feedback, advanced feedback, delayed feedback, comparative feedback, isolation feedback etc.

because it encompasses all other characterizations [29]. These characteristics are briefly explained in Table 1.

For developing an effective adaptive feedback framework, the characteristics of adaptation and feedback have to be taken into consideration. The pedagogy, domain and learner models have to be represented in a knowledge base where effective algorithms can be used to implement effective feedback strategies. The next section discusses a promising knowledge representation technique for modeling knowledge in an adaptive learning environment.

2.2 Knowledge Modeling in Cognitive Knowledge Base

Conventional technologies for knowledge base modeling and manipulation such as linguistic knowledge base, expert knowledge base and ontology are man-made rather than machine built. *The absence of thorough and sufficient operations on acquired knowledge, inflexible to learn knowledge synergy, and weak trans-formability among different knowledge bases gave rise to a novel approach, the cognitive knowledge base (CKB)* [30]. Based on the previous studies in cognitive science and neurophysiology [31, 32], *the foundations of human knowledge in the long-term memory can be represented by an object-attribute-relation model based on the synaptic structure of human memory, which represents the hierarchical and dynamic neural clusters of knowledge retained in memory as well as the logical model of knowledge bases* [33].

The cognitive knowledge base is a structure that manipulates knowledge as a dynamic concept network like the human knowledge processing [34, 35]. In CKB a concept is a cognitive unit which identifies and models real-world concrete entities and a perceived-world (abstract entity) [30]. The basic unit of knowledge in a CKB is a formal concept represented as an OAR model according to concept algebra [36]. While complex knowledge such as a theme are represented as multiple associate concepts, which forms a partial dynamic concept network (DCN).

The CKB structure as shown in Fig. 1 consist of the logical model, physical model, linguistic knowledge base and knowledge manipulation engine. The logical model of knowledge bases shared by humans and cognitive systems is known as the object-attribute-relation model [33]. The logical structure is modeled as a hierarchical network of concepts and themes. The logical knowledge base represents knowledge as Cartesian products of formal concepts. The physical knowledge base implements the memory structures of knowledge as a DCN. The linguistic knowledge base comprises of the initial words as modeled in WordNet and the representation of these words and their relation in form of the OAR model.

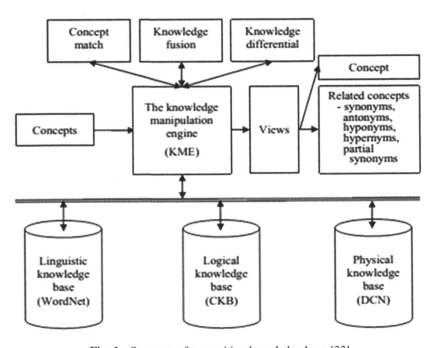

Fig. 1. Structure of a cognitive knowledge base [33]

The cognitive processes of concept memorization and knowledge fusion, similar to human and cognitive systems learning processes, are used for knowledge acquisition in CKB. Concept memorization involves acquiring concepts based on the formal structure model as shown in Fig. 2.

Concept|SM $\triangleq$ {< *ConceptID* : S | 1 < *ConceptID*|S < 100 >,   // ID of the concept
    < *A* : Ξ | *A*|Ξ = {A_1|SM, A_2|SM, ..., A_n|SM} >,   // Attributes
    < *O* : Ξ | *O*|Ξ = {O_1|SM, O_2|SM, ..., O_m|SM} >,   // Objects
    < *RC* : Ξ | *RC*|Ξ = *O*|Ξ × *A*|Ξ >,   // Internal relations
    < *RI* : Ξ | *RI*|Ξ = *C*|SM′ × *C*|SM > ,   // Input relations
    < *RO* : Ξ | *RO*|Ξ = *C*|SM′ × *C*|SM >   // Output relations
    }

Fig. 2. Structure of a formal concept [33].

Where:

A|Ξ- - set of attributes that denotes the intention of the concept.
O|Ξ- - set of objects that instantiated the extension of the concept.
RC|Ξ- - set of internal relations between objects and attributes.
RI|Ξ- - set of input relations between the concept and other concepts.
RO|Ξ- - set of output relations between the concept and other concepts denotes the intention of the concept.

The newly acquired item is retained as a formal concept based on the structure in Fig. 2. The concept memorization phase consists of two algorithms, the *EnterConcept* process model and the *ConceptMatch* process model. The former checks the condition of the knowledge base if available it indexes the current concept and its time stamp. The latter implements a content-addressed search mechanism to match a set of existing concepts in the knowledge base to the new concept.

The use of a cognitive knowledge base that models pedagogy, domain and learner knowledge and their connections can be used as a base for generating adaptive feedback in computer-based learning environment. This research focuses on designing such feedback framework. The next section discusses our proposed framework for adaptive feedback.

3   Proposed Framework for Adaptive Feedback

The relationships between pedagogical (knowledge and expertise of teaching), domain (Knowledge of subject taught) and learner (learner's knowledge, characteristics and interactions) models are quit complex. However, in other to provide effective feedback which is compliant with the characteristics of adaptation, a link has to be established between these three models. The cognitive-PDL model proposed in Fig. 3 shows the connections of these model using a cognitive knowledge base. The decision to use a CKB arises from the fact that concepts can be represented with both internal (relationship between concepts and attributes in a specific model) and output (relationships between concepts and attributes across models) relationships.

What distinguish our approach is the specification of the relationships between pedagogy, domain and learner knowledge. In practice, it means that instead of looking

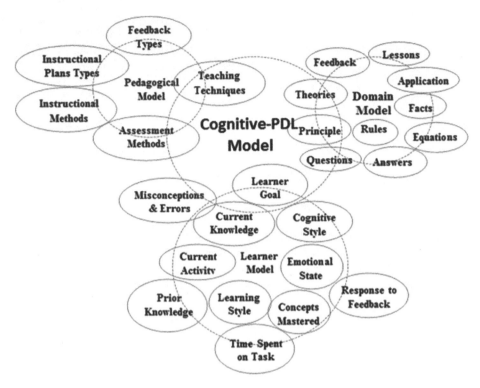

Fig. 3. Structure of a cognitive knowledge based pedagogical, domain and learner model

at all these components in isolation, we look at each concept as a 5-tuple. By definition each concept in a sub-model is represented as shown in Eq. 1.

$$c \triangleq (O, A, R^c, R^i, R^o) \tag{1}$$

Where:

- O is a nonempty set of an instance of the concept, $O = O_1, O_2, \dots O_m = \mathcal{P}U$ where $\mathcal{P}U\psi$ denotes a power set of ψU
- A is a nonempty set of attributes, $A = a_1, a_2, \dots a_n = \mathcal{P}M$
- $R^c \subseteq O \times A$ is a set of internal relations.
- $R^i \subseteq C' \times C$ is a set of input relations, where C' is a set of external concepts.
- $R^o \subseteq C' \times C$ is a set of output relations [33].

Knowledge acquisition in the Cognitive-PDL model will be through concept memorization and knowledge fusion. An algorithm is required to index a current concept to the Cognitive-PDL model after verifying its availability. Another algorithm searches the knowledge base to establish relationship between the new concept and a set of existing concepts in order to incorporate the new concept as acquired knowledge. A third algorithm decides the appropriate feedback to be provided based on the learners current state and pedagogical principles.

4   Conclusion

The need for adaptive feedback in learning environment cannot be over-emphasized. Learners have different personal characteristics such as prior knowledge, learning progress and learning preferences. Providing feedback based of an individual's characteristics is key to bridging the gap between students who excel most and those who excel less. Although, there are various tools that provide adaptive feedback in learning environments, some problems still exist. One of the problems we are looking into is How to design effective tutoring feedback strategies?. Most researchers have proposed methods that are not fully grounded in pedagogical principles. Also, the three main knowledge components of a learning environment (pedagogical, domain and learner model) are mostly treated in isolation. We propose a cognitive knowledge based framework for adaptive feedback, which combines the three facets of knowledge in a learning environment, using concept algebra. The decision to use a CKB arises from the fact that concepts can be represented with both internal (relationship between concepts and attributes in a specific model) and output (relationships between concepts and attributes across models) relationships.

Acknowledgment. This work was supported by the University of Malaya Research Grant [RP040B-15AET, 2015].

References

1. Stoyanov, S., Kirchner, P.: Expert concept mapping method for defining the characteristics of adaptive e-learning: ALFANET project case. Educ. Technol. Res. Dev. **52**, 41–54 (2004)
2. Brusilovsky, P.: Adaptive and intelligent technologies for web-based eduction. Ki **13**, 19–25 (1999)
3. Advisors, E.G.: Learning to adapt: a case for accelerating adaptive learning in higher education. In: Education Growth Advisors, March 2013
4. Lo, J.-J., Chan, Y.-C., Yeh, S.-W.: Designing an adaptive web-based learning system based on students' cognitive styles identified online. Comput. Educ. **58**, 209–222 (2012)
5. Narciss, S., Sosnovsky, S., Schnaubert, L., Andrès, E., Eichelmann, A., Goguadze, G., et al.: Exploring feedback and student characteristics relevant for personalizing feedback strategies. Comput. Educ. **71**, 56–76 (2014)
6. Thurlings, M., Vermeulen, M., Bastiaens, T., Stijnen, S.: Understanding feedback: a learning theory perspective. Educ. Res. Rev. **9**, 1–15 (2013)
7. Hattie, J., Gan, M.: Instruction based on feedback. In: Mayer, R., Alexander, P. (eds.) Handbook of Research on Learning and Instruction, pp. 249–271. Routledge, New York (2011)
8. Le, N.-T.: A classification of adaptive feedback in educational systems for programming. Systems **4**, 22 (2016)
9. Zhihong, T., Liu, W., Libing, L., Zeqing, Y.: The application of ontology model in intelligent tutoring system. In: Proceedings - International Conference on Computer Science and Software Engineering, CSSE 2008, pp. 1176–1179 (2008)
10. Anderson, J.R., Corbett, A.T., Koedinger, K.R., Pelletier, R.: Cognitive tutors: lessons learned. J. Learn. Sci. **4**, 167–207 (1995)

11. Bimba, A.T., Idris, N., Al-Hunaiyyan, A., Mahmud, R.B., Abdelaziz, A., Khan, S., et al.: Towards knowledge modeling and manipulation technologies: a survey. Int. J. Inf. Manag. **36**, 857–871 (2016)

12. Luckin, R., Holmes, W.: Intelligence unleashed: an argument for AI in education (2016)

13. van der Kleij, F.M., Eggen, T.J., Timmers, C.F., Veldkamp, B.P.: Effects of feedback in a computer-based assessment for learning. Comput. Educ. **58**, 263–272 (2012)

14. Narciss, S.: Designing and evaluating tutoring feedback strategies for digital learning. Digital Educ. Rev. **23**, 7–26 (2013)

15. Economides, A.A.: Adaptive feedback characteristics in CAT. Int. J. Instr. Technol. Distance Learn. **3**, 15–54 (2006)

16. Ana, G.S., Macario, D.-S.J.: Designing feedback to support language acquisition using the InGenio authoring tool. Procedia Soc. Behav. Sci. **1**, 1239–1243 (2009)

17. Prince, M.J., Felder, R.M.: Inductive teaching and learning methods: definitions, comparisons, and research bases. J. Eng. Educ. **95**, 123–138 (2006)

18. Driscoll, M.P.: Psychology of Learning for Instruction. Allyn & Bacon, Boston (2005)

19. Gillani, B.B.: Learning Theories and the Design of e-Learning Environments. University Press of America, Lanham (2003)

20. Yilmaz, K.: The cognitive perspective on learning: its theoretical underpinnings and implications for classroom practices. Clearing House J. Educ. Strat. Issues Ideas **84**, 204–212 (2011)

21. Scheuer, O., Loll, F., Pinkwart, N., McLaren, B.M.: Computer-supported argumentation: a review of the state of the art. Int. J. Comput. Support. Collaborative Learn. **5**, 43–102 (2010)

22. Kazanidis, I., Satratzemi, M.: Adaptivity in a SCORM compliant adaptive educational hypermedia system. In: Leung, H., Li, F., Lau, R., Li, Q. (eds.) ICWL 2007. LNCS, vol. 4823, pp. 196–206. Springer, Heidelberg (2008)

23. Virvou, M., Tsiriga, V.: Web passive voice tutor: an intelligent computer assisted language learning system over the WWW. In: Proceedings - IEEE International Conference on Advanced Learning Technologies, ICALT 2001, pp. 131–134 (2001)

24. Carmona, C., Conejo, R.: A learner model in a distributed environment. In: De Bra, P.M., Nejdl, W. (eds.) AH 2004. LNCS, vol. 3137, pp. 353–359. Springer, Heidelberg (2004)

25. Muñoz-Merino, P.J., Pardo, A., Scheffel, M., Niemann, K., Wolpers, M., Leony, D., et al.: An ontological framework for adaptive feedback to support students while programming. In: International Semantic Web Conference (2011)

26. Brusilovsky, P.: Methods and techniques of adaptive hypermedia. User Model. User-Adap. Inter. **6**, 87–129 (1996)

27. Chieu, V.M.: Constructivist learning: an operational approach for designing adaptive learning environments supporting cognitive flexibility. Universiteit Twente (2005)

28. Specht, M.: Adaptive Methoden in computerbasierten Lehr/Lernsystemen. GMD-Forschungszentrum Informationstechnik (1998)

29. Carter, J.: Instructional learner feedback: a literature review with implications for software development. Comput. Teach. **12**, 53–55 (1984)

30. Wang, Y.: Concept algebra: a denotational mathematics for formal knowledge representation and cognitive machine learning. J. Adv. Math. Appl. **4**, 1–26 (2015)

31. Hampton, J.A.: Psychological representation of concepts. In: Conway, M.A. (ed.) Cognitive Models of Memory, pp. 81–110. MIT Press, Cambridge (1997)

32. Leone, N., Pfeifer, G., Faber, W., Eiter, T., Gottlob, G., Perri, S., et al.: The DLV system for knowledge representation and reasoning. ACM Trans. Comput. Logic (TOCL) **7**, 499–562 (2006)

33. Wang, Y.: On a novel cognitive knowledge base (CKB) for cognitive robots and machine learning. Int. J. Softw. Sci. Comput. Intell. (IJSSCI) **6**, 41–62 (2014)

34. Wang, Y., Tian, Y., Hu, K.: The operational semantics of concept algebra for cognitive computing and machine learning. In: 2011 10th IEEE International Conference on Cognitive Informatics & Cognitive Computing (ICCI*CC), pp. 49–58 (2011)
35. Wang, Y.: On concept algebra: a denotational mathematical structure for knowledge and software modeling. Int. J. Cogn. Inf. Nat. Intell. (IJCINI) 2, 1–19 (2008)
36. Valipour, M., Yingxu, W.: Formal properties and rules of concept algebra. In: 2015 IEEE 14th International Conference on Cognitive Informatics & Cognitive Computing (ICCI*CC), pp. 128–135 (2015)

Creative Computing

Towards Developing a Therapeutic Serious Game Design Model for Stimulating Cognitive Abilities: A Case for Children with Speech and Language Delay

Nadia Akma Ahmad Zaki[1(✉)], Tengku Siti Meriam Tengku Wook[2], and Kartini Ahmad[3]

[1] Faculty of Art, Computing and Creative Industry,
Sultan Idris Education University, Tanjong Malim, Perak, Malaysia
nadiaakma@fskik.upsi.edu.my
[2] Faculty of Information Science and Technology,
National University of Malaysia, Bangi, Selangor, Malaysia
tsmeriam@ukm.edu.my
[3] Faculty of Health Sciences, National University of Malaysia,
Bangi, Selangor, Malaysia
kart@ukm.edu.my

Abstract. Recently, serious games aimed at cognitive therapy have been gaining increasing importance in general health applications, creating new possibilities for various groups, including children, to access new forms of treatment. However, to develop therapeutic serious games that stimulate the cognitive abilities of children with speech and language delay (CSLD), the needs of this group and a few other issues must be considered. Therefore, this paper aims to identify the problems affecting the cognitive functions of CSLD as well as their needs, to assist in the development of a serious game for CSLD. We conducted a preliminary study through a semi-structured interview with experts in the area of therapy. Our results indicate that CSLD indeed have major difficulties that affect the development of their cognitive abilities such as memory, attention, perception, problem solving, decision making, language, learning, and reasoning. In addition, CSLD also lack preverbal and motor skills. These findings reinforce the need to propose a model for therapeutic serious game design that stimulates the cognitive abilities of CSLD.

Keywords: Therapeutic · Serious games · Serious game design · Cognitive stimulation · Cognitive abilities · Children with speech and language delay

1 Introduction

Serious games have increasingly become a good option for therapeutic treatment and are considered non-pharmacological therapies to train, evaluate, enhance, and stimulate various cognitive functions. The cognitive stimulation concept is an intervention that affects most of the human cognitive domains [1]. Existing work has highlighted the potential of cognitive stimulation in enhancing the cognitive ability of children [2, 3].

© Springer International Publishing AG 2017
S. Phon-Amnuaisuk et al. (eds.), *Computational Intelligence in Information Systems*,
Advances in Intelligent Systems and Computing 532, DOI 10.1007/978-3-319-48517-1_23

The low cognitive ability of some children can be stimulated with the use of serious games because these types of games can boost brain function and improve well-being [4, 5]. Besides that, these games provide new solutions in the form of fun, repetitive exercises, which increase the motivation and engagement of the players, and are also attractive, user-friendly, pleasant, and tied to an objective [6–8].

Language difficulties affecting children are the most common developmental problems encountered by clinicians and the number one concern voiced by parents [9, 10]. Children with speech and language delay (CSLD) also face developmental problems. Previous studies have discovered that pre-school-level CSLD are at a high risk of having learning disabilities such as poor literacy development [11–14]. Hence, this can lead to lowered school performance, which may persist into young adulthood; and would in turn affect their psychosocial development, causing emotional and behavioural problems at school [12, 15–17]. Past studies have also shown that language difficulties often co-occur together with cognitive delays [18] and are associated with lower IQ (intelligence quotient) scores [19, 20].

In Malaysia, statistics from the Social Welfare Department showed an increase in children with learning difficulties, in which most suffer from delayed speech and language problems [21]. These children are recommended to undergo rehabilitation and therapy that focus on lingustic and cognitive ability, as the chances for them to heal would be much higher [14, 22, 23]. Therapists have established diverse, well-structured therapeutic procedures for CSLD. Overall, a typical clinical session for these patients is designed as a pair-wise playing and talking session between the therapist and the child. During the therapeutic process of CSLD, drawings, still diagrams, educational toys, and action pictures are widely used together with real objects representing daily concepts and events [24]. Therefore, therapists have to provide a variety of materials for different tasks in each therapy session to stimulate the cognitive and language ability of CSLD. Recently, Malaysia has been facing a shortage of speech therapists [25]. This could lead to difficulties for CSLD, who require further treatment, so as to be qualified for a therapeutic session.

Therefore in this study, we introduce new opportunities for serious games to stimulate cognitive abilities, with therapeutic implications for CSLD. The combination of therapeutic content and gaming elements with therapeutic objectives can create new, motivating, and engaging therapeutic environments [26]. In order to successfully stimulate the cognitive side of the mind through therapeutic serious games, these games must be well-designed to effectively and efficiently deliver training. Meanwhile, the user must feel at ease with the technology, and be willing and motivated in accepting new concepts and in learning new skills. Furthermore, serious game design must take into account the fundamental knowledge regarding the needs and preferences of the target population, as well as ensure that the therapeutic adherence would not be low and that the intervention will succeed [27]. However, the lack of suitable therapeutic serious games as a tool for stimulating the cognitive ability of children, especially CSLD, has given rise to a pressing need to develop a game that would meet the requirements of this target group. Therefore, this paper aims to identify the problems affecting the cognitive functions of CSLD as well as their needs, to assist in the development of therapeutic serious games for CSLD from 4 to 7 years old.

The remaining sections of the paper are organised as follows; Sect. 2 outlines the overview of the main themes related to this work; Sect. 3 describes the preliminary study made by experts in the area of therapy; Sect. 4 discusses the findings; and Sect. 5 lays out our conclusions and directions for future work.

2   Related Works

2.1   Serious Games for Therapeutic and Cognitive Stimulation

Serious games are digital applications with the primary purpose of more than just entertainment. Recently, therapeutic serious games have captured the attention of healthcare practitioners as a proven and efficient form of medical therapy, which could increase patient motivation and engagement in the therapeutic process [28]. Therapeutic serious games are games that integrate gaming elements with therapeutic objectives [26]. Most of these therapeutic serious games have been applied to treat autistic spectrum disorder, anxiety disorders, language learning impairments, cognitive behavioural therapy, and solution-focused therapy [26]. There are numerous evidences that prove the effectiveness of these therapies in improving the cognition, behaviour, mood, and activities of a patient's daily living, and in delaying institutionalisation, as well as in improving the quality of life of patients and caregivers alike [29]. One of these approaches, i.e. the cognition-focused intervention, is typically designed to promote cognitive stimulation and minimise cognitive impairment; with the direct or indirect aim of improving cognitive functioning [30, 31].

There are a few types of cognitive interventions such as cognitive training, cognitive stimulation, and cognitive rehabilitation [1, 29, 32]. Evidence suggests that cognitive stimulation treatment is the best treatment out of all the non-pharmacological therapies that yields positive outcomes, although this approach is labour-intensive, and requires further evaluations in regard to its cost-effectiveness [29]. Cognitive stimulation refers to interventions that promote increased patient engagement in mentally stimulating activities, where engagement levels are associated with the rates of cognitive decline i.e. the higher the involvement, the lower the rate of cognitive decline [32]. The premise is that practice that includes specific cognitive function tasks may improve, or at least maintain functioning in a given domain and that any effects of practice could be generalised and also induce a general improvement in cognitive and social functioning [29].

Serious games that are used for cognitive stimulation would typically involve a set of tasks designed in the form of activities that reflect cognitive functions such as attention, memory, problem-solving, language, reasoning, learning, perception, and decision making or acting in a particular context [33, 34]. The knowledge or cognitive aspect embedded in serious games has the potential to stimulate the ability of certain individuals to interpret, learn, and understand a situation [35]. This aspect of the game has been designed with a focus on engaging players in the game-play experience, whilst simultaneously addressing the cognitive component and optimising user performance [36, 37]. Despite a growing number of recent studies on the use of serious games as a form of therapy for stimulating the cognitive abilities of various groups, therapeutic serious games that focus on children are still scarce [38] and only focus on a few cognitive aspects. Therefore, the potential of therapeutic serious games for stimulating the cognitive abilities of CSLD is highly warranted.

2.2   User Characteristics of CSLD

Speech refers to the motor act of communicating by articulating verbal expressions; language is commonly thought of in its spoken form, which includes receptive language (understanding), expressive language (the ability to convey information, feelings, thoughts, and ideas), and visual forms, such as sign language [11, 13]. In general, a child is considered to have speech and language delay if his or her speech and language development is significantly below the norm for his or her age level [14, 39]. It has been reported that 5 to 8 per cent of preschool children have this disorder, which often persists into their school years [16]. This disorder is three times more common in boys than in girls [10, 39, 40].

CSLD include children who suffer from developmental delays of speech and language, expressive language disorder, receptive language disorders, hearing loss, intellectual disabilities, mental retardation, delayed growth, down-syndrome, autism, cerebral palsy, Attention Deficit Hyperactivity Disorder (ADHD), and physical speech problems [10, 11, 14, 16, 39]. The causes of risk factors for speech and language delay vary widely. Several studies on the potential risk factors for speech and language delay in children investigated heterogeneous populations with different individual and environmental characteristics. The most consistently reported risk factors were family history of speech and language delay, male gender, prematurity, and low birth weight. Other risk factors reported less consistently include parent education level, childhood illness, birth order, larger family size, low socioeconomic status, and the parents' higher than normal concerns with their children's language ability [11, 39, 40].

CSLD experience difficulties in understanding and expressing their interest and wants, which results in feelings of disappointment and low self-esteem, forcing them to socialise less [39, 41]. If this disorder continues up to adulthood, they are at risk of experiencing learning problems, stress, isolation from the community, and will have problems to secure jobs because of their communication impairment [16].

2.3   The Cognitive Development of a Child

Cognitive development refers to the progressive and continuous growth of attention, perception, memory, learning, language, decision making, and problem solving; where information is received, transformed, stored, and used to solve problems and process languages [42, 43]. Many theories have been proposed regarding how children learn or adapt to their environment and how cognitive development works. One of the significant theorists in this area is Jean Piaget, who used four basic concepts namely schema, assimilation, adaptation, and equilibrium to elaborate upon the activity process of an individual's cognitive structure [42]. At the centre of Piaget's theory is the principle that cognitive development occurs in a series of four distinct, universal stages, each characterised by increasingly sophisticated and abstract levels of thought [44, 45]. The first distinct stage is sensorimotor (from birth to 2 years), followed by preoperational (2 to 7 years), concrete operational (7 to 11 years), and lastly, formal operational (begins in adolescence and spans into adulthood). Since this study focuses on CSLD between 4 to 7 years old; they definitely will be counted as still in the preoperational stage (see [44, 45]). However, we also took into account the sensorimotor stage because even at age 4 (according to chronological age), the need to understand the development of individual psychology accurately no longer becomes relevant; this includes the use of predictive factors of health, and intelligence and mental capacity [46, 47].

3   Preliminary Study

In order to identify the problems affecting the cognitive functions of CSLD as well as their needs, so as to assist in the development of therapeutic serious games, a preliminary study with a group of experts was conducted. This is because it is crucial to consider the needs and preferences of real users before developing the design of therapeutic serious games. According to [48], at least 5 respondents are needed to conduct an interview, while [49] states that the number of respondents to be interviewed varies and could be only a few to many. The interview method was chosen because it is an effective method for obtaining in-depth data about a particular role or set of tasks and in determining what users want in the early design and user requirement phase [49, 50]. Therefore, a total of six experts were chosen as the respondents for the semi-structured interview in this study, which is based on a series of fixed questions with a scope that could be expanded based on their responses. The respondents consisted of five professionals' speech therapists from the Private and Public sectors and one professional occupational therapist from the Private sector.

The selection of respondents were based on the criteria that they would have much more experience in interacting with CSLD and in stimulating their cognitive ability during therapy sessions. Out of the 6 interviews, three were conducted face-to-face and the rest were done over the phone. The interview process took about three days, with an estimated time of 20–30 min per respondent. The interviews were audio-recorded so that they could be played back and annotated later at a more appropriate pace. The data obtained from the respondents were then transcribed into written form to be analysed via thematic analysis [51]. Thematic networks were subsequently used to organise the thematic analysis of qualitative data [52]. A thematic analysis seeks to unearth the themes relevant in a text at different levels, while thematic networks aim to facilitate the structuring and interpretation of these themes.

There are three levels of thematic networks; the lowest-order premises evident in the text are known as basic themes; the medium-order are categories of basic themes grouped together to summarise more abstract principles (organising themes); and super-ordinate themes or global themes encapsulate the core, principal metaphor, which in turn, encapsulates the main point in the text. These are then represented as web-like maps portraying the salient themes of each of the three levels, illustrating the relationship between them [52].

By adapting thematic analysis steps, we managed to derive 19 basic themes made up of repeating words mentioned in the interviews, as shown in Table 1. The 19 basic themes were then identified and matched according to 14 organising themes, as outlined in Table 2. The 14 organising themes were then grouped into three global themes, which have similar, coherent groupings, namely cognitive abilities, preverbal skills, and motor skills. The cognitive abilities group involve eight cognitive functions such as perception, learning, attention, memory, language, decision making, reasoning, and problem solving. The preverbal skills group includes five organising themes, namely eye contact, play, turn-taking, imitation, and attention. Attention is an organising theme that appeared in both the cognitive abilities and preverbal skills group. The motor skills group is made up of two organising themes, which are fine and gross motors.

Based on the salient themes of each of the three levels in Tables 1 and 2, we illustrated the relationship between them using a thematic analysis network. This thematic network is formed by working from the peripheral basic themes inwards and then to global themes. As a result, we produced a thematic network analysis, as shown in Fig. 1, in which the problems affecting the cognitive function of CSLD are identified. In addition, we also found that CSLD experience problems in preverbal and motor skill development.

Table 1. Identified Basic Themes

Basic themes	
1. Weak memory	2. Difficulty and delay in making decisions
3. Limited object search	4. Difficulty understanding abstract concepts
5. Doing things repeatedly	6. Lacking the ability to manipulate objects
7. Easily distracted	8. Poor imitation skills
9. Inappropriate play behaviours	10. Trouble with reasoning and logic skills
11. Short attention span	12. Delayed turn-taking skills
13. Poor expressive language	14. Lack of comprehension and understanding
15. Poor eye contact skills	16. Difficulty interpreting information
17. Poor hand-eye coordination	18. Difficulty and delay in solving problems
19. Delay in reading, spelling, and writing	

Table 2. Global and organising themes

Global Themes	Organising Themes		
Cognitive Abilities	1. Perception	2. Learning	
	4. Memory	5. Language	
	6. Decision making	7. Reasoning	
	8. Problem solving		3. Attention
Preverbal Skill	9. Eye contact	10. Play	
	11. Turn-taking	12. Imitation	
Motor Skills	13. Fine motor	14. Gross motor	

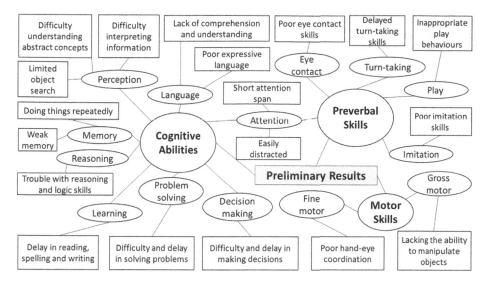

Fig. 1. Thematic network analysis

4 Discussion

Based on our preliminary study with the six experts in this study, our results using a thematic analysis has revealed that CSLD indeed have major difficulties in the development of their cognitive abilities such as memory, attention, perception, problem solving, decision making, language, learning, and reasoning. Besides that, CSLD have poor memory, as proven from sessions with them where the same thing needs to be done repeatedly until they can remember. According to [22, 53], working memory (short-term memory) plays an important role in phonologic development and language acquisition. The working memory is necessary for many cognitive tasks that we engage in through out our daily life and it is required for retaining and maintaining task-relevant information for problem solving and in maintaining attention [54]. Therefore, it is crucial that the working memory of CSLD be enhanced.

The results from our study reveal that CSLD face difficulty and delay in solving problems and making decisions. At the same time, they also have problems paying attention; they lack concentration, and are easily distracted with their surroundings, besides having a short attention span. Furthermore, they also experience learning difficulties, experiencing delays in reading, spelling, and writing. CSLD also have problems in language development either in the pre-speech, comprehension (lack of understanding), and/or pronunciation levels (poor expressive language).

Meanwhile, another problem identified by this study is that CSLD have difficulties with reasoning and logic such as the ability to sort, match, sequence, predict, and compare. In addition to having a weakness in matrix reasoning, they are also perceived as having weak perception such as difficulties interpreting information, difficulties understanding abstract concepts, and having limited object search. These issues reflect the weak visuospatial processing functions of CSLD. The findings of this study provide evidence of additional problems faced by children with speech and language difficulties, which are supported by recent well-documented studies [15, 39–41, 55].

In addition, CSLD also have difficulties in grasping preverbal skills; they cannot maintain appropriate eye contact, which is another important component of social interaction; and are lacking in joint attention skills such as sharing and shifting attention from one activity, person, or object to another. Furthermore, CSLD also experience delay in turn-taking skills, and lack the ability to imitate others. They also have poor play skills such as the inability to complete a puzzle, not retaining action on objects, and displaying disinterest towards age-appropriate toys. If these preverbal skills were not honed at this age, a child's speech and language development would be affected, which would in turn, disrupt his or her learning process. Meanwhile, some CSLD may have problems with their fine motor skills, i.e. poor hand-eye coordination, and in their gross motor skills. CSLD who have difficulty sitting still during an activity and maintaining a table-top posture (upper body support) would be affected in their ability to participate in activities involving fine motor skill such as writing, drawing, and cutting, all of which could potentially impact their learning. The ability to sit still during an activity is crucial for a child to learn speech and language. Good sitting skills help the child to pay attention to an activity. A child who is not able to sit still or is hyperactive would usually be hard to interact with. This shows that gross motor skills are important to enable children to perform every day functions, and without them, their other everyday functions would be adversely affected.

5   Conclusion

Therapeutic serious games for stimulating cognitive abilities signifies the increasing amount of cognitive processes that have been integrated within the overall game-playing experience. Despite a growing interest regarding the use of serious games as a form of therapy for stimulating cognitive abilities, therapeutic serious games focusing on children are still scarce. This has warranted the development of a serious game that really meets the needs of children, especially CSLD. To design a therapeutic serious game for the cognitive stimulation of CSLD in a systematic manner, we need to be guided by a model that clarifies the relationship between design and cognition. Currently, however, there is no such model in the literature. In addition, the game design implications are only based on the assumption that designers should identify cognitive difficulties in performing a certain task and then attempt to minimise them by careful design. Results from our preliminary study indicate that CSLD face major difficulties in developing cognitive abilities such as memory, attention, perception, problem solving, decision making, language, learning, and reasoning. Thus, regarding the direction for future work, we will propose a model for therapeutic serious game design that stimulates the cognitive abilities and addresses the cognitive difficulties of CSLD, whilst also taking into account the difficulties they face in developing their preverbal and motor skills.

References

1. Clare, L., Woods, R.T.: Cognitive training and cognitive rehabilitation for people with early-stage Alzheimer's disease: a review. Neuropsychol. Rehabil. **14**, 385–401 (2004)
2. Hackman, D.A., Farah, M.J., Meaney, M.J.: Socioeconomic status and the brain: mechanistic insights from human and animal research. Nat. Rev. Neurosci. **11**, 651–659 (2010)
3. Bonnier, C.: Evaluation of early stimulation programs for enhancing brain development. Acta Paediatr. Int. J. Paediatr. **97**, 853–858 (2008)
4. Bavelier, D., Green, C.S., Seidenberg, M.S.: Cognitive development: gaming your way out of dyslexia? Curr. Biol. **23**(7), R282–R283 (2013)
5. Wankoff, L.S.: Warning signs in the development of speech, language, and communication: when to refer to a speech-language pathologist. J. Child Adolesc. Psychiatr. Nurs. **24**, 175–184 (2011)
6. Hussaan, A.M., Sehaba, K.: Adaptive Serious Game for Rehabilitation of Persons with Cognitive Disabilities. 2013 IEEE 13th Int. Conf. Adv. Learn. Technol. 65–69 (2013)
7. Andrade, K. de O., Fernandes, G., Martins J., J., Roma, V.C., Joaquim, R.C., Caurin, G.A.P.: Rehabilitation robotics and serious games: an initial architecture for simultaneous players (BRC 2013), ISSNIP, pp. 1–6 (2013)
8. Dores, A., Carvalho, I., Barbosa, F., Almeida, I., Guerreiro, S., Leitão, M., de Sousa, L., Castro-Caldas, A.: Serious games: are they part of the solution in the domain of cognitive rehabilitation? In: Ma, M., Fradinho Oliveira, M., Madeiras Pereira, J. (eds.) Serious Games Development and Applications SE - 9, pp. 95–105. Springer, Berlin Heidelberg (2011)
9. Buschmann, A., Jooss, B., Rupp, A., Dockter, S., Blaschtikowitz, H., Heggen, I.: Children with developmental language delay at 24 months of age: results of a diagnostic work-up. Dev. Med. Child Neurol. **50**, 223–229 (2008)
10. Parakh, M., Parakh, P., Bhansali, S., Gurjar, A.: A clinico-epidemiologic study of neurologic associations and factors related to speech and language delay. Natl. J. Community Med. **3**, 518–523 (2012)
11. McLaughlin, M.R.: Speech and language delay in children. Am. Fam. Physician **83**, 1183–1188 (2011)
12. Lafferty, A.E., Gray, S., Wilcox, M.J.: Teaching alphabetic knowledge to pre-school children with developmental language delay and with typical language development. Child Lang. Teach. Ther. **21**, 263–277 (2005)
13. Shetty, P.: Speech and language delay in children: a review and the role of a pediatric dentist. J. Indian Soc. Pedod. Prev. Dent. **30**, 103–108 (2012)
14. Lawrence, R., Bateman, N.: 12 minute consultation: an evidence-based approach to the management of a child with speech and language delay. Clin. Otolaryngol. **38**, 148–153 (2013)
15. Hay, I., Elias, G., Fielding-Barnsley, R., Homel, R., Freiberg, K.: Language delays, reading delays, and learning difficulties: interactive elements requiring multidimensional programming. J. Learn. Disabil. **40**, 400–409 (2007)
16. Nelson, H., Nygren, P., Walker, M., Panoscha, R.: Screening for speech and language delay in preschool children: systematic evidence review for the US preventive services task force. Pediatrics **117**(12), e298–e319 (2006)
17. Hagberg, B.S., Miniscalco, C., Gillberg, C.: Clinic attenders with autism or attention-deficit/hyperactivity disorder: cognitive profile at school age and its relationship to preschool indicators of language delay. Res. Dev. Disabil. **31**, 1–8 (2010)

18. Jansen, R., Ceulemans, E., Grauwels, J., Maljaars, J., Zink, I., Steyaert, J., Noens, I.: Young children with language difficulties: a dimensional approach to subgrouping. Res. Dev. Disabil. **34**, 4115–4124 (2013)

19. Liao, S.F., Liu, J.C., Hsu, C.L., Chang, M.Y., Chang, T.M., Cheng, H.: Cognitive development in children with language impairment, and correlation between language and intelligence development in kindergarten children with developmental delay. J. Child Neurol. (2014)

20. Lundervold, A.J., Posserud, M., Sørensen, L., Gillberg, C.: Intellectual function in children with teacher reported language problems. Scand. J. Psychol. **49**, 187–193 (2008)

21. Jabatan Kebajikan Masyarakat: Laporan Statistik Jabatan Kebajikan Masyarakat (2014)

22. van der Schuit, M., Segers, E., van Balkom, H., Verhoeven, L.: How cognitive factors affect language development in children with intellectual disabilities. Res. Dev. Disabil. **32**, 1884–1894 (2011)

23. Law, J., Garrett, Z., Nye, C.: Speech and language therapy interventions for children with primary speech and language delay or disorder. Cochrane Database Syst. Rev. **1**, 1–62 (2003)

24. Cagatay, M., Ege, P., Tokdemir, G., Cagiltay, N.E.: A serious game for speech disorder children therapy. In: 2012 7th International Symposium on Health Informatics and Bioinformatics, HIBIT 2012, pp. 18–23 (2012)

25. Bernama: UKM Study Shows Malaysia Needs More Speech Therapists. http://www.themalaymailonline.com/malaysia/article/ukm-study-shows-malaysia-needs-more-speech-therapists

26. Wrzesien, M., Raya, M.A., Botella, C., Burkhardt, J.-M., Breton-López, J., Ortega, A.R.: A pilot evaluation of a therapeutic game applied to small animal phobia treatment. In: 5th International Conference on Serious Games Development and Applications, pp. 10–20 (2014)

27. Boot, W.R., Champion, M., Blakely, D.P., Wright, T., Souders, D.J., Charness, N.: Video games as a means to reduce age-related cognitive decline: attitudes, compliance, and effectiveness. Front. Psychol. **4**, 31 (2013)

28. Mader, S., Natkin, S., Levieux, G.: How to analyse therapeutic games: the player/game/therapy model. Entertain. Comput. **2012**, 193–206 (2012)

29. Mapelli, D., Di Rosa, E., Nocita, R., Sava, D.: Cognitive stimulation in patients with dementia: randomized controlled trial. Dement. Geriatr. Cogn. Dis. Extra. **3**, 263–271 (2013)

30. Muscio, C., Tiraboschi, P., Guerra, U.P., Defanti, C.A., Frisoni, G.: Clinical trial design of serious gaming in mild cognitive impairment. Front. Aging Neurosci. **7**, 26 (2015)

31. Dreyer, B.P.: Early childhood stimulation in the developing and developed world: if not now, when? Pediatrics **127**, 975–977 (2011)

32. Kelly, M.E., Loughrey, D., Lawlor, B.A., Robertson, I.H., Walsh, C., Brennan, S.: The impact of cognitive training and mental stimulation on cognitive and everyday functioning of healthy older adults: a systematic review and meta-analysis. Ageing Res. Rev. **15**, 28–43 (2011)

33. Wouters, P.: Game design: the mapping of cognitive task analysis and game discourse analysis in creating effective and entertaining serious games, pp. 25–27 (2010)

34. Slootmaker, A., Netherlands, T., Kurvers, H., Hummel, H., Koper, R., Learning, T.E., Learning, B.: Developing scenario-based serious games for complex cognitive skills acquisition: design, development and evaluation of the EMERGO platform. J. Univers. Comput. Sci. **20**, 561–582 (2014)

35. Torrente, J., del Blanco, A., Moreno-Ger, P., Fernández-Manjón, B.: Designing serious games for adult students with cognitive disabilities. In: Huang, T., Zeng, Z., Li, C., Leung, C.S. (eds.) ICONIP 2012, Part IV. LNCS, vol. 7666, pp. 603–610. Springer, Heidelberg (2012)

36. Sedig, K., Haworth, R.: Interaction design and cognitive gameplay: role of activation time. In: Proceedings of the First ACM SIGCHI Annual Symposium on Computer-human Interaction in Play, pp. 247–256. ACM, New York (2014)

37. Guía, E. De, Penichet, V.M.R., de la Guía, E., Lozana, M.D., Penichet, V.M.R.: Tangible user interfaces applied to cognitive therapies. In: Proceedings of IUI 2014 Workshop: Interacting with Smart Objects (2014)

38. Ahmad Zaki, N.A., Tengku Wook, T.S.M., Ahmad, K.: Analysis and classification of serious games for cognitive stimulation. In: The 5th International Conference on Electrical Engineering and Informatics 2015, pp. 685–690 (2015)

39. Hotonu, A., Aldous, A., Schafer-Dreyer, R.: Including Children with Speech and Language Delay in the Foundation Stage. A&C Black, London (2011)

40. Hawa, V.V., Spanoudis, G.: Toddlers with delayed expressive language: an overview of the characteristics, risk factors and language outcomes. Res. Dev. Disabil. **35**, 400–407 (2014)

41. Hauner, K.K.Y., Shriberg, L.D., Kwiatkowski, J., Allen, C.T.: A subtype of speech delay associated with developmental psychosocial involvement. J. Speech. Lang. Hear. Res. **48**, 635–650 (2005)

42. Singleton, N.C., Bartolotta, T.E., Shulman, B.B., Birtcher, K., Parker, K.E.: Language development: foundations, processes, and clinical applications. William Brottmiller (2014)

43. Herr, J.: Working with Young Children. Goodheart-Willcox (2008)

44. Shaffer, D.R., Kipp, K.: Developmental Psychology: Childhood and Adolescence. Cengage Learning, Belmont (2014)

45. Santrock, J.W.: Essentials of Life-Span Development. Mc Graw Hill, New York (2014)

46. Hoyer, W., Roodin, P.: Adult Development and Aging. McGraw-Hill, Boston (2009)

47. Barak, B., Schiffman, L.G.: Cognitive age: a nonchronological age variable. Adv. Consum. Res. **8**, 602–606 (1981)

48. Nielsen, J.: Usability Engineering (1993)

49. Devi, K.R., Sen, A., Hemachandran, K.: A working framework for the user-centered design approach and a survey of the available methods. Int. J. Sci. Res. Publ. **2**, 1–8 (2012)

50. Maguire, M.: Methods to support human-centred design. Int. J. Hum. Comput. Stud. **55**, 587–634 (2001)

51. Braun, V., Clarke, V.: Using thematic analysis in psychology. Qual. Res. Psychol. **3**, 77–101 (2006)

52. Attride-Stirling, J.: Thematic networks: an analytic tool for qualitative research. Qual. Res. **1**, 385–405 (2001)

53. Davies, P., Shanks, B., Davies, K.: Improving narrative skills in young children with delayed language development. Educ. Rev. **56**, 271–286 (2004)

54. Lee, S., Baik, Y., Nam, K., Ahn, J., Lee, Y., Oh, S., Kim, K.: Developing a cognitive evaluation method for serious game engineers. Cluster Comput., 1–10 (2013)

55. Rescorla, L., Ross, G.S., McClure, S.: Language delay and behavioral/emotional problems in toddlers: findings from two developmental clinics. J. Speech Lang. Hear. Res. **50**, 1063–1078 (2007)

3D Facial Expressions from Performance Data

Fadhil Hamdan$^{(\boxtimes)}$, Haslanul Matussin, Saadah Serjuddin,
Somnuk Phon-Amnuaisuk, and Peter David Shannon

Media Informatics Special Interest Group, School of Computing and Information
Technology, Universiti Teknologi Brunei, Gadong, Brunei
fadhil.hamdan105@gmail.com, shinumakei@hotmail.com,
saadahserjuddin@gmail.com,
{somnuk.phonamnuaisuk,peter.shannon}@utb.edu.bn

Abstract. Advances in facial animation technology have been extended into various creative applications such as computer games, 3D animations and interactive multimedia. In this work, we explore performance-driven facial animations using a rigged bone approach. A 3D face model is rigged with bones positioned at the desired control points. The performance information for controlling these control points are extracted using the marker-based technique where color dots are painted on a performer's face at desired positions. Facial expression performances are recorded using a low cost camera. The information from the recorded performances is extracted and mapped onto a targeted face. Hence facial expression performances of the actor can be applied to different 3D model faces, provided that they share the same control points. This approach affords reusability of the facial performances without expensive equipment. We present the creative process of our approach, discuss the outcome and the prospect of future works.

Keywords: Face rigged bone · Performance driven facial animation · MEL script

1 Introduction

Decades ago, the only way to make a 2D or 3D animation was to manually alter the content frame by frame - a tedious job. Increments in computing power and advances in 2D/3D animation authoring tools have improved animation quality and sped up the process greatly. Now, animating a 3D character is a simple process of plugging in motion capture (MOCAP) data from required animation clips.

In the character animation field, facial animation requires great attention to create nuance of expressions. This is a difficult task if one is looking for realistic animation output. Contemporary techniques employed for computerised facial animation include bone-driven animation, blend shape (or shape interpolation) and physiological model. The rigged bone facial animation technique provides a simple setup but realistic muscle movement might not be obtained since each

© Springer International Publishing AG 2017
S. Phon-Amnuaisuk et al. (eds.), *Computational Intelligence in Information Systems*,
Advances in Intelligent Systems and Computing 532, DOI 10.1007/978-3-319-48517-1_24

bone may control many vertices. The blend shape technique demands intensive manual labour as many key target shapes must be prepared. The physiological model could provide a realistic animation if the physiological model is accurately prepared. However, the approach is computationally expensive. Weighting the pros and cons of these approaches, the rigged bone approach appears to be a good candidate for our interest of a low cost setup.

In this work, we use a simple low-end camera to record facial expression performances derived from markers on a performer's face. The movements of control markers are tracked. The coordinates of the tracking points are then applied to 3D models through scripting.

The rest of the paper is organized into the following sections: Sect. 2 discusses the background of adaptive behaviors in a 3D environment; Sect. 3 discusses our approach and gives the details of the techniques behind it; Sect. 4 provides a critical discussion of the output from the proposed approach; and finally, the conclusion and further research are presented in Sect. 5.

2   Background

Human facial expressions are controlled by muscles which may be voluntarily or involuntarily controlled [1]. Facial expressions convey emotions and are essential in human interactions. Computer-generated facial expressions are sought in creative industries such as game publishers, film producers, production houses, etc. Realistic and caricature facial expressions [2] have been explored by researchers in the creative computing community [3,4].

Intuitively, simulating anatomical characteristics of bones, tissues and skin should give a realistic animation output although the approach requires intensive computing power. Early pioneers in computerized facial animations attempted to describe facial movements. *Facial Action Coding System (FACS)* defines 46 action units (AUs), various facial expressions can be generated using combinations of these AUs.

Computerised facial animation has been explored by pioneers such as [5,6]. Parke introduced an interpolation technique which can be seen as the forefather of the popular *blend shape* animation technique [7,8]. The blend shape technique morphs the original facial pose into the target facial expression pose. This is a kind of *key frame* animation where *in-betweens* are filled in between two morph targets. With knowledge of the face model, the in-between can be calculated using a linear interpolation function [9,10].

The advantage of the blend shape approach is its fast playback since computing linear interpolation of vertices is simple. However, manually setting key facial poses is tedious and the quality of animation depends on these key poses. If the key poses are set too far apart, the interpolation could introduce distortion to the face. The number of key expression poses also directly limits the number of possible expressions.

Another popular alternative approach is the skeleton animation approach [11] where parts of the polygon surface are associated with bones. In [12], the whole face is rigged with bones and each bone controls a set of vertices on the facial model. Various facial expressions are obtained by manipulating these bones using the performance data (i.e., coordinates of tracking points) [13]. The approach offers a convenient means to animate various expressions with a rigged face model.

Facial expressions has also become a challenge in ICT when there is a need to maintain live video stream over a low bit rate communication channel. Instead of sending the whole video sequence, which is a huge amount of data, if both sides of the communication channel maintain their own facial models then only control information instructing those facial models can be sent over with a fraction of the data size. The *Moving Pictures Experts Group* has designed a *Face Animation Parameter* (FAP) component for the MPEG-4 *Face and Body Animation* (FBA) international standard. FAP defines 84 facial feature points in its facial model [14] as well as face animation parameters which define animation behaviours such as mouth closed, gaze and head orientation, etc. This is still an open research topic.

3   Performance Driven Facial Animation

In this study, six emotional categories are explored: happiness, sadness, anger, disgust, fear and surprise. To demonstrate how the facial performances obtained from a human's face can be translated into different 3D face models, three types of face models are used in this study: human, cat and alien[1].

Figure 1 shows how the bones are positioned at important facial components: eye, eye lids, mouth, nose, etc. Every bone has its own functions and control area. Their movements could deform the facial polygons' surface and give illusion of different facial expressions.

We recorded 6 basic expressions with a digital single-lens reflex camera (DSLR) mounted on a tripod to stabilize the camera. The performer's face was painted with color markers. The green markers cover four important muscle movement areas: the eyes, the nose, the mouth and the eyebrow areas (markers 1–22). Two reference points were marked with orange markers - one at the performer's forehead and the other at the performer's chin (markers 23, 24). All performances were recorded with a frame size of 1080*720 pixels resolution, at 25 frames per second. Figure 2 illustrates the facial marker setup and the six emotional expressions used in this study.

Once the facial performances were recorded, all markers' positions were extracted using image processing techniques [15]. All clips were stabilised to remove unwanted head movement (using information extracted from the orange stabilised markers), then the (x, y) coordinate of all the markers were extracted for each performance. The object tracking algorithm tracks each individual marker separately. The differences between the original pose and the performance poses yield a time series of coordinates for all the bone control points (see Fig. 3). Only (x, y) coordinates are extracted, the z coordinate is available from the model.

---

[1] The three facial models (see Fig. 1) used in this study are obtained from http:// tf3dm.com/.

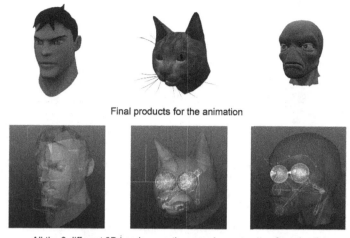

Final products for the animation

All the 3 different 3D heades use the same bone structure for animation

Fig. 1. Three 3D head models and their bones' positions

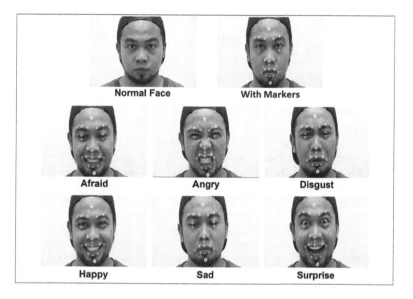

Fig. 2. Prepare a subject with markers at desired control points before facial performance. A total of 24 markers are used in this experiment. (Color figure online)

4 Creative Process and Results

Let d_m denote the distance between the marker m and a fixed reference point from the face model, at frame f; the differences between any two frame f_1 and f_2 is the movement Δd. If a performance has f frames, then the facial expression

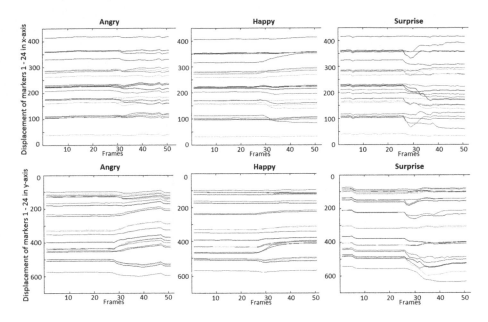

Fig. 3. Displacement plots of all 24 markers in x-axis and y-axis (examples of angry, happy and surprise categories).

performance can be represented using a matrix P of size $m \times f$, where each entry denotes Δd_m of the marker m at frame f.

Control information extracted from the facial performance is applied to the 3D model via a script. The control script moves the bones over time according to the facial performance information. Portions on the model surface may be associated with one or more bones. Hence their movements deform and modify the model surface according to the performance. Figure 4 shows track points for two arbitrary frames taken from angry, happy, sad and surprise emotional expressions. Coordinate information is stored in a MEL script by setting the time and the keyframes of each bone position:

```
currentTime 1 ;
setKeyframe"cc_r_browA01.tx";
setKeyframe "cc_r_browA01.ty";
setKeyframe "cc_r_browB01.tx";
setKeyframe "cc_r_browB01.ty";
...
setKeyframe "cc_eye01.RightUpperLid";
setKeyframe "cc_eye01.RightLowerLid";
...
setKeyframe "cc_upperJaw01.rx";
setKeyframe "cc_r_lipConner01.tx";
setKeyframe "cc_r_lipConner01.ty";
setKeyframe "cc_r_upperLip01.tx";
... and so on
```

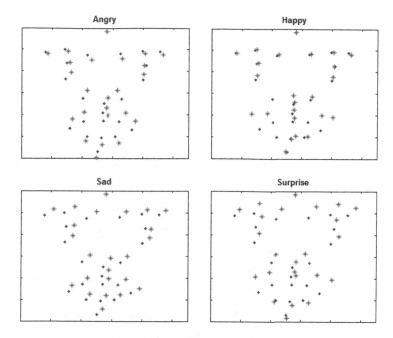

Fig. 4. Facial expressions of various expressions - each with two sets of markers' positions: '.' and '*'

Since the rigged-bone approach controls animation through bone movements, it naturally supports the reuse of performance data. Given a new 3D model with the same control structure (the same set of markers), the reference performance can be applied to the new model using the following mapping:

$$\Delta d_m^{new} = \Delta d_m^{ref} \frac{d_m^{new}}{d_m^{ref}}$$

Figure 5 shows snapshots of the performances of six emotions applied to the three head models.

We have successfully implemented the approach and have shown that the same reference performance can be applied to different head models using a rigged-bone approach. We have also experienced various issues associated with this approach, for example, although it is conceptually logical that animated performances should be applicable to different face models; an alien model does not have a nose and therefore the code used for the nose movement will not be implemented to the alien model. Quite a bit of manual adjustment work was also required for different face models which have dissimilar polygon mesh structure.

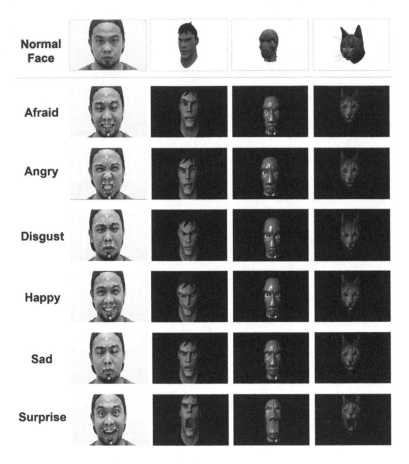

Fig. 5. Snapshots of six facial performances from the three models

5 Conclusion and Future Direction

In this paper, we investigate performance-driven facial animation process. The 3D face model was rigged with bones which each controls a set of vertices on a face model. Animating these bones using movement information extracted from facial performances deform the model's facial surface according to the performer's expressions.

We employed a low cost camera to record facial performances and extract the performances (as a time series of control data) corresponding to the bone position on the 3D face model. This approach has three advantages; (i) it offers comparatively, good quality facial animation in less production time; (ii) the recorded performance can be reused with different models; and (iii) since tracking can be done in real-time, the approach can be extended to support real-time performances.

The focal point in this area is its realistic facial expressions as it engages and communicates with the audience at a deeper level. However, this is not a simple task due to the *uncanny valley*, the psychological phenomenon argued in [16]. In future works, we plan to increase the level of realism by adding more markers to the face so that the system can detect nuance of expressions and produce better animations. Adding additional components such as hair, eye flashes and half-body movements associated with the facial expressions will increase the level of realism.

Acknowledgments. We wish to thank anonymous reviewers for their comments that have helped improve this paper. We would like to thank the GSR office for their partial financial support given to this research.

References

1. Hjorztsjo, C.H.: Man's face and mimic language. Accessed from Carl-Herman Hjortsjö, Man's face and mimic language on June 2016 diglib.uibk.at.at/ (1969)
2. Phon-Amnuaisuk, S.: Exploring particle-based caricature generations. In: Abd Manaf, A., Zeki, A., Zamani, M., Chuprat, S., El-Qawasmeh, E. (eds.) ICIEIS 2011. CCIS, vol. 252, pp. 37–46. Springer, Heidelberg (2011). doi:10.1007/978-3-642-25453-6_4
3. Arya, A., DiPaola, S.: Multispace behavioural model for face-based affective social agents. EURASIP J. Image Video Process. Spec. Issue Facial Image Process. **2007**, 048757 (2007)
4. Byun, M., Badler, N.: FacEMOTE: qualitative parametric modifiers for facial animations. In: Proceedings of the 2002 SIGGRAPH/Eurographics Symposium on Computer Animation (SCA 2002), pp. 65–71 (2002)
5. Parke, F.I.: Computer generated animation of faces. In: Proceedings of the ACM Annual Conference (ACM 1972), vol. 1, pp. 451–457 (1972)
6. Waller, B.M., Lembeck, M., Kuchenbuch, P., Burrows, A.M., Liebal, K.: Gibbon-FACS: a muscle-based facial movement coding system for hylobatids. Int. J. Primatol. **33**(4), 809. doi:10.1007/s10764-012-9611-6
7. Chuang, E., Bregler, C.: Performance driven facial animation using blendshape interpolation. Technical report CS-TR-2002-02. Stanford University (2002)
8. Deng, Z., Bulut, M., Neumann, U., Narayanan, S.S.: Automatic dynamic expression synthesis for speech animation. In: Proceedings of IEEE Computer Animation and Social Agents (CASA 2004), pp. 267–274 (2004)
9. Pighin, F., Hecker, J.: Synthesizing realistic facial expression from photographs. In: Proceedings of the 25th Annual Conference on Computer Graphics and Interactive Techniques (SIGGRAPH 1998), pp. 75–84. ACM, New York (1998)
10. Ekman, P., Friesen, W.V.: Measuring facial movement. Environ. Psychol. Nonverbal Behav. **1**(1), 56–75 (1976)
11. Magnenat-Thalmann, N., Laperrière, R., Thalmann, D.: Joint-dependent local deformations for hand animation and object grasping. In: Proceedings of Graphics Interface 1988, Edmonton, pp. 26–33 (1988)
12. Schleifer, J.: Character setup from rig mechanics to skin deformations: a practical approach. In: Proceedings of the 29th International Conference on Computer Graphics and Interactive Techniques (SIGGRAPH 2002). A note from short course (2002)

13. Willaims, L.: Performance-driven facial animation. In: Proceedings of the 17th Annual Conference on Computer Graphics and Interactive Techniques SIGGRAPH 1990, pp. 235–242 (1990)
14. Pandzic, I.S., Forchheimer, R.: MPEG-4: Facial Animation: The Standard, Implementation and Applications. Wiley, New York (2002)
15. Bouguet, J.Y.: Pyramidal implementation of the affine Lucas Kanade feature tracker: description of the algorithm. Technical report Intel Corporation. robots.standford.edu/cs223b04/algo_affine_tracking.pdf. Accessed June 2016
16. Mori, M.: The uncanny valley. IEEE Robot. Autom. **19**(2), 98–100 (2012)

Computational Complexity
and Algorithms

Constrained Generalized Delaunay Graphs are Plane Spanners

Prosenjit Bose[1], Jean-Lou De Carufel[2], and André van Renssen[3,4(✉)]

[1] School of Computer Science, Carleton University, Ottawa, Canada
jit@scs.carleton.ca
[2] School of Electrical Engineering and Computer Science,
University of Ottawa, Ottawa, Canada
jdecaruf@uottawa.ca
[3] National Institute of Informatics, Tokyo, Japan
[4] JST, ERATO, Kawarabayashi Large Graph Project, Tokyo, Japan
andre@nii.ac.jp

Abstract. We look at generalized Delaunay graphs in the constrained setting by introducing line segments which the edges of the graph are not allowed to cross. Given an arbitrary convex shape C, a constrained Delaunay graph is constructed by adding an edge between two vertices p and q if and only if there exists a homothet of C with p and q on its boundary that does not contain any other vertices visible to p and q. We show that, regardless of the convex shape C used to construct the constrained Delaunay graph, there exists a constant t (that depends on C) such that it is a plane t-spanner of the visibility graph.

1 Introduction

A geometric graph G is a graph whose vertices are points in the plane and whose edges are line segments between pairs of vertices. A graph G is called plane if no two edges intersect properly. Every edge is weighted by the Euclidean distance between its endpoints. The distance between two vertices u and v in G, denoted by $\delta_G(u, v)$, or simply $\delta(u, v)$ when G is clear from the context, is defined as the sum of the weights of the edges along the shortest path between u and v in G. A subgraph H of G is a t-spanner of G (for $t \geq 1$) if for each pair of vertices u and v, $\delta_H(u, v) \leq t \cdot \delta_G(u, v)$. The smallest value t for which H is a t-spanner is the *spanning ratio* or *stretch factor* of H. The graph G is referred to as the *underlying graph* of H. The spanning properties of various geometric graphs have been studied extensively in the literature (see [1,2] for an overview of the topic).

Most of the research has focused on constructing spanners where the underlying graph is the complete Euclidean geometric graph. We study this problem in a more general setting with the introduction of line segment *constraints*. Specifically, let P be a set of points in the plane and let S be a set of line segments

Research supported in part by FQRNT, NSERC, and Carleton University's President's 2010 Doctoral Fellowship.

© Springer International Publishing AG 2017
S. Phon-Amnuaisuk et al. (eds.), *Computational Intelligence in Information Systems*,
Advances in Intelligent Systems and Computing 532, DOI 10.1007/978-3-319-48517-1_25

with endpoints in P, with no two line segments intersecting properly. The line segments of S are called *constraints*. Two vertices u and v can *see each other* or *are visible to each other* if and only if either the line segment uv does not properly intersect any constraint or uv is itself a constraint. If two vertices u and v can see each other, the line segment uv is a *visibility edge*. The *visibility graph* of P with respect to a set of constraints S, denoted $Vis(P, S)$, has P as vertex set and all visibility edges as edge set. In other words, it is the complete graph on P minus all edges that properly intersect one or more constraints in S.

This setting has been studied extensively within the context of motion planning amid obstacles. Clarkson [3] was one of the first to study this problem and showed how to construct a linear-sized $(1+\epsilon)$-spanner of $Vis(P, S)$. Subsequently, Das [4] showed how to construct a spanner of $Vis(P, S)$ with constant spanning ratio and constant degree. Bose and Keil [5] showed that the Constrained Delaunay Triangulation is a $4\pi\sqrt{3}/9 \approx 2.419$-spanner of $Vis(P, S)$. The constrained Delaunay graph where the empty convex shape is an equilateral triangle was shown to be a 2-spanner of $Vis(P, S)$ [6]. In the case of rectangles, the spanning ratio is at most $\sqrt{2} \cdot (2l/s + 1)$, where l and s are the length of the long and short side of the rectangle [7]. We look at the constrained generalized Delaunay graph, where the empty convex shape can be any convex polygon.

In the unconstrained setting, it is known that generalized Delaunay graphs are spanners [8], regardless of the convex shape used to construct them. These bounds are very general, but unfortunately not tight. In special cases, better bounds are known. For example, when the empty convex shape is a circle, Dobkin *et al.* [9] showed that the spanning ratio is at most $\pi(1 + \sqrt{5})/2 \approx 5.09$. Improving on this, Keil and Gutwin [10] reduced the spanning ratio to $4\pi/3\sqrt{3} \approx 2.42$. Recently, Xia showed that the spanning ratio is at most 1.998 [11]. On the other hand, Bose *et al.* [12] showed a lower bound of 1.58, which is greater than $\pi/2$, which was conjectured to be the tight spanning ratio up to that point. Later, Xia and Zhang [13] improved this to 1.59.

Chew [14] showed that if an equilateral triangle is used instead, the spanning ratio is 2 and this ratio is tight. In the case of squares, Chew [15] showed that the spanning ratio is at most $\sqrt{10} \approx 3.16$. This was later improved by Bonichon *et al.* [16], who showed a tight spanning ratio of $\sqrt{4 + 2\sqrt{2}} \approx 2.61$.

In this paper, we show that the constrained generalized Delaunay graph G is a spanner of $Vis(P, S)$ whose spanning ratio depends solely on the properties of the empty convex shape C used to create it: We show that G satisfies the α_C-diamond property and the visible-pair κ_C-spanner property (defined in Sect. 3.2), which implies that it is a t-spanner for:

$$
t = \begin{cases} 2\kappa_C \cdot \max\left(\frac{3}{\sin(\alpha_C/2)}, \kappa_C\right), & \textit{if } G \textit{ is a triangulation} \\ 2\kappa_C^2 \cdot \max\left(\frac{3}{\sin(\alpha_C/2)}, \kappa_C\right), & \textit{otherwise.} \end{cases}
$$

This proof is not a straightforward adaptation from the work by Bose *et al.* [8] due to the presence of constraints. For example, showing that a region contains no vertices that are visible to some specific vertex v requires more work than showing that this same region contains no vertices, since we allow vertices in the region that are not visible to v. Induction cannot be applied in a straightforward manner as in the unconstrained case because not all pairs of vertices are visible to each other. Moreover, we prove a slightly stronger result, where constraints are not necessarily edges of the graph. We elaborate on this point in more detail in Sect. 2.

2   Preliminaries

Throughout this paper, we fix a convex shape C. We assume without loss of generality that the origin lies in the interior of C. A *homothet* of C is obtained by scaling C with respect to the origin, followed by a translation. Thus, a homothet of C can be written as $x + \lambda C = \{x + \lambda z : z \in C\}$, for some scaling factor $\lambda > 0$ and some point x in the interior of C after translation.

For a given set of vertices P and a set of constraints S, the constrained generalized Delaunay graph is usually defined as follows. Given any two visible vertices p and q, let $C(p, q)$ be any homothet of C with p and q on its boundary. The constrained generalized Delaunay graph contains an edge between p and q if and only if pq is a constraint or there exists a $C(p, q)$ such that there are no vertices of P in the interior of $C(p, q)$ visible to both p and q. We assume that no four vertices lie on the boundary of any homothet of C.

Now, slightly modify this definition such that there is an edge between two visible points p and q if and only if there exists a $C(p, q)$ such that there are no vertices of P in the interior of $C(p, q)$ visible to both p and q. Note that this modified definition implies that constraints are not necessarily edges of the graph, since constraints may not necessarily adhere to the visibility property. Indeed, our modified graph is always a subgraph of the constrained generalized Delaunay graph. Therefore, any result proven on our modified graph also holds for the graph that includes all the constraints. As such, we prove the stronger result on our modified graph. For simplicity, in the remainder of the paper, when we refer to the constrained generalized Delaunay graph, we mean our modified subgraph of the constrained generalized Delaunay graph.

2.1   Auxiliary Lemmas

Next, we present three auxiliary lemmas that are needed to prove our main results. First, we reformulate a lemma that appears in [17].

Lemma 2.1. *Let C be a closed convex curve in the plane. The intersection of two distinct homothets of C is the union of two sets, each of which is either a segment, a single point, or empty.*

Though the following lemma was applied to constrained θ-graphs in [6], the property holds for any visibility graph. We say that a region R *contains* a vertex v if v lies in the interior or on the boundary of R. We call a region *empty* if it does not contain any vertex of P in its interior. We also note that we distinguish between *vertices* and *points*. A *point* is any point in $\mathbb{R}^2$, while a *vertex* is part of the input.

Lemma 2.2. *Let u, v, and w be three arbitrary points in the plane such that uw and vw are visibility edges and w is not the endpoint of a constraint intersecting the interior of triangle uvw. Then there exists a convex chain of visibility edges from u to v in triangle uvw, such that the polygon defined by uw, wv and the convex chain is empty and does not contain any constraints.*

Let p and q be two vertices that can see each other and let $C(p, q)$ be a convex polygon with p and q on its boundary. We look at the constraints that have p as an endpoint and the edge (or edges) of $C(p, q)$ on which p lies, and extend them to half-lines that have p as an endpoint (see Fig. 1a). Given the cyclic order of these half-lines around p and the line segment pq, we define the clockwise neighbor of pq to be the half-line that minimizes the strictly positive clockwise angle with pq. Analogously, we define the counterclockwise neighbor of pq to be the half-line that minimizes the strictly positive counterclockwise angle with pq. We define the *cone* C_q^p that contains q to be the region between the clockwise and counterclockwise neighbor of pq. Finally, let $C(p, q)_q^p$, the *region of $C(p, q)$ that contains q with respect to p*, be the intersection of $C(p, q)$ and C_q^p (see Fig. 1b).

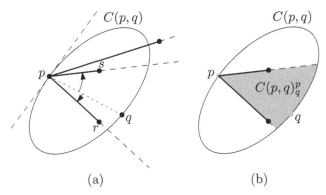

(a)   (b)

Fig. 1. Defining the region of $C(p, q)$ that contains q with respect to p: (a) The clockwise and counterclockwise neighbor of pq are the half-lines through pr and ps, (b) $C(p, q)_q^p$ is marked in gray

Lemma 2.3. *Let p and q be two vertices that can see each other and let $C(p, q)$ be any convex polygon with p and q on its boundary. If there is a vertex x in $C(p, q)_q^p$ (other than p and q) that is visible to p, then there is a vertex y (other than p and q) that is visible to both p and q and such that triangle pyq is empty.*

Proof. We have two visibility edges, namely pq and px. Since x lies in $C(p, q)_q^p$, p is not the endpoint of a constraint such that q and x lie on opposite sides of the line through this constraint. Hence, we can apply Lemma 2.2 and we obtain a convex chain of visibility edges from x to q and the polygon defined by pq, px and the convex chain is empty and does not contain any constraints. Furthermore, since the convex chain is contained in triangle pxq, which in turn is contained in $C(p, q)$, every vertex along the convex chain is contained in $C(p, q)$ (see Fig. 2).

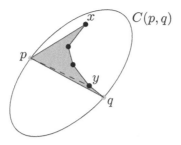

Fig. 2. Vertex y lies in $C(p, q)$ and is visible to both p and q

Let y be the neighbor of q along this convex chain. Hence, y is visible to q and contained in $C(p, q)$. Furthermore, p can see y, since the line segment py is contained in the polygon defined by pq, px and the convex chain, which is empty and does not contain any constraints. This also implies that triangle pyq is empty. □

3   The Constrained Generalized Delaunay Graph

Before we show that every constrained generalized Delaunay graph is a spanner, we first show that they are plane.

3.1   Planarity

In order to show that the constrained generalized Delaunay graph is plane, we first observe that no edge of the graph can contain a vertex, as this vertex would lie in $C(p, q)$ and be visible to both endpoints of the edge.

Observation 3.1. *Let pq be an edge of the constrained generalized Delaunay graph. The line segment pq does not contain any vertices other than p and q.*

Lemma 3.2. *The constrained generalized Delaunay graph is plane.*

Proof. We prove this by contradiction, so assume that there exist two edges pq and rs that intersect. It follows from Observation 3.1 that neither p nor q lies on rs and that neither r nor s lies on pq, so the edges intersect properly. Since pq is contained in $C(p, q)$ and rs is contained in $C(r, s)$, the boundaries of $C(p, q)$ and $C(r, s)$ intersect or one of $C(p, q)$ and $C(r, s)$ contains the other.

We first show that this implies that $p \in C(r, s)$, $q \in C(r, s)$, $r \in C(p, q)$, or $s \in C(p, q)$. If one of $C(p, q)$ and $C(r, s)$ contains the other, this holds trivially. If the two homothets intersect and either $p \in C(r, s)$ or $q \in C(r, s)$, we are done, so assume that neither p nor q lies in $C(r, s)$. Lemma 2.1 states that the boundaries of $C(p, q)$ and $C(r, s)$ intersect each other at most twice. These intersections split the boundary of $C(p, q)$ into two parts: one that is contained in $C(r, s)$ and one

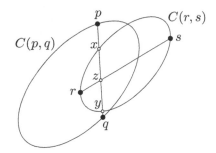

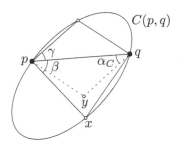

Fig. 3. $C(p, q)$ and $C(r, s)$ intersect and pq intersects $C(r, s)$ at x and y

Fig. 4. The constrained generalized Delaunay graph satisfies the α_C-diamond property

that is not. Since $p \notin C(r, s)$ and $q \notin C(r, s)$, p and q lie on the arc of $C(p, q)$ that is not contained in $C(r, s)$ (see Fig. 3). However, pq intersects $C(r, s)$, since otherwise pq cannot intersect rs. Let x and y be the two intersections of pq with the boundary of $C(r, s)$ (if the boundary of $C(r, s)$ is parallel to pq, x and y are the two endpoints of the interval of this intersection). We note that x and y split $C(r, s)$ into two parts, one of which is contained in $C(p, q)$, and that r and s cannot lie on the same part. In particular, one of r and s lies on the part that is contained in $C(p, q)$, proving that $r \in C(p, q)$, or $s \in C(p, q)$.

In the remainder of the proof, we assume without loss of generality that $r \in C(p, q)$ (see Fig. 3). Let z be the intersection of pq and rs. Hence, z can see both p and r. Also, z is not the endpoint of a constraint intersecting the interior of triangle pzr. Therefore, it follows from Lemma 2.2 that there exists a convex chain of visibility edges from p to r. Let v be the neighbor of p along this convex chain. Since v is part of the convex chain, which is contained in pzr, which in turn is contained in $C(p, q)$, it follows that v is a vertex visible to p contained in $C(p, q)$. Furthermore, since the polygon defined by pz, zr and the convex chain does not contain any constraints, v lies in $C(p, q)_q^p$. Thus, it follows from Lemma 2.3 that there exists a vertex in $C(p, q)$ that is visible to both p and q, contradicting that pq is an edge of the graph.   □

3.2   Spanning Ratio

Let x and y be two distinct points on the boundary ∂C of C. These two points split ∂C into two parts. For each of these parts, there exists an isosceles triangle with base xy such that the third vertex lies on that part of ∂C. We denote the base angles of these two triangles by $\alpha_{x,y}$ and $\alpha'_{x,y}$. We define α_C as follows:

$$\alpha_C = \min\{\max(\alpha_{x,y}, \alpha'_{x,y}) : x, y \in \partial C, x \neq y\}. \tag{1}$$

Note that since this function is defined on a compact set, the minimum and maximum exist and this function is well-defined.

Given a graph G and an angle $0 < \alpha < \pi/2$, we say that an edge pq of G satisfies the α-*diamond property*, when at least one of the two isosceles triangles with base pq and base angle α does not contain any vertex visible to both p and q. A graph G satisfies the α-diamond property when all of its edges satisfy this property [18].

Lemma 3.3. *Let C be any convex polygon. The constrained generalized Delaunay graph satisfies the α_C-diamond property.*

Proof. Let pq be any edge of the constrained generalized Delaunay graph. Since pq is an edge, there exists a $C(p,q)$ such that $C(p,q)$ does not contain any vertices that are visible to both p and q. The vertices p and q split the boundary $\partial C(p,q)$ of $C(p,q)$ into two parts and each of these parts defines an isosceles triangle with base pq. Let β and γ be the base angles of these two isosceles triangles and assume without loss of generality that $\beta \geq \gamma$ (see Fig. 4). Let x be the third vertex of the isosceles triangle having base angle β.

Translate and scale $C(p,q)$ such that it corresponds to C. This transformation does not affect the angles β and γ. Hence, since $p \neq q$ and both lie on the boundary of $C(p,q)$, the pair $\{\beta, \gamma\}$ is one of the pairs considered when determining α_C in Eq. 1. Hence, since $\beta \geq \gamma$, it follows that $\alpha_C \leq \beta$. Let y be the third point of the isosceles triangle having base pq and base angle α_C that lies on the same side of pq as triangle pxq (see Fig. 4). Since $\alpha_C \leq \beta$, triangle pyq is contained in triangle pxq. By convexity of $C(p,q)$, pxq is contained in $C(p,q)$. Hence, since $C(p,q)$ does not contain any vertices visible to both p and q, triangle pyq does not contain any vertices visible to both p and q either. Hence, pq satisfies the α_C-diamond property. □

For the next property, fix O to be a point in the interior of C. Let x and y be two distinct points on ∂C, such that x, y, and O are collinear. Again, x and y split ∂C into two parts. Let $\ell_{x,y}$ and $\ell'_{x,y}$ denote the lengths of these two parts. We define $\kappa_{C,O}$ as follows:

$$\kappa_{C,O} = \max \left\{ \frac{\max(\ell_{x,y}, \ell'_{x,y})}{|xy|} : x, y \in \partial C, x \neq y, \text{and } x, y, \text{and } O \text{ are collinear} \right\}.$$

We note that the constrained generalized Delaunay graph does not depend on the location of O inside C, as the presence of any edge pq is defined in terms of $C(p,q)$, which does not depend on the location of O. Therefore, we define κ_C as follows: $\kappa_C = \min\{\kappa_{C,O} : O \text{ is in the interior of } C\}$. Throughout the remainder of this section, we assume that O is picked such that $\kappa_C = \kappa_{C,O}$. We refer to this O as the *center* of C.

Given a constrained generalized Delaunay graph G, let p and q be two vertices on the boundary of a face f of the constrained generalized Delaunay graph, such that p can see q and the line segment pq does not intersect the exterior of f. If for every such pair p and q on every face f, there exists a path in G of length at most $\kappa \cdot |pq|$, then G satisfies the *visible-pair κ-spanner property*. We show that the constrained generalized Delaunay graph satisfies the visible-pair κ_C-spanner property. However, before we do this, we bound the length of the union of the boundary of a sequence of homothets that have their centers on a line.

Let a set of $k+1$ vertices $v_1, ..., v_{k+1}$ be given, such that all vertices lie on one side of the line through v_1 and v_{k+1}. For ease of exposition, assume the line through v_1 and v_{k+1} is the x-axis and all vertices lie on or above this line. We consider only point sets for which there exists $C_1, ..., C_k$, a set of homothets of C, such that the center of each homothet lies on the x-axis, C_i has v_i and v_{i+1} on its boundary, and no C_i contains any vertices other than v_i and v_{i+1}. Let ∂C be the boundary of C above the x-axis and let $\partial(v_i, v_{i+1})$ be the part of the boundary of C_i between v_i and v_{i+1} that lies above the x-axis.

Lemma 3.4. *Let $C(v_1, v_{k+1})$ be the homothet with v_1 and v_{k+1} on its boundary and its center on the x-axis. It holds that $\sum_{i=1}^{k} |\partial(v_i, v_{i+1})| \leq |\partial C(v_1, v_{k+1})|$.*

Proof. We prove the lemma by induction on k, the number of homothets. If $k = 1$, $\partial(v_1, v_2)$ is the same as $\partial C(v_1, v_2)$, so the lemma holds.

If $k > 1$, we assume that the induction hypothesis holds for all sets of at most $k - 1$ homothets. Since homothet C_i does not contain any vertices other than v_i and v_{i+1}, it follows that none of the homothets are fully contained in the union of the other homothets.

Let C_{k-1} be the homothet that defines the rightmost intersection r with the x-axis when C_k is not part of the set of homothets. Let l be the leftmost intersection of C_k and the x-axis (see Fig. 5). Let $\partial(v_1, v_k) = \bigcup_{i=1}^{k-1} \partial(v_i, v_{i+1})$ and let $\partial(v_k, r)$ be the part of ∂C_{k-1} between v_k and r above the x-axis. Let $\partial(l, v_k)$ be the part of ∂C_k between l and v_k above the x-axis. Since $\sum_{i=1}^{k} |\partial(v_i, v_{i+1})| = |\partial(v_1, v_k)| + |\partial(v_k, v_{k+1})|$, to prove the lemma, we show that $|\partial(v_1, v_k)| + |\partial(v_k, v_{k+1})| \leq |\partial C(v_1, v_{k+1})|$.

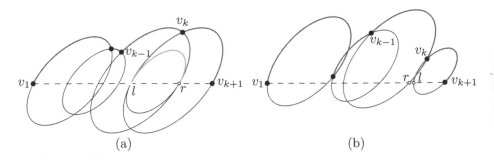

Fig. 5. The partial boundaries $\partial(v_1, v_k)$ and $\partial(v_k, v_{k+1})$ (blue), $\partial(l, v_k)$ and $\partial(v_k, r)$ (red), and $\partial C(l, r)$ (orange): (a) l lies to the left of r, (b) l lies on or to the right of r (Color figure online)

Let $c = |\partial C(v_1, v_{k+1})| / |v_1 v_{k+1}|$, so $|\partial C(v_1, v_{k+1})| = c \cdot |v_1 v_{k+1}|$. Since $|\partial(v_1, v_k)| + |\partial(v_k, r)| = |\partial(v_1, v_{k-1})| + |\partial(v_{k-1}, r)|$, it follows from the induction hypothesis that $|\partial(v_1, v_k)| + |\partial(v_k, r)| = |\partial(v_1, v_{k-1})| + |\partial(v_{k-1}, r)| \leq$

$|\partial C(v_1, r)| = c \cdot |v_1 r|$. Since the center of C_k lies on the x-axis, it follows that $|\partial C_k| = |\partial(l, v_k)| + |\partial(v_k, v_{k+1})| = c \cdot |lv_{k+1}|$. We consider two cases: (a) l lies to the left of r, (b) l lies on or to the right of r.

Case (a): If l lies to the left of r, let $C(l, r)$ be the homothet centered on the x-axis with l and r on its boundary (see Fig. 5a). Hence, it follows that $|\partial C(l, r)| = c \cdot |lr|$. Since $C(l, r)$ has l and on its left boundary, it is contained in C_k, and since it has r on its right boundary, it is contained in C_{k-1}. Hence, $C(l, r)$ is contained in the intersection of C_{k-1} and C_k. Since the length of the boundary of this intersection above the x-axis is $|\partial(l, v_k)| + |\partial(v_k, r)|$ and $C(l, r)$ is convex, it follows that $|\partial C(l, r)| \leq |\partial(l, v_k)| + |\partial(v_k, r)|$. Hence, we have that

$$\sum_{i=1}^{k} |\partial(v_i, v_{i+1})| = |\partial(v_1, v_k)| + |\partial(v_k, v_{k+1})|$$

$$= |\partial(v_1, v_k)| + |\partial(v_k, r)| - |\partial(v_k, r)| + |\partial C_k| - |\partial(l, v_k)|$$
$$\leq c \cdot |v_1 r| - |\partial(v_k, r)| + c \cdot |lv_{k+1}| - |\partial(l, v_k)|$$
$$= c \cdot |v_1 v_{k+1}| + c \cdot |lr| - |\partial(l, v_k)| - |\partial(v_k, r)|$$
$$= |\partial C(v_1, v_{k+1})| + |\partial C(l, r)| - |\partial(l, v_k)| - |\partial(v_k, r)|$$
$$\leq |\partial C(v_1, v_{k+1})|.$$

Case (b): If l lies on or to the right of r (see Fig. 5b), we have that

$$\sum_{i=1}^{k} |\partial(v_i, v_{i+1})| = |\partial(v_1, v_k)| + |\partial(v_k, v_{k+1})|$$

$$\leq |\partial(v_1, v_k)| + |\partial(v_k, r)| + |\partial(l, v_k)| + |\partial(v_k, v_{k+1})|$$
$$\leq c \cdot |v_1 r| + c \cdot |lv_{k+1}|$$
$$\leq c \cdot |v_1 v_{k+1}|$$
$$= |\partial C(v_1, v_{k+1})|,$$

completing the proof.   □

Lemma 3.5. *The constrained generalized Delaunay graph satisfies the visible-pair κ_C-spanner property.*

Proof. Let p and q be two vertices on the boundary of a face f of the constrained generalized Delaunay graph, such that p can see q and pq does not intersect the exterior of f. Assume without loss of generality that pq lies on the x-axis. Let $C(p, q)$ be the homothet of C with p and q on its boundary and its center on pq. We show that there exists a path between p and q of length at most $\kappa_C \cdot |pq|$. Since by definition κ_C is at least $|\partial C(p, q)|/|pq|$ (where $\partial C(p, q)$ is the boundary of $C(p, q)$ above the x-axis), showing that there exists a path between p and q of length at most $|\partial C(p, q)|$ completes the proof. If pq is an edge of the graph, this follows from the triangle inequality, so assume this is not the case.

We grow a homothet C' with its center on pq by moving its center from p to q, while maintaining that p lies on the boundary of C' (see Fig. 6a). Let v_1 be the first vertex hit by C' that is visible to p and lies in $C(p,q)_q^p$. We assume without loss of generality that v_1 lies above pq. Since v_1 is the first vertex satisfying these conditions, pv_1 is either an edge or a constraint: Since v_1 is the first visible vertex we hit in $C(p,q)_q^p$, we have that $C(p,q)_q^p \cap C'$ contains no vertices visible to p. Hence, there is no vertex visible to both p and v_1. Therefore, Lemma 2.3 implies that $C(p,q)_q^p \cap C'$ does not contain any vertices visible to v_1. Finally, if pv_1 is not a constraint, $C(p,q)_q^p \cap C'$ contains the region that is visible to both p and v_1. Hence, if pv_1 is not a constraint, the region that is visible to both p and v_1 does not contain any vertices and pv_1 is an edge.

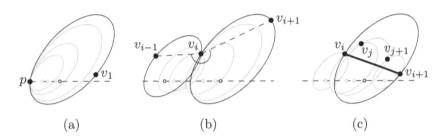

Fig. 6. Constructing a path from p to q: (a) growing C' from p, (b) growing C' while maintaining that v_i lies on its boundary, (c) refining when $v_i v_{i+1}$ is a constraint

We continue constructing a sequence of vertices $p, v_1, v_2, ..., v_k, q$ until we hit q by moving the center of C' along pq towards q and each time we hit a vertex v_i, we require that it lies on the boundary of C' until we hit the next vertex v_{i+1} that is visible to v_i and v_i is not the endpoint of a constraint that lies in the counterclockwise angle $\angle v_{i-1} v_i v_{i+1}$ (see Fig. 6b). Since v_{i+1} is the first vertex satisfying these conditions starting from v_i, $v_i v_{i+1}$ is either an edge or a constraint. This in turn implies that these vertices all lie above pq, since pq is visible and does not intersect the exterior of f.

Unfortunately, we cannot assume that there exists an edge between every pair of consecutive vertices: If $v_i v_{i+1}$ is a constraint, there can be vertices visible to both v_i and v_{i+1} on the opposite side of the constraint. For pairs of vertices v_i, v_{i+1} that do not form an edge, we refine the construction of the sequence between them: We start with C' such that it does not cross $v_i v_{i+1}$ and v_i lies on its boundary. We construct a sequence of vertices from v_i to v_{i+1} by moving the center of C' along pq towards q, maintaining that v_i lies on its boundary (see Fig. 6c). For the first vertex we hit, we require that it is visible to v_i and lies in $C'^{v_i}_{v_{i+1}}$.

We continue moving the center of C' along pq towards q, but we now maintain that v_i' lies on the boundary of C'. Each time we hit a vertex v_j, we require that it lies on the boundary of C' until we hit the next vertex v_{j+1}' that is visible to v_j and v_j is not the endpoint of a constraint that lies in the counterclockwise angle $\angle v_{j-1} v_j v_{j+1}$. In other words, we construct a more fine-grained sequence when consecutive vertices define a constraint and there is no edge between them. Note that we may need to repeat this process a number of times, since there need not be edges between the vertices of the finer grained sequence either. However, since the point set is finite, this process terminates.

This way, we end up with a path $p, v_1', v_2', ..., v_l', q$ from p to q that lies above pq. Furthermore, since C is convex, we can upper bound the length of each edge $v_i v_{i+1}$ by the part of the boundary of $C(v_i, v_{i+1})$, the homothet with v_i and v_{i+1} on its boundary and its center on pq, that does not intersect pq. Hence, the total length of the path is upper bounded by the length of the union of the boundaries of these homothets above pq. By construction, none of the homothets corresponding to consecutive vertices along the path contain any of the other vertices along the path. Hence, we can apply Lemma 3.4 and it follows that the total length of the path is at most $|\partial C(p, q)|$, completing the proof.   □

We are now ready to prove that the constrained generalized Delaunay graph is a spanner. Das and Joseph [18] showed that any plane graph that satisfies the diamond property and the good polygon property (similar to the visible-pair κ-spanner property) is a spanner. Subsequently, Bose et al. [19] improved slightly on the spanning ratio. They showed that a geometric (constrained) graph G is a spanner of the visibility graph when it satisfies the following properties:

1. G is plane.
2. G satisfies the α-diamond property.
3. The spanning ratio of any one-sided path in G is at most κ.
4. G satisfies the visible-pair κ'-spanner property.

In particular, G is a t-spanner for $t = 2\kappa\kappa' \cdot \max\left(\frac{3}{\sin(\alpha/2)}, \kappa\right)$.

It follows from Lemmas 3.2, 3.3, and 3.5 that the constrained generalized Delaunay graph satisfies these four properties. Moreover, even though in general the constrained generalized Delaunay graph is not a triangulation, if for a specific convex shape it is, it satisfies the visible-pair 1-spanner property: Since every face consists of three vertices that are pairwise connected by an edge, the shortest path between two vertices p and q on this face has length $1 \cdot |pq|$. Therefore, we obtain the following theorem:

Theorem 3.6. *The constrained generalized Delaunay graph G is a t-spanner of Vis(P, S) for*

$$
t = \begin{cases} 2\kappa_C \cdot \max\left(\frac{3}{\sin(\alpha_C/2)}, \kappa_C\right), & \text{if } G \text{ is a triangulation} \\ 2\kappa_C^2 \cdot \max\left(\frac{3}{\sin(\alpha_C/2)}, \kappa_C\right), & \text{otherwise.} \end{cases}
$$

4   Conclusion

In light of other recent results in the constrained setting, such as the fact that Yao- and θ-graphs with sufficiently many cones are spanners, the result presented in this paper raises a tantalizing question: What conditions need to hold for a graph to be a spanner in the constrained setting? In particular, these and previous results show a number of sufficient conditions, but do not immediately give rise to a set of necessary conditions.

References

1. Bose, P., Smid, M.: On plane geometric spanners: A survey and open problems. Comput. Geom. Theory Appl. **46**(7), 818–830 (2013)
2. Narasimhan, G., Smid, M.: Geometric Spanner Networks. Cambridge University Press, New York (2007)
3. Clarkson, K.: Approximation algorithms for shortest path motion planning. In: Proceedings of the 19th Annual ACM Symposium on Theory of Computing (STOC 1987), pp. 56–65 (1987)
4. Das, G.: The visibility graph contains a bounded-degree spanner. In: Proceedings of the 9th Canadian Conference on Computational Geometry (CCCG 1997), pp. 70–75 (1997)
5. Bose, P., Keil, J.M.: On the stretch factor of the constrained Delaunay triangulation. In: Proceedings of the 3rd International Symposium on Voronoi Diagrams in Science and Engineering (ISVD 2006), pp. 25–31 (2006)
6. Bose, P., Fagerberg, R., Renssen, A., Verdonschot, S.: On plane constrained bounded-degree spanners. In: Fernández-Baca, D. (ed.) LATIN 2012. LNCS, vol. 7256, pp. 85–96. Springer, Heidelberg (2012). doi:10.1007/978-3-642-29344-3_8
7. Bose, P., De Carufel, J.L., van Renssen, A.: Constrained empty-rectangle Delaunay graphs. In: Proceedings of the 27th Canadian Conference on Computational Geometry (CCCG 2015), pp. 57–62 (2015)
8. Bose, P., Carmi, P., Collette, S., Smid, M.: On the stretch factor of convex Delaunay graphs. J. Comput. Geom. **1**(1), 41–56 (2010)
9. Dobkin, D.P., Friedman, S.J., Supowit, K.J.: Delaunay graphs are almost as good as complete graphs. Discrete Comput. Geom. **5**(1), 399–407 (1990)
10. Keil, J.M., Gutwin, C.A.: Classes of graphs which approximate the complete Euclidean graph. Discrete Comput. Geom. **7**(1), 13–28 (1992)
11. Xia, G.: The stretch factor of the Delaunay triangulation is less than 1.998. SIAM J. Comput. **42**(4), 1620–1659 (2013)
12. Bose, P., Devroye, L., Löffler, M., Snoeyink, J., Verma, V.: Almost all Delaunay triangulations have stretch factor greater than $\pi/2$. Comput. Geom. Theory Appl. (CGTA) **44**(2), 121–127 (2011)
13. Xia, G., Zhang, L.: Toward the tight bound of the stretch factor of Delaunay triangulations. In: Proceedings of the 23rd Canadian Conference on Computational Geometry (CCCG 2011), pp. 175–180 (2011)
14. Chew, L.P.: There are planar graphs almost as good as the complete graph. J. Comput. Syst. Sci. **39**(2), 205–219 (1989)
15. Chew, L.P.: There is a planar graph almost as good as the complete graph. In: Proceedings of the 2nd Annual Symposium on Computational Geometry (SoCG 1986), pp. 169–177 (1986)
16. Bonichon, N., Gavoille, C., Hanusse, N., Perković, L.: The stretch factor of L_1 and L_∞-Delaunay triangulations. In: Epstein, L., Ferragina, P. (eds.) ESA 2012. LNCS, vol. 7501, pp. 205–216. Springer, Heidelberg (2012). doi:10.1007/978-3-642-33090-2_19
17. Swanepoel, K.: Helly-type theorems for homothets of planar convex curves. Proc. Am. Math. Soc. **131**(3), 921–932 (2003)

18. Das, G., Joseph, D.: Which triangulations approximate the complete graph? In: Djidjev, H. (ed.) Optimal Algorithms. LNCS, vol. 401, pp. 168–192. Springer, Heidelberg (1989). doi:10.1007/3-540-51859-2_15
19. Bose, P., Lee, A., Smid, M.: On generalized diamond spanners. In: Dehne, F., Sack, J.-R., Zeh, N. (eds.) WADS 2007. LNCS, vol. 4619, pp. 325–336. Springer, Heidelberg (2007). doi:10.1007/978-3-540-73951-7_29

Solving the Longest Oneway-Ticket Problem and Enumerating Letter Graphs by Augmenting the Two Representative Approaches with ZDDs

Jun Kawahara[1], Toshiki Saitoh[2(✉)], Hirofumi Suzuki[3], and Ryo Yoshinaka[4]

[1] Graduate School of Information Science, NAIST, Ikoma, Japan
[2] Graduate School of Engineering, Kobe University, Kobe, Japan
saitoh@eedept.kobe-u.ac.jp
[3] Graduate School of Information Science and Technology,
Hokkaido University, Sapporo, Japan
[4] Graduate School of Information Sciences, Tohoku University, Sendai, Japan

Abstract. Several researchers have studied subgraph enumeration algorithms that use a compressed expression for a family of sets, called a zero-suppressed binary decision diagram (ZDD), to solve subgraph optimization problems. We have two representative approaches to manipulate ZDDs effectively. One is fundamental mathematical operations on families of sets over ZDDs. The other is a direct construction method of a ZDD that represents desired subgraphs of a graph and is called frontier-based search. In this research, we augment the approaches by proposing two new operations, called disjoint join and joint join, on family algebra over ZDDs and extending the frontier-based search to enumerate subgraphs that have a given number of vertices of specified degrees. Employing the new approaches, we present enumeration algorithms for alphabet letter graphs on a given graph. Moreover, we solve a variant of the longest path problem, called the Longest Oneway-ticket Problem (LOP), that requires computing the longest trip on the railway network of the Japan Railways Group using a oneway ticket. Numerical experiments show that our algorithm solves the LOP and is faster than the existing integer programming approach for some instances.

1 Introduction

Several researchers have studied enumeration algorithms that use a compact data structure for a family of sets, called a zero-suppressed binary decision diagram (ZDD) [9], to solve optimization problems. We can represent and store feasible solutions of an optimization problem as a ZDD, which can be exponentially smaller than the number of solutions stored in the ZDD. The ZDD representation of feasible solutions is of higher value than just compactly storing them. ZDDs support fundamental mathematical operations on families of sets such as the union and the intersection of two families based on the recursive structure of ZDDs, called *family algebra* [6]. Moreover, we can count the number of solutions represented by a ZDD, compute the solution with the maximum/minimum total

© Springer International Publishing AG 2017
S. Phon-Amnuaisuk et al. (eds.), *Computational Intelligence in Information Systems*,
Advances in Intelligent Systems and Computing 532, DOI 10.1007/978-3-319-48517-1_26

weight, and sample a solution uniformly and randomly [6] in time proportional to the size of the ZDD. These operations can be done without extracting elements of sets one by one. Combining these features, we can solve optimization problems under various constraint conditions.

For graph optimization problems, we can represent a set of subgraphs of a given graph as a ZDD by identifying a subgraph with the set of its edges. A framework for directly constructing a ZDD that represents desired subgraphs has been proposed, called the *frontier-based search* (FBS) [5,6,11]. This framework enables us to efficiently construct ZDDs representing spanning trees [11], *s-t* paths [6], and various subgraphs [5]. By storing some information into nodes of a ZDD in the process of the construction, the FBS can flexibly impose various conditions on subgraphs such as the number of edges, the number of connected components, existence or nonexistence of loops, the connectivity of specified vertices, and so forth. Using this feature, Inoue et al. [3] proposed algorithms to optimize power loss on a grid network, Yoshinaka et al. [13] presented algorithms to solve and enumerate some pencil-and-paper puzzles on graphs, and Takizawa et al. [12] applied the FBS to evacuation planning.

The contributions of the paper are threefold. The first is to design the FBS for subgraphs whose degrees are given as an input to show the flexibility of the FBS. Specifically, we specify the number of vertices whose degree is ℓ for each integer ℓ and construct the ZDD representing all the subgraphs satisfying the condition. The second is to develop some ZDD operations for family algebra, called disjoint join and joint join, to show the versatility of ZDD operations. The disjoint (resp. joint) join operation constructs the family of sets that are the unions of disjoint (resp. overlapping) sets from two given families. The other is an FBS-based algorithm for converting the ZDD representing a set of edge sets into the one representing the set of the sets of edges and involved vertices. This algorithm enables us to compute the set of the vertex-disjoint unions of two subgraphs included in two subgraph sets represented by ZDDs.

By the proposed algorithms, we solve a variant of the longest path problem, called the Longest Oneway-ticket Problem (LOP) proposed by Miyashiro et al. [10], that requires computing the longest trip on the railway network of the Japan Railways Group using a oneway ticket. This problem is NP-hard (see Sect. 7.1) and is equivalent to computing the heaviest path (called type-L), cycle (type-O), and path connected with a cycle (called type-P). So far, to the authors' best knowledge, no other algorithm is proposed for the problem except for their integer programming approach. We construct the ZDDs representing the type-L, type-O, and type-P subgraphs of a given graph. Numerical experiments show that our algorithm can solve the LOP for some instances within several minutes, while an integer programming approach cannot solve it within one day. Moreover, we create ZDDs for subgraphs that look like multiple Roman alphabet letters by combining the three algorithms. We confirm the performance of our algorithms by numerical experiments.

This paper is organized as follows. Sect. 2 presents brief summaries of ZDDs, family algebra of ZDDs, and the FBS. In Sect. 3, we propose the FBS for graphs whose degrees are specified. Sect. 4 gives a proposal of new operations on ZDDs, and Sect. 5 provides an algorithm to obtain a set of sets of edges and vertices. In Sect. 6, we construct the ZDD representing a set of (multiple) letter graphs using the algorithms. We show the results of numerical experiments to show the effectiveness of our algorithms in Sect. 7.

2   Preliminaries

2.1   Zero-Suppressed Binary Decision Diagrams

A ZDD [9] represents a family of sets over a finite universal set $U = \{x_1, \ldots, x_{|U|}\}$. We assume that U is totally ordered and write $x_i < x_j$ if and only if $i < j$. A ZDD is defined to be a labeled directed acyclic graph that satisfies the following properties: (i) There is only one node with indegree 0, called the *root node* and denoted by n_{root}, (ii) there are exactly two nodes **0** and **1** with outdegree 0, called the *0-terminal* and the *1-terminal*, (iii) each non-terminal node has exactly two outgoing arcs, labeled by 0 and 1 and called the *0-arc* and the *1-arc*, respectively, (iv) for $j \in \{0, 1\}$, we call the node pointed by the j-arc of a node n the j-*child* of n and denote by $c_j(n)$, (v) each non-terminal node n is labeled by an element of U, and (vi) the label of a non-terminal node is strictly smaller than those of its children.

We say that a subset U' of U *corresponds* to a path P from n to n' in a ZDD if and only if there exists a node n'' labeled with x whose 1-arc is in P for all and only $x \in U'$. A ZDD Z represents a family $\mathcal{F}$ of sets on U as follows: A subset U' of U is in $\mathcal{F}$ if and only if there is a path from the root to **1** in Z such that U' corresponds to the path.

A ZDD is said to be *reduced* if it satisfies the following two conditions: (i) There are no distinct nodes that have the same label, 0-child and 1-child, and (ii) there is no node whose 1-child is **0**. It is known that for any family $\mathcal{A}$ of sets, a unique reduced ZDD exists for $\mathcal{A}$ and it has the smallest number of nodes of ZDDs for $\mathcal{A}$ [8]. A linear-time algorithm that reduces a ZDD can be found in [6, Sect. 7.1.4, Algorithm R]. The reduced ZDD representing a family $\mathcal{F}$ of sets is denoted by $Z^{\mathcal{F}}$.

ZDDs have important features for combinatorial optimization. Suppose that the solution space of a combinatorial optimization problem is 2^U, and the weight of each element in U is given. We also suppose that the set $\mathcal{F}$ of all feasible solutions is represented as $Z^{\mathcal{F}}$. We can compute the maximum and minimum total weights of sets in $\mathcal{F}$ from $Z^{\mathcal{F}}$ and count the number of the feasible solutions in $\mathcal{F}$ in time proportional to the number of nodes in $Z^{\mathcal{F}}$ [6, Sect. 7.1.4, Algorithm B and C].

One can perform fundamental mathematical operations on families of sets over ZDDs, e.g., the union of two families (see Table 1). Let $|Z|$ denote the number of nodes of a ZDD Z, which we call the *size* of Z. Based on the recursive structure of ZDDs, for two families of sets $\mathcal{A}$ and $\mathcal{B}$, we can compute the ZDDs $Z^{\mathcal{A} \cup \mathcal{B}}$ and $Z^{\mathcal{A} \cap \mathcal{B}}$ from the ZDDs $Z^{\mathcal{A}}$ and $Z^{\mathcal{B}}$ in $\mathrm{O}(|Z^{\mathcal{A}}||Z^{\mathcal{B}}|)$ time [1,8]. By contrast, the size of the resulting ZDD of the join $\mathcal{A} \bowtie \mathcal{B}$ can be exponential in

Table 1. Examples of operations of family algebra

$\mathcal{A} \cup \mathcal{B} = \{S \mid S \in \mathcal{A} \text{ or } S \in \mathcal{B}\}$	$\mathcal{A} \bowtie \mathcal{B} = \{A \cup B \mid A \in \mathcal{A} \text{ and } B \in \mathcal{B}\}$
$\mathcal{A} \cap \mathcal{B} = \{S \mid S \in \mathcal{A} \text{ and } S \in \mathcal{B}\}$	$\mathcal{A}/\mathcal{B} = \{S \mid S \cup B \in \mathcal{A} \text{ and } S \cap B = \emptyset \text{ for all } B \in \mathcal{B}\}$
$\mathcal{A} \setminus \mathcal{B} = \{S \mid S \in \mathcal{A} \text{ and } S \notin \mathcal{B}\}$	$\mathcal{A} \bmod \mathcal{B} = \mathcal{A} \setminus (\mathcal{B} \bowtie (\mathcal{A}/\mathcal{B}))$

the input size $|Z^{\mathcal{A}}| + |Z^{\mathcal{B}}|$ in the worst case, where the join operation is defined as $\mathcal{A} \bowtie \mathcal{B} = \{A \cup B \mid A \in \mathcal{A} \text{ and } B \in \mathcal{B}\}$. For any binary operation $\diamond$, we write $Z^{\mathcal{A}} \diamond Z^{\mathcal{B}}$ for representing $Z^{\mathcal{A} \diamond \mathcal{B}}$.

2.2 Frontier-Based Search

Throughout this paper, we fix a given undirected edge-weighted graph $G = (V, E)$ and let $m = |E|$ and $E = \{e_1, \ldots, e_m\}$. We assume that G is a simple and connected graph. We use the term subgraph only as a subgraph of G and consider only subgraphs $(\bigcup_{e \in E'} e, E')$ for $E' \subseteq E$. That is, subgraphs contain no isolated (degree zero) vertex, and thus we identify a subgraph with its edge set.

In this subsection, we introduce the *frontier-based search*, FBS for short, for constructing a ZDD representing a set of subgraphs on a given graph. The universe set of the ZDD is E, which is totally ordered as $e_1 < \cdots < e_m$. We describe the FBS for multiple cycle graphs and single cycle graphs as examples [5, 6]. In this paper, a multiple (resp. single) cycle graph is defined as a graph that consists of at least one (resp. exactly one) cycle.

We describe the FBS for multiple cycle graphs, also proposed by Knuth [6]. The algorithm constructs the ZDD representing the set of all the multiple cycle graphs on G. It first creates the root node n_{root} and labels it with e_1. Then, it constructs the ZDD in a breadth-first manner, that is, for $i = 1, \ldots, m - 1$, it creates nodes with label e_{i+1} after all the nodes with label e_i are created. The destinations of the arcs of each non-terminal node with label e_i must be nodes labeled e_{i+1} or the 0/1-terminal.

We describe only an outline of the FBS. (See the details in [5].) For each node n in the ZDD under construction by the FBS, we store an array into n. The array $n.\mathbf{deg}$ represents a map from a certain subset of V in concern to natural numbers, where every path from the root to the node n corresponds to a subgraph in which those vertices have the degree specified by the array. Conversely, for a node n, if there exists a node n' with the same label as n such that $n.\mathbf{deg}$ is equal to $n'.\mathbf{deg}$, we merge n and n'. We call this operation *node sharing*.

For a vertex v, suppose that e_i has the largest index of all the edges incident to v. Since edges $e_{i+1}, \ldots, e_m$ do not affect the degree of v, we no longer refer to $\deg[v]$ after e_i is processed. In this case, we say that v is *fixed*. More formally, for $i = 1, \ldots, m$ we define $F_i = \left(\bigcup_{j=1}^{i} e_j\right) \cap \left(\bigcup_{j=i+1}^{m} e_j\right)$, called the *$i$-th frontier*.

We also define $F_0 = \emptyset$. For a node n with label e_i, we store $n.\mathbf{deg}[v]$ only for $v \in F_{i-1}$ into the node.

Using the array $\mathbf{deg}$, we can prune some nodes. For example, if the degree of a vertex in a subgraph is three, the corresponding node can be pruned due to the violation of the degree condition of cycles. By pruning and node sharing, we can construct the ZDD representing the set of all the multiple cycles on G.

We can also prune a node corresponding to the set of subgraphs each of which has a connected component isolated from another one. We can construct the ZDD representing the set of all the single cycles on G by combining constructing the ZDD for multiple cycles with this pruning technique. We refer the reader to [5] for further details.

3   FBS for Degree Specified Graphs

In this section, we propose an algorithm for constructing ZDDs representing subgraphs whose degrees are specified. Before explaining the algorithm, we present some definitions. Let δ be a partial function mapping from $\{1, \ldots, |V| - 1\}$ to $\{0, \ldots, |V|\}$. For any integer h, $\delta(h)$ means that there are $\delta(h)$ vertices whose degrees are h in the subgraph. If $\delta(h)$ is undefined, it means that there exist an arbitrary number of vertices whose degrees are h in the subgraph. Note that since we ignore degree zero vertices in subgraphs, we do not define $\delta(0)$. Let $\mathrm{dom}(\delta)$ be the domain of definition, i.e., $\mathrm{dom}(\delta) = \{h \mid \delta(h) \text{ is defined}\}$. We define a graph class $\mathcal{M}(\delta)$ as

$$\mathcal{M}(\delta) = \{G' \mid G' = (\bigcup_{e \in E'} e, E'), E' \subseteq E, E' \neq \emptyset, G' \text{ is connected,}$$
$$|\{v \mid d_{G'}(v) = h\}| = \delta(h) \text{ for any } h \in \mathrm{dom}(\delta)\},$$

where $d_{G'}(v)$ is the degree of v on a subgraph G'. Our goal is to construct the ZDD representing all the subgraphs in $\mathcal{M}(\delta)$ for a given δ. Note that $\mathcal{M}(\delta)$ is the class of single cycle graphs if $\delta(2)$ is undefined and $\delta(h) = 0$ for $h \in \{1\} \cup \{3, \ldots, |V| - 1\}$. Therefore, we can say that the algorithm in this section is a generalization of the algorithm in the previous section.

For a node n with label e_i, we store into n another array, say $n.\mathbf{dn}$, such that for each $h \in \mathrm{dom}(\delta)$, $n.\mathbf{dn}[h]$ is the number of fixed vertices whose degrees are h. We consider a situation in which we are creating a node labeled e_{i+1} as the destination of the x-arc of a node n_i with label $e_i = \{v, w\}$, where $x = 0$ or 1. When $x = 1$, $\mathbf{deg}[v]$ and $\mathbf{deg}[w]$ are incremented by one. Suppose that e_i has the largest index of all the edges incident to v. Let $h = .\mathbf{deg}[v]$. If $h \in \mathrm{dom}(\delta)$, $.\mathbf{dn}[h] \leftarrow n_i.\mathbf{dn}[h] + 1$, which means that the number of fixed vertices with degree h increases by one. Then, if $.\mathbf{dn}[h]$ exceeds $\delta(h)$, can be pruned because it violates the condition of $\mathcal{M}(\delta)$. When $i = m$, i.e., the last edge is processed, we also confirm whether there exists $h \in \mathrm{dom}(\delta)$ such that $.\mathbf{dn}[h] \neq \delta(h)$. If so, it is decided that the destination of the x-arc is the 0-terminal. Otherwise, it is the 1-terminal. We can assure that the subgraph set represented by the constructed ZDD includes only connected subgraphs in the same way.

4   New ZDD Operations

One of our goals is to obtain the family of sets that is the unions of disjoint (edge) sets from two given families of (edge) sets. For this purpose, we propose the following new ZDD operations for family algebra:

- (disjoint join) $\mathcal{A} \bowtie \mathcal{B} = \{A \cup B \mid A \in \mathcal{A},\ B \in \mathcal{B},\ \text{and}\ A \cap B = \emptyset\}$,
- (joint join) $\mathcal{A} \bowtie \mathcal{B} = \{A \cup B \mid A \in \mathcal{A},\ B \in \mathcal{B},\ \text{and}\ A \cap B \neq \emptyset\}$,

where $\mathcal{A}$ and $\mathcal{B}$ are families of sets. For a family $\mathcal{F}$ of sets over U and $x \in U$, we define $\mathcal{F}_0 = \mathcal{F} \bmod \{\{x\}\} = \{S \mid S \in \mathcal{F} \text{ and } x \notin S\}$ and $\mathcal{F}_1 = \mathcal{F}/\{\{x\}\} = \{S \setminus \{x\} \mid S \in \mathcal{F} \text{ and } x \in S\}$. By definition, $\mathcal{F} = (\mathcal{F}_1 \bowtie \{\{x\}\}) \cup \mathcal{F}_0$ (the definitions of operators $\bowtie$, $/$, and $\bmod$ are shown in Table 1).

Lemma 1. *Let $\mathcal{A}$ and $\mathcal{B}$ be families of sets over U and $x \in U$.*
The disjoint join and the joint join can be described as follows.

$$
\mathcal{A} \bowtie \mathcal{B} = \begin{cases} \{\emptyset\} & \text{if } \mathcal{A} = \{\emptyset\} \text{ and } \mathcal{B} = \{\emptyset\}, \\ \emptyset & \text{if } \mathcal{A} = \emptyset \text{ or } \mathcal{B} = \emptyset, \\ (((\mathcal{A}_0 \bowtie \mathcal{B}_1) \cup (\mathcal{A}_1 \bowtie \mathcal{B}_0)) \bowtie \{\{x\}\}) \cup (\mathcal{A}_0 \bowtie \mathcal{B}_0) & \text{otherwise}, \end{cases}
\tag{1}
$$

and

$$
\mathcal{A} \bowtie \mathcal{B} = \begin{cases} \emptyset & \text{if } \mathcal{A} = \emptyset, \mathcal{A} = \{\emptyset\}, \mathcal{B} = \emptyset, \text{ or } \mathcal{B} = \{\emptyset\}, \\ (((\mathcal{A}_1 \bowtie \mathcal{B}_1) \cup (\mathcal{A}_1 \bowtie \mathcal{B}_0) \cup (\mathcal{A}_0 \bowtie \mathcal{B}_1)) \bowtie \{\{x\}\}) \cup (\mathcal{A}_0 \bowtie \mathcal{B}_0) & \text{otherwise}. \end{cases}
\tag{2}
$$

We can show the lemma by induction but it is omitted due to the page limitation.

Lemma 1 indicates that the two operations can be computed recursively. Suppose that $\mathcal{A}$ and $\mathcal{B}$ are represented as ZDDs. By choosing the smallest element in $(\bigcup_{A \in \mathcal{A}} A) \cup (\bigcup_{B \in \mathcal{B}} B)$ as x in the lemma, we obtain the ZDDs representing $\mathcal{A}_0$, $\mathcal{A}_1$, $\mathcal{B}_0$ and $\mathcal{B}_1$ in a constant time because they are just (the pointers of) 0-children and 1-children of the root nodes of $\mathcal{A}$ and $\mathcal{B}$, respectively. (Note that the root node has the smallest label, and that we often identify a ZDD node with the ZDD consisting of the reachable nodes from the node [9].) Moreover, for two ZDDs $\mathcal{A}'$ and $\mathcal{B}'$ and a variable x which is smaller than any variables occurring in $\mathcal{A}'$ or $\mathcal{B}'$, it takes a constant time to compute $(\mathcal{A}' \bowtie \{\{x\}\}) \cup \mathcal{B}'$. Let $Z_i^{\mathcal{F}} = Z^{\mathcal{F}_i}$ for $i = 0, 1$. We describe the disjoint join operation in Algorithm 1. We can design the joint join operation in a similar way.

In the same way as the join operation, for ZDDs Z and Z', the size of the resulting ZDDs of $Z \bowtie Z'$ and $Z \bowtie Z'$ can be exponential in the input size $|Z|+|Z'|$ in the worst case.

Algorithm 1. $Z^{\mathcal{A}} \bowtie Z^{\mathcal{B}}$

---

1 Let r^A and r^B be the root nodes of $Z^{\mathcal{A}}$ and $Z^{\mathcal{B}}$, respectively;

2 **if** r^A *is 1-terminal and* r^B *is 1-terminal* **then** **return** 1-terminal;

3 **if** r^A *is 0-terminal or* r^B *is 0-terminal* **then** **return** 0-terminal;

4 Let x^A and x^B be the labels of r^A and r^B, respectively;

5 **if** $x^A < x^B$ **then** **return** $((Z_1^A \bowtie Z^{\mathcal{B}}) \bowtie Z^{\{\{x^A\}\}}) \cup (Z_0^A \bowtie Z^{\mathcal{B}})$;

6 **if** $x^B < x^A$ **then** **return** $((Z_1^B \bowtie Z^{\mathcal{A}}) \bowtie Z^{\{\{x^B\}\}}) \cup (Z_0^B \bowtie Z^{\mathcal{A}})$;

7 **if** $x^A = x^B$ **then** **return** $(((Z_0^A \bowtie Z_1^B) \cup (Z_1^A \bowtie Z_0^B)) \bowtie Z^{\{\{x^B\}\}}) \cup (Z_0^A \bowtie Z_0^B)$;

---

5 ZDDs over the Set of Edges and Vertices

In this section, we propose an algorithm for converting the ZDD representing the set of edge sets into the one representing the set of sets of edges and involved vertices. More precisely, given the ZDD Z representing the set $\mathcal{E}$ of edge sets (e.g., constructed by the FBS), we construct the ZDD Z' representing the set $\mathcal{S} = \{ E' \cup (\bigcup_{e \in E'} e) \mid E' \in \mathcal{E} \}$. The added information of vertices is redundant but it will be useful for computing vertex-disjoint graphs in the next section.

In principle we will insert vertex variables into appropriate places of the original ZDD Z by tracing paths of Z to obtain Z', possibly making copies of nodes of Z if necessary. We insert a node labeled v after a node labeled by e_i where i is the largest such that $v \in e_i$. If Z has a node n labeled with e such that $\mathcal{G}(n)$ has two subgraphs one of which includes a vertex v with $v > e$ and the other does not, the node n will have copies in Z' so that one must pass the 1-arc of a v node to reach the 1-terminal and another must pass the 0-arc of a v node. This conversion is realized by a variant of the FBS, where the information that each node n' labeled with e_i of Z' under construction stores is a pair of (the pointer of) a node of Z and a subset of the $(i-1)$-th frontier

$$F_{i-1} = \left(\bigcup_{j<i} e_j\right) \cap \left(\bigcup_{j \geq i} e_j\right).$$

We have the following theorem on the complexity of this conversion, where $f_{\max} = \max_i |F_i|$.

Theorem 1. *We have* $|Z'| \leq (2f_{\max} + 3)2^{f_{\max}}|Z|$. *One can construct* Z' *from* Z *in* $\mathrm{O}(f_{\max}^2 2^{f_{\max}}|Z|)$ *time.*

6 Constructing ZDDs for Letter Graphs

We construct ZDDs for letter graphs using FBS and ZDD operations described in Sects. 3 and 4. Let us define letter graphs we treat in this paper. Let $G' = (V', E')$ be a graph. If a graph G'' is obtained from G' by removing an edge $\{u, v\}$ and adding a new vertex w and two edges $\{u, w\}$ and $\{w, v\}$, we call G'' a subdivision of G'. Moreover, a subdivision of a subdivision of G' is also a subdivision of G'. A vertex is *branching* if its degree is greater than or equal to three and called a *pendant* if its degree is one.

Table 2. Types of letters and their equivalent letters. Column "$\delta(1,3,4)$" describes the values $\delta(1)$, $\delta(3)$, and $\delta(4)$ for each type. Note that $\delta(2) = \perp$ and $\delta(h) = 0$ for any $h \geq 5$ for each type.

Type	Eq	$\delta(1,3,4)$	Type	Eq	$\delta(1,3,4)$	Type	Eq	$\delta(1,3,4)$
L	C I J L M N	(2, 0, 0)	H	H K	(4, 2, 0)	B	B	(0, 2, 0)
	S U V W Z					☌	☌	(0, 2, 0)
O	D O	(0, 0, 0)	Q	Q	(1, 0, 1)	A	A R	(2, 2, 0)
P	P	(1, 1, 0)	X	X	(4, 0, 1)	♂	♂	(2, 2, 0)
E	E F G T Y	(3, 1, 0)	8	8	(0, 0, 1)	♀	♀	(3, 1, 1)

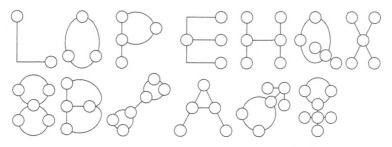

Fig. 1. Graph expressions of L, O, P, E, H, Q, X (top half) and 8, B, ☌, A, ♂, ♀ (bottom half).

López [7] topologically classifies the capital letters in the Roman alphabet in the Sans Serif font, that are shown in Table 2. We choose a representative for each group and express them as graphs shown in Fig. 1. For subsequent usage and experiments, in addition to letters in the Roman alphabet, we treat digit 8 of the Sans Serif font and the male, female, and opposition symbols, ♂, ♀, and ☌, in the Waldi symbol font. For □ = L, O, P, E, H, Q, X, 8, B, ☌, A, ♂, ♀, we define the type-□ class by the set consisting of the corresponding graph in Fig. 1 and its subdivisions and denote by $Z^{\square}$ the ZDD representing all the type-□ subgraphs. We call a graph in the type-□ class for some □ a *letter graph*. A *multi-letter graph* is a graph such that each connected component of the graph is a letter graph.

Enumerating Letter Graphs: We first design the FBS for the type-L, O, P, E, H, Q, X, and 8 classes. We here consider only the type-E class. We define the degree function δ_E as $\delta_E(1) = 3$, $\delta_E(2) = \perp$, $\delta_E(3) = 1$, and $\delta_E(h) = 0$ for $h \in \{4, \ldots, |V| - 1\}$. It can be shown that $\mathcal{M}(\delta_E)$ is equivalent to the type-E class. Therefore, we construct the ZDD for $\mathcal{M}(\delta_E)$ by the FBS. We describe the degree function δ for each type in Table 2.

Next, we treat the types-B, ☌, A, ♂, and ♀. They are not straightforward because there is no δ such that each of these type classes is equivalent to $\mathcal{M}(\delta)$. To solve the problems, we propose ZDD construction algorithms for those types by combining the FBS in Sect. 3 with the new ZDD operations in Sect. 4. Our algorithms for the types consist of two steps: We first construct the ZDD for $\mathcal{M}(\cdot)$ and then exclude the subgraphs of the other types by ZDD operations. In the following, we consider only the types-B and $\male^\wp$.

Both a type-B graph and a type-$\male^\wp$ graph have two vertices of degree three and the other vertices have degree two or zero in the graphs. We define a degree function $\delta_{B,\male^\wp}$ as $\delta_{B,\male^\wp}(2) = \bot$, $\delta_{B,\male^\wp}(3) = 2$, and $\delta_{B,\male^\wp}(h) = 0$ for $h \in \{1\} \cup \{4, \ldots, |V| - 1\}$.

In our algorithm, we first construct the ZDD Z for the $\mathcal{M}(\delta_{B,\male^\wp})$ class by using the FBS in Sect. 3. Note that the union of the type-B and $\male^\wp$ classes is equivalent to the $\mathcal{M}(\delta_{B,\male^\wp})$ class. Next, we exclude all the subgraphs in the type-$\male^\wp$ class from Z for enumerating type-B graphs. To exclude the graphs, we first construct the ZDD Z^O for the $\mathcal{M}(\delta_O)$ class and then compute the joint join of Z^O and Z^O, i.e., $Z^O \bowtie Z^O$. The set of subgraphs represented by $Z^O \bowtie Z^O$ includes all the type-B subgraphs but no type-$\male^\wp$ subgraphs. Thus, we extract the type-B class by $Z \cap (Z^O \bowtie Z^O)$, which is the ZDD for the type-B class. In contrast, the ZDD for the type-$\male^\wp$ class is obtained by $Z \setminus (Z^O \bowtie Z^O)$.

Similarly, we can construct ZDDs by applying the new FBS algorithm and using the proposed operations for the other types A, $\male$, and $\female$.

Enumerating Multi-letter Graphs: Given a multiset L of letters, we show a ZDD construction algorithm for multi-letter graphs. Our algorithm first constructs a ZDD $Z^\square$ for each letter $\square$ in L. Next, for each letter $\square$, we convert the ZDD $Z^\square$ to the ZDD that represents a family of sets of edges and vertices using Sect. 5. Finally, we carry out the disjoint join operation for the ZDDs in turn and obtain the ZDD representing all the multi-letter graphs for L.

7   Experiments

In this section, we show the results of some computer experiments. We implement the algorithms described above in C++ language, employing the TdZdd library [4][1] for the FBS and the SAPPOROBDD library[2] for handling ZDDs and carrying out ZDD operations. We compile the program using g++ 4.9.3 with the -O3 optimization option and execute it on a machine with the following specification: OS: Linux, CentOS 6.7; CPU: Intel(R) Xeon E5-2650 @ 2.00 GHz; Memory: 128 GB.

7.1   Finding the Longest Oneway-Ticket

In this subsection, we solve the LOP. In this problem, a railway network is given, and the rule for oneway tickets of the network (e.g., the Japan Railways Group, JR for short) is that a passenger can take a route from a departure station until the passenger reaches a station that has already been visited once.

---

[1] It is available at https://github.com/kunisura/TdZdd.

[2] The SAPPOROBDD library has not been officially published but is available at https://github.com/takemaru/graphillion/tree/master/src/SAPPOROBDD.

The goal station may have been visited twice. Miyashiro et al. [10] remarked that the passenger's trajectory looks like an L (path), O (cycle), or P (cycle plus a path connected with the cycle), considered the problem that requires computing the maximum weighted subgraph shaped like an L, O, or P for a given weighted graph, and called it the LOP. The NP-hardness of the LOP is proven by a reduction from the longest s-t path problem [2]. In the reduction, given an instance graph G' of the longest s-t path problem, we obtain the instance graph of the LOP by adding four enormously heavy edges $\{s, u\}$, $\{u, v\}$, $\{v, s\}$ and $\{t, w\}$ to G', where u, v and w are new vertices.

We use the JR train map in Sep. 2016 as G in this experiment. In the preliminary experiment we conducted, it was impossible to construct the ZDD for the type-P class on account of the graph size. Therefore, we consider reducing the input graph size as described in the following. This technique is also used by Miyashiro et al. [10]. We remove vertices whose degree is two. More precisely, for each vertex v with degree two, when v is adjacent to vertices u and w, we remove v, $\{u, v\}$, and $\{v, w\}$ from the input graph and add edge $\{u, w\}$ to the graph. Let the weight of $\{u, w\}$ be the sum of those of $\{u, v\}$ and $\{v, w\}$. If removing a degree two vertex generates parallel edges, we do not do so. We call the resulting graph the *simplified graph* and denote it by G'. We can show that it is sufficient to construct the ZDDs for type-L, O, P, 8 and $\mathcal{M}(\delta_{B, \wp})$ classes to solve the LOP.

We conducted the following experiment. The simplified graph G' has 348 vertices and 484 edges. We constructed the ZDD for the type-L, O, P, 8, and $\mathcal{M}(\delta_{B, \wp})$ classes by the FBS described in Sect. 3. Table 3 shows the result of constructing the ZDDs for G'. All the maximum type-$\wp$, B, and 8 graphs are lighter than the maximum type-P graph. Therefore, we have proven that the maximum solution of the LOP for G is a type-P graph and found that its length is 11140.1 km. The solution is shown in Fig. 2 and https://github.com/junkawahara/LOP.

Table 3. Constructing ZDDs for solving the LOP.

Class	Time	# of solutions	Weight (km)
type-L ($\mathcal{M}(\delta_L)$)	429.99	2.53e+34	11110.9
type-O ($\mathcal{M}(\delta_O)$)	28.17	9.25e+24	8141.9
type-P ($\mathcal{M}(\delta_P)$)	1053.58	1.13e+34	11140.1
type-8 ($\mathcal{M}(\delta_8)$)	107.89	1.21e+27	9142.3
$\mathcal{M}(\delta_{B, \wp})$	819.29	1.26e+33	11013.7

Fig. 2. The solution of the LOP for the JR map in Sep. 2016.

We implemented and ran the integer programming method proposed by Miyashiro et al. [10] and tried to compute the maximum weighted type-P subgraph on G' by Gurobi ver.6.50. The formulae of the method cannot ensure the connectivity of a computed subgraph, that is, redundant loops may be generated. Therefore, we need to repeatedly solve the instances of the integer programming problem by adding the constraints to prohibit the loops having been generated

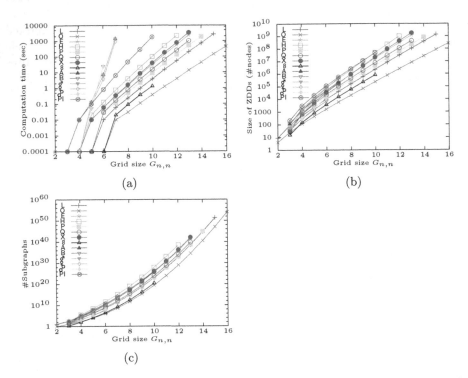

Fig. 3. (a) Computation time, (b) ZDD size, (c) the number of subgraphs for grids. The Y-axes of all the graphs are log scale.

until no loop is generated. In our experiment, we solved them more than 500 times but could not obtain the solution within one day. Note that Miyashiro et al. succeeded in solving the LOP by contracting some components of G' and reducing the graph size. This technique does not affect the correctness of their solution due to a special property of the JR train map but cannot be applied to general graphs.

7.2   Enumerating Letter and Multi-letter Graphs

In this subsection, we enumerate Roman letter graphs by implementing the algorithms proposed in Sect. 6. For each type, we record the computation time setting timeout at one hour, the number of nodes in a ZDD, and the number of subgraphs of the type in $n \times n$ grids $G_{n,n}$. The results are shown in Fig. 3.

From Fig. 3(a) and (b), we can see that the computation time and the number of nodes in the ZDD strongly correlate for all types. Fig. 3(b) and (c) show that if the number of subgraphs of some type is larger than that of another type, this tendency is also shown in the number of nodes in the ZDDs. From the experiment results, the number of type-O graphs is the smallest and that of type-E graphs is the largest on grids. Moreover, the slope of the number of nodes in the ZDDs is gentler than that of the number of the graphs. This means that the ZDDs effectively represent the many graphs in a compressed way for all types. Comparing to ZDDs that are constructed by FBS only, Z^O and Z^L for example, the ones that require both FBS and ZDD operations like Z^A and Z^B demand much more time to compute, because the ZDDs constructed by the operations in progress become huge even if the resulting ZDDs are small.

We also enumerate multi-letter graphs for "PI" in grids and the results are also shown in Fig. 3. The number of "PI" is 140,592,018,849,858,624,628,322,744 in $G_{10,10}$ and it takes 1,760 s to enumerate "PI" in $G_{10,10}$. Most of the execution time was consumed by the disjoint join operation.

References

1. Bryant, R.E.: Graph-based algorithms for boolean function manipulation. IEEE Trans. Comput. **C–35**(8), 677–691 (1986)
2. Garey, M.R., Johnson, D.S.: Computers and Intractability: A Guide to the Theory of NP-Completeness. W. H. Freeman and Company, New York (1979)
3. Inoue, T., Takano, K., Watanabe, T., Kawahara, J., Yoshinaka, R., Kishimoto, A., Tsuda, K., Minato, S., Hayashi, Y.: Distribution loss minimization with guaranteed error bound. IEEE Trans. Smart Grid **5**(1), 102–111 (2014)
4. Iwashita, H., Minato, S.: Efficient top-down ZDD construction techniques using recursive specifications. Hokkaido University, Division of Computer Science, TCS Technical reports TCS-TR-A-13-69 (2013)
5. Kawahara, J., Inoue, T., Iwashita, H., Minato, S.: Frontier-based search for enumerating all constrained subgraphs with compressed representation. Hokkaido University, Division of Computer Science, TCS Technical reports TCS-TR-A-14-76 (2014)
6. Knuth, D.E.: The Art of Computer Programming. Combinatorial Algorithms, Part 1, vol. 4A. Addison-Wesley, Upper Saddle River (2011)
7. López, R.: How does a topologist classify the letters of the alphabet? CoRR abs/1410.3364 (2014). http://arXiv.org/abs/1410.3364
8. Minato, S.: Zero-suppressed BDDs for set manipulation in combinatorial problems. In: The 30th ACM/IEEE Design Automation Conference, pp. 272–277 (1993)
9. Minato, S.: Zero-suppressed BDDs and their applications. Int. J. Softw. Tools Technol. Transfer **3**(2), 156–170 (2001)
10. Miyashiro, R., Kasai, T., Matsui, T.: Saicho katamichi kippu no genmitsukai wo motomeru (Strictly solving the longest oneway-ticket problem) (in Japanese). In: Proceedings of the 2000 Fall National Conference of Operations Research Society of Japan, pp. 24–25 (2000)
11. Sekine, K., Imai, H., Tani, S.: Computing the Tutte polynomial of a graph of moderate size. In: Proceedings of the 6th International Symposium on Algorithms and Computation (ISAAC), pp. 224–233 (1995)
12. Takizawa, A., Takechi, Y., Ohta, A., Katoh, N., Inoue, T., Horiyama, T., Kawahara, J., Minato, S.: Enumeration of region partitioning for evacuation planning based on ZDD. In: Proceedings of 11th International Symposium on Operations Research and its Applications in Engineering, Technology and Management 2013 (ISORA 2013), pp. 1–8 (2013)
13. Yoshinaka, R., Saitoh, T., Kawahara, J., Tsuruma, K., Iwashita, H., Minato, S.: Finding all solutions and instances of numberlink and slitherlink by ZDDs. Algorithms **5**(2), 176–213 (2012). http://dx.doi.org/10.3390/a5020176

Author Index

© Springer International Publishing AG 2017
S. Phon-Amnuaisuk et al. (eds.), *Computational Intelligence in Information Systems*,
Advances in Intelligent Systems and Computing 532, DOI 10.1007/978-3-319-48517-1

Printed in the United States
By Bookmasters